QUANTITATIVE ANALYSIS IN ARCHAEOLOGY

QUANTITATIVE ANALYSIS IN ARCHAEOLOGY

Todd L. VanPool & Robert D. Leonard

WILEY-BLACKWELL

A John Wiley & Sons, Ltd., Publication

This edition first published 2011
© 2011 Todd L. VanPool and Robert D. Leonard

Blackwell Publishing was acquired by John Wiley & Sons in February 2007. Blackwell's publishing program has been merged with Wiley's global Scientific, Technical, and Medical business to form Wiley-Blackwell.

Registered Office
John Wiley & Sons Ltd, The Atrium, Southern Gate, Chichester, West Sussex, PO19 8SQ, United Kingdom

Editorial Offices
350 Main Street, Malden, MA 02148-5020, USA
9600 Garsington Road, Oxford, OX4 2DQ, UK
The Atrium, Southern Gate, Chichester, West Sussex, PO19 8SQ, UK

For details of our global editorial offices, for customer services, and for information about how to apply for permission to reuse the copyright material in this book please see our website at www.wiley.com/wiley-blackwell.

The right of Todd L. VanPool and Robert D. Leonard to be identified as the author of this work has been asserted in accordance with the UK Copyright, Designs and Patents Act 1988.

Wiley also publishes its books in a variety of electronic formats. Some content that appears in print may not be available in electronic books.

Designations used by companies to distinguish their products are often claimed as trademarks. All brand names and product names used in this book are trade names, service marks, trademarks or registered trademarks of their respective owners. The publisher is not associated with any product or vendor mentioned in this book. This publication is designed to provide accurate and authoritative information in regard to the subject matter covered. It is sold on the understanding that the publisher is not engaged in rendering professional services. If professional advice or other expert assistance is required, the services of a competent professional should be sought.

Library of Congress Cataloging-in-Publication Data

VanPool, Todd L., 1968-
    Quantitative analysis in archaeology / Todd L. VanPool, Robert D. Leonard.
        p.   cm.
    ISBN 978-1-4051-8951-4 (hardback : alk. paper) – ISBN 978-1-4051-8950-7 (pbk. : alk. paper)
    1. Archaeology–Methodology.   2. Quantitative research.   3. Archaeology–Research.   I. Leonard, Robert D.   II. Title.
    CC75.7.V36 2010
    930.1072–dc22

                                                            2010030199

A catalogue record for this book is available from the British Library.

Set in 10.5/13pt Minion by Toppan Best-set Premedia Limited
Printed in Malaysia

01   2011

To Connie Woebke and Glenn McCoy, two teachers who taught TVP what counts

To Grandmas Maggie and Ferne, for helping RDL learn his numbers

Contents

Contents

Tables

Figures

Equations

Acknowledgments

First and foremost, we express our gratitude to the many students who have contributed to this volume through their participation in our Quantitative Methods in Anthropology classes. Our interaction with them has been the inspiration for both the structure and contents of this book. Robert R. Sokal's and F. James Rohlf's excellent text *Biometry*, has, no doubt, influenced this volume, given that we have taught from it for years. We also gratefully acknowledge the helpful comments of Drs. R. Lee Lyman and Gordon F.M. Rakita, as well as two anonymous reviewers. They read drafts of this volume and helped strengthen it tremendously. We thank Drs. Christine S. VanPool and Marcus Hamilton for their useful comments and suggestions through the years. Finally, Robert Leonard thanks Drs. Donald Grayson and Loveday Conquest, and Todd VanPool thanks Dr. Stephen R. Durand for introducing us to quantitative methods. Thank you all so very much!

1

Quantifying Archaeology

If archaeologists do anything, it is count. We count stones, bones, potsherds, seeds, buildings, settlements, and even particles of earth – virtually everything that constitutes the archaeological record. We also measure essentially everything that we touch. Length, weight, thickness, depth, volume, area, color, and height are only some of the simplest measurements taken. We are exaggerating only slightly when we state that our predilection for counting and measuring ensures fame (if not fortune) to anyone who brings to our attention some forgotten or never known aspect of the archaeological record that archaeologists should be counting and/or measuring.

Most archaeologists are in the counting and measuring business not for its own sake, but to help us fashion a meaningful perspective on the past. Quantification isn't required to back up every proposition that is made about the archaeological record, but for some propositions it is absolutely essential. For example, suppose we proposed an idea about differences in Hallstatt assemblages in Central Europe that could be evaluated by examining ceramic variation. Having observed hundreds of the pots, we could merely assert what we felt the major differences and similarities to be, and draw our conclusions about the validity of our original idea based upon our simple observations. We might be correct, but no one would take our conclusions seriously unless we actually took the relevant measurements and demonstrated that the differences and/or similarities were meaningful in a way that everyone agreed upon and understood. Quantification and statistics serve this end, providing us with a common language and set of instructions about how to make meaningful observations of the world, how to reduce our infinite database to an accurate and understandable set of characterizations, and how to evaluate differences and similarities. Importantly, statistics do this by using a framework that

Quantitative Analysis in Archaeology, Todd L. VanPool and Robert D. Leonard
© 2011 Todd L. VanPool and Robert D. Leonard

allows us to specify the ways in which we can be wrong, and the likelihood that we are mistaken. Statistics consequently provide archaeologists with a means to make arguments about cause that will ultimately help us construct explanations.

Statistical thinking plays an important role in archaeological analysis because archaeologists rely so heavily on samples. The archaeological record contains only the material remains of human activity that time and the vagaries of the environment (including human activity) have allowed to be preserved. The artifacts, features, and other material manifestations of human behavior that enter the archaeological record are only a small subset of those originally produced. Funding constraints, time limits, and our emphasis on conserving the archaeological record further dictate that archaeologists generally recover only a small subset of those materials that have been preserved. Thus, we have a sample of the material remains that have been preserved, which is only a sample of all of the materials that entered the archaeological record, which is only a sample of all of the materials that humans have produced.

Archaeologists are consequently forced to understand and explain the past using imperfect and limited data. Connecting our sample to a meaningful understanding of the past necessitates the application of a statistical framework, even when quantitative methods are avoided as part of a purportedly humanistic approach. It is only through statistical reasoning, no matter how implicit, that any form of general conclusion can be formed from the specifics of the archaeological record. Regardless of whether an archaeologist studies the social differentiation of Cahokia's residents, subsistence shifts during the Mexican colonial occupation of New Mexico, or the religious systems of Upper Paleolithic cave dwellers, they are going to employ a statistical approach, even if they don't acknowledge it. Quantitative methods allow us to make this approach explicit and make our arguments logically coherent and thereby facilitate their evaluation. Even the most ardent humanist should appreciate this.

As important as statistics are, we must remember that they are only tools, and subservient to theory. Our theoretical perspectives tell us which observations are important to make and how explanations are constructed. Statistics are useful only within this larger context, and it is important to remember their appropriate role. It is also important to recognize that the use of statistics does not equal science. The historical confluence of events that brought statistics, computers, the hypothetico-deductive method, and the theoretical advances of the New Archaeology to our discipline in a relatively brief span of time in the 1960s make it appear that they are inseparable. Nothing could be farther from the truth. While this might seem self-evident, at least one quite popular introductory archaeology textbook overstates the relationship, as a discussion of the role of science in archaeology begins with a brief discussion of statistics. Not the role of theory, not the scientific method, but statistics! Statistics do not a science make, and statistical analyses conducted in the absence of theory are merely vacuous description.

This book approaches quantification and statistics from the perspective that they are a simple set of tools that all competent archaeologists must know. Most readers will use statistics innumerable times throughout their career. Others may never calculate a mean or standard deviation willingly, but at least they will know the basics of the statistical tool kit. Choosing not to use a tool is fine. Remaining ignorant is unfortunate and unnecessary. At the very least, knowledge of statistics is needed to evaluate the work of others who do use them.

So, why should two archaeologists write a book about statistics when there are thousands of excellent statistics books in existence? Here are our reasons in no particular order. First, few of us entered archaeology because we wanted to be mathematicians. In fact, many archaeologists became interested in archaeology for very humanistic (or even romantic) reasons, and many avoided math in school like the plague. There definitely needs to be a book that is sympathetic to those coming from a non-quantitative background. We seek to achieve this goal by presenting the clearest description of techniques possible, with math no more complicated than simple algebra, but with enough detail that the reader will be able to actually understand how each technique operates.

Second, most statistics textbooks use examples that are not anthropological, and are very hard to relate to the archaeological record. While knowledge of dice examples is useful when playing craps in Las Vegas, the implications of these examples for archaeological studies are often difficult to decipher. Our examples are almost exclusively archaeological, and we hope that they provide good illustrations of how you might approach various archaeological data sets from a statistical perspective.

Third, archaeologists do not always need the standard set of statistics that are presented in popular textbooks. Some techniques of limited importance to archaeology are overemphasized in these texts, while other extremely important statistical methods are underemphasized or do not appear at all.

Fourth, it is our observation that many degree-granting programs in archaeology focus solely on computer instruction in quantitative methods rather than on the tried and true pencil and paper method. We have nothing against the use of computers and statistical software, as long as it is done by people who first learn statistical techniques by putting pencil to paper. However, our experience has shown us that when all training is focused on using a statistical package instead of learning a statistical method, the computer becomes a magic black box that produces "results" that students who don't know what actually happened inside the box are (hopefully) trained to interpret. This lack of understanding causes confusion and, more importantly, embarrassment when insupportable or erroneous conclusions are drawn. These students need a friendly text to which they can refer to help clarify how the quantitative methods work and how their results should be understood.

Finally, many disciplines use samples, but few are as wholly reliant on them as is archaeology. This in turn means that the application of quantitative reasoning has special significance in archaeological research that needs to be explored if we are to produce the best archaeological analyses we can. This consideration is absent from statistical texts written for general audiences, but should be central to those specifically for archaeologists. It certainly will be central to the discussions that follow this chapter.

Ultimately, our goal is to illustrate the utility and structure of a quantitative framework to the reader (i.e., you), and to provide a full understanding of each statistical method so that you will understand how to calculate a statistical measure, why you would want to do so, and how the statistical method works mathematically. If you understand these issues, you will find each method to be intuitively meaningful and will appreciate the significance of its assumptions, limitations, and strengths. If you don't understand these factors, your applications will be prone to error and misinterpretations, and, as a result, archaeology as a discipline will suffer. Hopefully, this text will serve to aid you, gentle reader, as we all work to accomplish our collective goals as a discipline.

Practice Exercises

1 Identify five attributes of artifacts or features that archaeologists routinely measure. Why do archaeologists find these attributes important? What information do they hope to gain from them?

2 Identify an archaeological problem that might interest you. What attributes of archaeological materials might be useful for your research problem? Why would you select these attributes as opposed to any others that might be possible?

2

Data

Quantitative methods and statistics are applied to *data*. Data are observations, not things. Data are not artifacts. They are not pots or stones or bones or any other component of the phenomenological world. We build data by making systematic observations on pots and stones and bones. What constitutes data is determined by our research questions and theoretical perspective. We create data to serve a purpose defined by a pre-existing intellectual framework. Most certainly, the real world exists in terms of various arrangements of matter and energy, but that real world is not to be confused with the observations that we make about it.

In addition to the theoretical perspective that we bring to bear and the research question we address, the tools with which we look at the world also influence what our data look like. As Gulliver's travels taught us, the world looks very different to Lilliputians and Brobdingnags, and the view is very different when the instrument we hold to our eye switches from a telescope to a microscope. There is no "high court" of archaeologists that makes the rules about what kind of observations we are restricted to make or what tools we use to make them. Data are what we determine them to be. Certainly, there is a range of observations that many archaeologists agree are useful for addressing particular problems. Michael Shott (1994: 79–81), for example, outlines a "minimum attribute set" for flaked stone debitage that archaeologists have found to be consistently useful for answering the questions they frequently ask. His list includes the "usual suspects" of weight, cortex, platform angle and raw material, among others. Despite the utility of these attributes for addressing certain questions, we, as archaeologists, are by no means restricted to looking at the world from a single perspective or using only these attributes. Shott (1994) in fact discusses how scholars have employed these and other attributes using a variety of perspectives to answer many different questions.

Quantitative Analysis in Archaeology, Todd L. VanPool and Robert D. Leonard
© 2011 Todd L. VanPool and Robert D. Leonard

Because we create our data by making observations of the world, we must ensure that we build our data in a systematic manner. Otherwise our data do not measure the same thing(s). All measurements and observations must have unambiguous definitions and they must be consistently recorded in the same way in order to eliminate measurement bias, if the resulting data are to be analytically useful (Lyman and VanPool, 2009). The first step to making systematic observations about the world is specifying what *variables* our data measure. A variable is any quality of the real world that can be characterized as having alternative states. The color of soil is a variable, as is projectile point length. With the former, the color spectrum is partitioned arbitrarily into segments with labels. The labels can be commonsensical (e.g., reddish brown) or we can use a standardized labeling system with more rigorous definitions for each label; archaeologists often use Munsell color charts, with standardized labels such as 10YR 4/1. With respect to projectile point length, archaeologists typically use the metric system where length is arbitrarily partitioned into usefully sized segments. How the measurement used to characterize a variable is partitioned depends on our theoretical perspective, methodology, and research problem. For example, millimeters might be the perfect sized segment with which to measure the height of a ceramic vessel, but likely are inappropriate for measuring the size of a settlement or a grain of corn pollen. Furthermore, a different analytic framework might not consider ceramic vessels, settlement size, or corn pollen worthy of measurement for addressing the same research problem.

An individual observation of a variable is called a *variate*. If we measure the rim angles of 10 ceramic vessels, we have manufactured 10 variates. These variates constitute data. Again, note that data are not variables. They are the totality of our observations, or variates. Also note that the word "data" is plural. A single observation is a variate and multiple variates are data. It is consequently improper to refer to data as a singular object; your data "are".

Scales of Measurement

A factor that is central to properly constructing data is identifying at what scale we should measure data. When we create data our observations can be recorded in one of four measurement scales: *nominal*; *ordinal*; *interval*; and *ratio*. The statistical tools available for analysis differ depending upon which measurement scale is being used.

Nominal level measurement

Nominal levels of measurement exist either when we use numbers as labels (hence the term "nominal"), or when we use numbers to represent the abundance of a class of phenomena (i.e., counts). This can be confusing, but the literature refers to both uses of numbers as being nominal level measurement, so it is best to understand the distinction.

Using numbers as labels is common in the world; we only need to turn the TV channel to ESPN to see numbers as labels on volleyball players, football players, field hockey players, horses, racecars and innumerable other rapidly moving objects. These numbers constitute labels only, and anyone performing arithmetic operations on football players' jersey numbers is wasting his or her time. The numbers simply aren't useful *except* as arbitrary names. The differences among numbers don't reflect an increase or decrease in the specific variable. Archaeologists use numbers or letters quite frequently as labels to code data. A ceramic analyst might use the label "1" to designate partially oxidized sherd cross-sections (sherds that exhibit oxidation on their edges but not in the center of the paste), "2" to designate sherds that exhibit no oxidation at all, and "3" for sherds that are fully oxidized. These labels have no mathematical significance, and to perform meaningful arithmetic operations on them is impossible. Our fictitious analyst could have used labels such as Sally and Harry almost as easily.

There is a distinction, however, between how our analyst used numbers as labels and the use of numbers to label racecars. While there is no analytic utility to considering the abundance and distribution of racecars labeled "1" through "8" across racetracks, a researcher might be interested in the abundance and distribution of eight classes of ceramic firing attributes across the landscape. This information can be used to evaluate ideas about differences in technology and site function. Additionally, if variation is present across sites or site components in terms of class abundance, most analysts will ultimately seek to explain those differences.

A number of good statistical tests exist to determine if differences in such abundances are meaningful, including the chi-square test, which is one of the first statistical tests used in archaeology (Spaulding, 1953). In archaeology, nominal level data are typically attributes of qualitative variables. These kinds of variables are called *discrete* variables (they are also referred to as *discontinuous* variables, or *meristic* variables) because they reflect differences that are fixed in the sense that there are only a limited number of mutually exclusive possible outcomes. The analyst then determines which of these possible outcomes is applicable for each variate (e.g., the species of animal from which a bone originated). Counting is the only appropriate arithmetic operation to be used on discrete variables, and the values assigned are always whole numbers. For example, we cannot have 5.5 pieces of Edwards Plateau Chert on a site, nor can we have 20.3 bison bones. Common discrete variables for different kinds of artifacts include:

- Bone: species, skeletal element (e.g., tibia), presence or absence of burning, type of butchering marks.
- Ceramics: temper type, extent of paste oxidation, type of decoration, type of paint, presence or absence of design elements, cultural-historical type.
- Ground stone: type of abrasion, direction of abrasion, artifact shape, number of used surfaces.
- Flaked stone: raw material type, the presence or absence of edge wear, type of edge wear, number of flake scars, presence or absence of heat treatment.

Ordinal level measurement

Ordinal levels of measurement imply an observable order or degree along some defined dimension of a variable. Ordinal measurements are only possible when measuring variables that occur along some continuum. For example, length, size, and floor area are variables that can be measured using a continuum ranging from small to large. Ordinal measures divide this underlying continuum into segments that are assumed to be asymmetrical. That is, we cannot assume equal values for our units of measure. Ordinal measures come with labels like small, medium, large, and extra large, where we cannot assume that the difference between large and extra large is the same as the difference between small and medium. We either do not have the necessary instrument with which to measure the variable in equal increments or the phenomena we are measuring cannot be (or does not need to be for our research question) measured in such a manner. This happens often in archaeology, particularly when we are using one kind of data to infer past behavior, or another set of data.

For example, Grayson (1984: 96–111) has argued that counts of faunal specimens recovered from archeological sites are ordinal counts, at best. His arguments rest on the fact that the abundance of bones is often the only means we have of studying the relative contributions of different species of animals to subsistence practices. Only in rare circumstances can archaeologists recover all of the bones of all of the animals or even a single bone from each animal utilized at a settlement. As a result, the analysis of the faunal material can at best only indicate which animals did and did not constitute a large portion of the diet, but cannot indicate the absolute number of any animal species consumed. Grayson concludes that both commonly used procedures for estimating the number of animals from a faunal assemblage – the minimum number of individuals (MNI) and the number of identified specimens (NISP) – produce ordinal measures of abundance. The same holds true for both palynological and ethnobotanical measures as well.

Settlement hierarchies are another example of ordinal measurement in archaeology. For example, Di Peso *et al.* (1974) argued that Medio period (AD 1200–1450) sites in northern Mexico could be divided into three tiers reflecting their relative size (Cruz *et al.*, 2004). At the top of the settlement hierarchy was the large regional center, Paquimé, which was characterized by multistory construction, public and ceremonial architecture, platform and effigy mounds, and craft specialization. Di Peso and his colleagues estimated more than 2000 people inhabited Paquimé. Subordinate to Paquimé were "second tier sites" that were roughly one-half to one-third the size of Paquimé that Di Peso believed were occupied by 500 to 1000 people. Di Peso further argued that smaller "third tier" sites, which he believed were occupied by 100 to 500 people, were subordinate to both Paquimé and the larger sites. The difference between the first, second, and third tier is not uniform, and each tier does not reflect a consistent unit of size.

In a similar manner, many researchers use rank orders to measure social complexity. Sahlins and Service's (1960) typology of bands, tribes, chiefdoms, and states is one such example. The difference between bands and tribes is not the same as the difference between chiefdoms and states. The frequencies of variates (cultures) in each class can be counted, and we can orient the classes from "More complex" to "Less complex". We cannot, however, add, subtract, multiply, or divide the classes (e.g., a band and a chiefdom cannot be added together to make a state).

Interval level measurement

With interval level measurements our continuum is partitioned symmetrically into even increments but with an arbitrary zero value. As we wrote the first draft of this paragraph we were in the middle of conducting fieldwork in the Chihuahuan desert near Galeana, Mexico. When the senior author stumbled out of his room in search of breakfast he noted that it felt hotter than the previous morning. This was an ordinal level observation, as having no thermometer he sensed that it was hotter, but he had no idea how much hotter it was. In contrast, a thermometer provides an interval level of measurement that is much more useful than the author's impression of relative coolness. The increments are symmetrical, and the difference between 40 degrees and 50 degrees is the same as the difference between 60 degrees and 70 degrees. The amount of difference among measurements can thus be specified in a way that is not possible when using measurements of an ordinal scale such as warmer and colder.

Common interval level measurements used in archaeology include:

- Direction using a compass.
- Years AD/BC. Each year reflects the passage of the same amount of time, thereby allowing time to be measured using a consistent measure, but year AD 1 is arbitrary in regards to time, and doesn't imply the beginning of time.
- Temperature.

Ratio level measurement

Ratio level measurement is the same as interval level measurement, with one exception: the addition of a true zero value. In the temperature example presented above, zero is an arbitrarily chosen value that does not mean that there is an absence of temperature, as illustrated by the differences between zero degrees in the Fahrenheit and Centigrade scales. Likewise, zero degrees on a compass does not mean an absence of direction. The compass's zero point is arbitrary and could just as easily be due south or northwest instead of north. In contrast, zero length of a piece of lithic debitage means just that, no length at all. Obviously, any piece of debitage measured must have a length value no matter how minute. Zero length represents the complete absence of length, though. The presence of a true zero value opens up our analysis to a broader assortment of mathematical and statistical procedures.

Consider for example the statement that, "a 4 cm long projectile point is twice as long as a 2 cm long projectile point." This observation is so obvious that it is trite. Replace point length with compass degrees (an interval measurement) so that the statement now reads, "a compass measurement of 4 degrees is twice as large as a compass reading of 2 degrees," and it becomes so obviously incorrect that it is laughable. The lack of an absolute zero value prevents interval data from being used to reflect any sort of "absolute" value or amount, but the presence of a zero value that actually reflects the absence of a trait (e.g., no weight) allows for interval data to be compared in such a manner. Nearly all measurements taken using the metric system are ratio measurements. Common interval measurements used in archaeology include:

- Artifact length, width, thickness, volume, and weight.
- Height and thickness of walls and other architectural features.
- Distance from a datum or other fixed point.
- Settlement size.

The relationship among the scales of measurement

We stress that the scale of measurement relates to our observations, not the variables *per se*. We can measure the same variable in different ways at different scales of measurement. For example, temperature can be measured on an ordinal scale (cold, temperate, hot), an interval scale (degrees Centigrade), and a ratio scale (degrees Kelvin, in which 0 refers to "absolute zero" and indicates the complete absence of molecular movement, which is what the variable "temperature" actually measures). Likewise, temporal relationships can be measured using ordinal scale measurements such as prehistoric and historic, interval levels of measurement such as years AD/BC, and at a ratio level using years BP (before present, defined as the number of years before 1950). Further, data can be measured at one scale but then organized for the purposes of analysis at other scales. Faunal data are typically initially recorded using nominal categories (species/genus present), but counts within each category can be used to create an ordinal measure (called a rank order) of species abundance of each species relative to each other. As mentioned in our discussion of Grayson (1984), these rank orders are typically ordinal measures, although archaeologists do sometimes (incorrectly) treat them as ratio scale data. Thus, determining the appropriate scale of measurement for a specific analysis requires the careful consideration of analytic goals, research methodology, disciplinary practices, and data.

There is no "right" scale of measurement, and we encourage you to not think of ratio measurements as "better" than ordinal level measures. The appropriate scale of measurement is determined by the research design. Sometimes we may only need to determine whether a site is prehistoric or historic. Other times, we may need ratio levels of measurement using years BP. Note though that ratio and interval levels of measurement can generally be transformed into ordinal and nominal levels of measurement, but it is rarely possible to transform a nominal or ordinal level of measurement into interval or ratio measures. It is possible for example to transform degree readings of directions taken with a compass into ordinal categories such as north and northwest. However, transforming ordinal measurements recorded as north, southeast, and so forth into compass degrees isn't possible. Thus it is generally preferable to take measurements at interval or ratio scales if not prohibitively expense or methodologically difficult.

The importance of the scales of measurement may not be intuitively obvious, but it is mathematically (and hence statistically) profound.

- Nominal variables can be counted, but no additional mathematical operations can be performed on them. For example, we might be able to count the number of known Zuni and Navaho sites in an area, but we cannot subtract, add, divide, or multiply categories such as Zuni and Navaho (e.g., a mathematic statement such as 3 Zuni – 2 Navaho = 1 Hopi doesn't make sense).
- Ordinal measures can be used to discuss the rank order reflected in data. In other words, we can compare the abundances of the various ordinal categories (e.g., a comparison of the number of small and large sites in an area), but, again, it is not possible to add, subtract, divide, or multiply the categories. We cannot reliably state that a small settlement is half as large as a medium-sized settlement and a quarter the size of a large settlement. The scale of measurement doesn't allow for such specificity.
- Interval levels of measurement do allow us to directly compare the magnitude of specific values, but only in limited ways. The difference between daytime temperatures of 100 degrees Fahrenheit and 50 degrees Fahrenheit is 50 degrees, but a 100-degree day is not twice as warm as a 50-degree day. We can consequently compare magnitudes of differences/similarities using a set scale, but not in terms of absolute magnitudes of an attribute.
- Like interval measures, ratio levels of measurement allow the direct comparison of the magnitude of difference among variates using a consistent scale, but the addition of a true zero value allows even more specificity. A pot that has a diameter of 10 cm is truly twice as wide as one with a diameter of 5 cm and five times as wide as a 2 cm diameter pot.

Many statistical applications assume interval and ratio level measurements. Unfortunately, archaeologists sometimes ignore the fact that many of their data are constructed in terms of ordinal or nominal level measurements, and consequently perform inappropriate analyses. This is the point of Grayson's (1984) discussion that we previously cited. Such analyses are unfortunately flawed and often meaningless, which relates to our next topic: *validity*.

Validity

Validity in measurement is ascertained by determining if we are measuring what we think we are, which is of course central to producing useful data. One dictionary provides this definition for valid: "well grounded or justifiable; being at once relevant and meaningful." This seems fairly straightforward; we are either measuring the volume of a pot or we are not.

Yet measuring the validity of data is frequently one of the hardest tasks facing archaeologists, because we are often using one variable as a proxy of another (e.g., scraper edge angle as a measure of scraper use). Most of our measurements of artifacts, features, and other aspects of the archaeological record are really taken to tell us about past human behavior, as opposed to being a study of the materials themselves. For example, arguments for the presence of social differentiation at the Copper Age cemetery at Varna, Bulgaria, are based on variation in grave goods, especially the presence of several burials with comparatively large amounts of grave goods, some containing hundreds of gold artifacts. Underlying these arguments is the premise that grave goods reflect social status in some way. The amount and type of grave goods is consequently being used as a measure of social status. Obviously, the validity of this measure must be evaluated, which is not an easy or straightforward theoretical and methodological issue (see Rakita and Buikstra, 2008).

Problems with validity can be evident even when trying to study attributes of artifacts. A researcher studying the importance of food storage in the Jomon culture of Japan might consider using pot frequency and volume as a measure of the importance of storage, based on the premise that the increased frequency of large pots reflects increased storage. This would require a measurement of changes in the number of pots and their volumes used during different time periods and at different sites. One way to detect such changes might be to compare the number and volumes of all of the whole pots recovered from various time periods, which perhaps will tell us during which periods people made and used large ceramic pots, presumably for more storage. But, then again, maybe not. Perhaps the large vessels are weaker than smaller vessels, and therefore are more likely to break. Given that pots from earlier time periods have had longer to break, the frequency of large pots in older assemblages may systematically be underrepresented relative to more recent periods. And what if the time periods are of different lengths, which might create differences in the numbers of whole pots even if there isn't any change in storage behavior? And what if some sites are in plowed fields while others aren't? Would this impact the frequencies of whole pots that the archaeologist would recover? What are the implications of all of these considerations?

Our point is that determining validity is not quite as simple as it seems. Can pot frequency and volume be used as a measure of the importance of storage? That is, is it "well grounded or justifiable; being at once relevant and meaningful?" Many archaeologists think so, and they are probably right, but a logical argument has to be made for the validity of the measurement and its application in a specific archaeological context. Importantly, the establishment of validity is a logical argument, not a mathematical one, or one made by reference to authority. In many instances, especially where we are restricted to surrogate measures of what interests us, we have to draw the conclusion that our measures are correlated sufficiently with the variable that really interests us so that we are willing to proceed. A couple of more examples may illustrate this issue further.

A good example where validity has been carefully evaluated comes from dendroclimatology and its use in reconstructing past temperature and rainfall patterns. There is a well-established correlation in many tree species between tree-ring width and specific environmental factors such as temperature and the amount of water available. Archaeologists can therefore use the widths of the rings of trees recovered in archaeological sites and elsewhere as surrogate measures for the environmental variables we want to measure. Despite a number of criticisms, enough arguments have been presented regarding the validity of dendroclimatology that most archaeologists are quite comfortable with it.

However, some issues of validity have not been so easy to address. The validity of measures used to estimate population size is one such intractable case. Archaeologists often wish to know population sizes of the occupants of houses, settlements, regions, or even continents, but our populations are generally long gone (although archaeologists studying the historic and modern records can sometimes rely on census data or direct counts). What surrogate measures can we use? A wide variety has been offered including room numbers, site numbers, size of habitation areas, and site areas (Paine, 1997). Many arguments have been made for and against the validity of each of these measures, which is very healthy for our discipline; the continued assessment of validity is one of the functions of the scientific enterprise. However, no single method has been universally accepted. When estimating population size, and indeed when measuring any archaeologically relevant variable, the validity of the selected measure must be properly established.

Accuracy and Precision

The battle isn't over, though, just because validity has been ascertained . We must focus on the *accuracy* and *precision* of our instruments (Lyman and VanPool, 2009). We define these terms following Sokal and Rohlf (1981: 13); "Accuracy is the closeness of a measured or computed value to its true value; precision is the closeness of repeated measurements of the same quantity." The terms thus are not synonymous. A biased instrument (e.g., a scale that adds 3 g to every measurement) could produce precise but inaccurate values. A device that produces random values could, at least occasionally, give accurate results, but they will not be replicated through repeated measurements.

In general, we wish to attain both accuracy and precision at a level that is consistent with our desire for useful data. For example, a well-functioning electronic distance measurer (EDM) provides accurate and precise measures of distance and elevation from a given point (a datum). Further, an EDM is both more accurate and precise than an optical transit, but we are still quite comfortable with maps generated by optical transits, suggesting that error isn't as important here as convenience. The EDM simply makes it easier to survey across long distances and uneven ground than the optical transit, and automatically calculates true distance without putting the users through the horror of remembering high-school trigonometry. It is consequently more precise and accurate than the optical transit, which is prone to user error. This implies that the EDM supplies accuracy and precision greater than we actually need, which is wonderful unless the greater accuracy and precision costs more than it is worth.

But how do we ascertain the appropriate level of accuracy and precision? In other words, how consistent should our measurements be to be adequately precise? And how close do our measurements need to be to the "true value" to be accurate? The answers to these questions depend again on our theoretical and analytic structure. Returning to our example of the size of a settlement, trying to get accurate and precise measurements of the surface area of a site to the nearest square millimeter is probably excessive and costly. Instead, measurements to the nearest square meter, or, for extremely large settlements, the nearest square kilometer are adequate. In contrast, measurements of the diameter of decorative glass beads used in jewelry will require a much smaller unit of measurement to be accurate and precise.

There is no "hard and fast" rule about measurement units that allow accurate and precise data. The analyst must chose a unit of measurement that will allow meaningful variation in the data to be evident, but that is not so detailed as to require excessive time and money (which are often the same thing in archaeology) to measure. A common rule of thumb is to create measurements in which the number of unit steps from the smallest to the largest measurement is between 30 and 300. For example, imagine we are measuring the weight of projectile points, with the heaviest and lightest points weighing 6.3 g and 1.6 g. Measuring the weight of the points to the nearest gram will result in a total of five measurements (2 g, 3 g, 4 g, 5 g, and 6 g), which likely will not give us an adequate understanding of the variation we seek to explore. However, if we measure the projectile points to the nearest .1 g, we will have a total of 47 unit steps (possible measurements), which is *likely* to be more useful for our analysis. When measuring settlement size, square millimeters is an inappropriate unit of measurement because it would likely result in differences of tens of thousands of square millimeters between sites. These differences aren't going to be analytically useful at such a fine scale. Likewise, too large a scale will result in the loss of meaningful variation within our data. Choosing some intermediate value will allow us to efficiently gather analytically useful data, relative to extremely small or excessively large units. We stress, though, that the 30 to 300 unit "rule of thumb" is exactly that, and should never be slavishly followed when research questions or theoretical structures demand more or less exact measurements.

Populations and Samples

An additional distinction that is significant when evaluating data is whether they represent a population or a sample of a population. A *population* is the totality of phenomena about which we wish to draw conclusions. A *sample* is some subset of that population.

Populations and samples are both finite in time and space, and are defined by the investigator. Ideally, we would like to measure the totality of the phenomena about which we wish to draw inferences (i.e., ideally we would like to work with a population). This is typically impossible for archaeologists, and, in those rare cases where it might be possible, is often expensive and unnecessary. As a result, archaeologists typically analyze samples in order to understand the characteristics of the population about which they are interested. For example, a researcher may wish to draw conclusions about the physical characteristics of Achulean handaxes without making observations on the totality ever excavated. We may consider the total excavated as the population, and look at only a small subset of them to draw conclusions about the population as a whole. We use statistics to analyze information gathered on samples to make inferences about the population. Note that we defined the population as all Achulean handaxes ever excavated. Our imaginary researcher, however, is probably interested in more than those excavated. His or her population of interest is probably really all Acheulean handaxes ever made, or at least made within a region or during a given time period. All of the Acheulean handaxes ever excavated is actually a sample of all of the Acheulean handaxes ever made because it is unlikely that we have recovered every single Acheulean handaxe. The same is true when analyzing houses, pots, or gold earrings. The likelihood that archaeologists recovered *everything* of *anything* is remote. Therefore, we sample. Thankfully, statistical techniques are available that allow us to make useful inferences about the population from our sample. Drawing inferences from a sample to characterize a population is actually one of the main purposes of statistics.

What constitutes an adequate sample size is case-by-case specific, and procedures will be introduced later on in the book that will help you to make such determinations. In general, however, the more homogeneous the population is, the smaller your sample size can be. A physician can draw inferences concerning your blood chemistry from a single blood sample because blood is a homogeneous fluid that does not differ from body part to body part. Much of the archaeological record is similarly redundant, whereas other segments are extremely heterogeneous.

So, how do we work from a sample to a population? This is the subject matter of the remainder of the text.

Practice Exercises

1 Define and differentiate the following terms:

   (a) population and sample.

   (b) data, variable, and variate.

   (c) validity, accuracy, and precision.

2 Briefly define and differentiate between the four scales of measurement (nominal, ordinal, interval, and ratio). Identify potential measurement scales for the following commonly measured archaeological variables: site age, site size, ceramic type, type and amount of grave goods, number of habitation rooms, ceramic vessel form, distance to nearest permanent water source, bone length, pot morphology, blade length, obsidian source, age at death.

3 Determine the commonly used measurement scales for the five attributes of artifacts or features you identified for Question 1 from Chapter 1's practice exercises. Can any of the measurements be recorded at more than one measurement scale? If so, what are the implications of analyses of the attributes based on the use of different measurement scales?

4 Determine potentially valid attributes for measuring the characteristics you identified as likely useful for addressing an interesting archaeological problem in Question 2 of Chapter 1's practice exercises.

   (a) To the best of your ability, logically defend the validity of these attributes in regard to addressing your research topic.

   (b) Identify possibly useful scales of measurement for the potentially useful attributes. What implications does the scale of measurement have on the attributes' validity, if any? (Would the attribute be a valid measure at one scale but less useful if measured at another scale?)

   (c) What units would be useful for measuring the potentially valid attributes at an appropriate measurement scale?

   (d) How would you go about evaluating the accuracy and precision of your measurements of the attributes? Do you know of any common problems with accuracy and precision that archaeologists have recognized with the attributes you selected?

5 Find an article from an archaeological journal such as *American Antiquity, Antiquity, Journal of Field Archaeology*, or *Journal of Archaeological Science* that utilizes summary tables and statistical analysis of numeric data. Briefly answer the following questions to the best of your ability:

   (a) What are the variables being measured by the author(s)?

   (b) At what scale are the data measured? Are the data continuous or meristic?

   (c) Are the data a valid measure of the phenomenon that the author(s) wishes to study?

   (d) Are the techniques used to gather the data precise and accurate?

6 How many unit steps would most likely be useful for measuring the cranial capacity of skeletal remains if the smallest specimen has a cranial capacity of 1120 cc and the largest specimen has a cranial capacity of 1420 cc?

3

Characterizing Data Visually

A picture is worth a thousand words.

Data are important because they allow ideas about the world to be proven wrong. In science, hypotheses are constructed and evaluated using data. While Thomas Henry Huxley actually wrote the apt words, "The great tragedy of Science – the slaying of a beautiful hypothesis by an ugly fact," he could have just as easily written that, "the most elaborate of hypotheses can be brought down by the most modest of data." Equally important is the fact that new research questions and hypotheses often follow from a careful analysis of data.

The acquisition of data begins with a question, and is followed by a series of observations that provide us with the information necessary to address the question within our theoretical structure. For example, if we wish to know if education influences voting patterns, we must gather information on voting patterns and education. The same holds true if we wish to know what characteristics of the landscape influenced people's subsistence strategies, and as a consequence, the artifacts we encounter as archaeologists; we must gather data on characteristics of the landscape and artifacts, features, ecofacts, and other data sources useful in reconstructing subsistence strategies from various locations. The immediate product of our problem-driven observations is a simple list of numbers or qualities (i.e., variates) that are data. Often, especially with larger sets of data, it is impossible to fully identify trends or patterns from the raw data themselves. We must instead characterize them so that we can identify trends without having to remember every single variate.

The summary or description of the data should take two forms that we see as sequential steps to the effective use of data in analysis. The first step is *always* the construction of a visual representation of the data, which is nothing more than the creation of an effective graphical summary. The second step is a numerical description of the data, which is the topic of Chapter 4. Here we focus on the first of these steps, the visual characterization of data, because visual representations of the data provide information that is useful during the numerical summary, as well as aiding our intuitive understanding of the patterns that structure our data. They are also a great way to identify simple coding and measurement errors that cause spurious data (e.g., failing to add the decimal to a measurement of 1.05 cm) (Lyman and VanPool, 2009).

Let us consider first a situation where data are of a variable that is continuously measured, providing us with a simple list of numbers such as those presented in Table 3.1. The data are the depth of features observed on the Carrier Mills archaeological project (Jefferies and Butler, 1982), which was recorded to help determine feature function.

Do the data make much sense? Can you identify general trends of any sort? Probably not. It would be a truly gifted individual who could discern a pattern from this jumbled bunch of numbers. As fine an instrument as the human brain is, it has a difficult time organizing these numbers in any meaningful way without help. The difficulty of dealing with data sets is especially illustrated by the fact that there are only 91 variates here, many fewer than constitute many archaeological data sets.

Some individuals might be tempted to simply run head on into an advanced statistical analysis of these data without a second thought. That would be a mistake. Each statistical method is developed for specific situations and requires its own assumptions about the characteristics of the data. While the specific assumptions associated with each statistical method will be discussed in future chapters, it is important to understand that the patterns in the data themselves, in conjunction

Table 3.1 Carrier Mills feature depths

Original measurements (cm)						
6	5	22	8	23	11	10
8	4	10	6	13	20	24
16	13	13	16	19	21	22
6	16	3	9	13	6	10
5	6	13	9	11	20	11
12	11	10	18	24	11	14
7	9	9	19	15	22	15
6	17	7	4	24	17	10
20	9	9	23	10	7	12
10	13	10	16	14	20	12
14	20	19	15	8	16	17
5	22	16	12	11	8	7
13	3	8	18	9	9	19

with the questions we ask, determine which statistical procedures are useful. Statistics are like a fine piece of clothing that must be tailor fitted for an individual. How can a suit or dress be altered if you don't know the shape of the individual who will wear it? Therefore, the best place to start any statistical analysis is with a simple visual description.

Frequency Distributions

A good starting point for visually representing archaeological data sets measured at the ratio or interval scales of measurement is a *frequency distribution*. Figure 3.1 is a frequency distribution of the data presented in Table 3.1. The continuous variable of feature depth has been divided into 22 *classes* (represented in the column with the heading "*Y*") with a *class interval* of 1 cm.

In Figure 3.1, *Y* represents the variable "feature depth". Scanning down the column, we see that the values range from 3 to 24 cm. Each of these possible values

Y	Implied limit	Tally mark	f
3	2.5–3.5	‖	2
4	3.5–4.5	‖	2
5	4.5–5.5	‖‖	3
6	5.5–6.5	‖‖‖ ‖	6
7	6.5–7.5	‖‖‖	4
8	7.5–8.5	‖‖‖‖	5
9	8.5–9.5	‖‖‖‖ ‖‖‖‖	8
10	9.5–10.5	‖‖‖‖ ‖‖‖	8
11	10.5–11.5	‖‖‖‖ ‖	6
12	11.5–12.5	‖‖‖	4
13	12.5–13.5	‖‖‖‖ ‖‖	7
14	13.5–14.5	‖‖	3
15	14.5–15.5	‖‖	3
16	15.5–16.5	‖‖‖‖ ‖	6
17	16.5–17.5	‖‖	3
18	17.5–18.5	‖	2
19	18.5–19.5	‖‖‖	4
20	19.5–20.5	‖‖‖‖	5
21	20.5–21.5	‖	1
22	21.5–22.5	‖‖‖	4
23	22.5–23.5	‖‖	3
24	23.5–24.5	‖	2
			$\Sigma f = 91$

Figure 3.1   Frequency distribution of Carrier Mills feature depths (cm)

is a class. These classes represent the possible values within which the variates that comprise our data vary. The values in the column titled *Implied limit* represent the resolutions of our measurements. For example, a variate with a value of 3 cm has a true depth that is between 2.5 and 3.5 cm. The *Tally marks* represent the number of times each value was observed and noted in Table 3.1. For example, the value 3 was observed two times. The column under the symbol *f* represents the *frequency* with which each value was observed and is calculated by summing the tally marks. *f* contains the same information as the tally marks, but in a different form. You have no doubt noticed the menacing presence of the figure Σf in the lower left corner of Figure 3.1. While the symbol looks intimidating and complicated, Σ is merely the Greek capital letter sigma, and represents the term for summation. This benign symbol will be used throughout the rest of the book. Σf simply represents the summation of all of our frequencies for the variable *Y*, or 91 $(2 + 2 + 3 + 6 \ldots + 2)$.

The implied limits raise an important point for data collection: when collecting data, we round to the closest value we have defined. Thus when we say that a feature has a depth of 5 cm, we are really saying that its true depth is closer to 5 cm than it is to 4 cm or 6 cm. Having a feature that is *exactly* 5 cm thick, not a micrometer more or less, is unlikely. The implied limits consequently identify the range of actual values represented by the data.

Understanding implied limits clarifies the understanding of our data, and provides analytic flexibility to represent data at a resolution that is analytically appropriate. Sometimes we measure data at a scale that is more exact than we really need, and we wish to reduce the numbers of classes considered (e.g., measuring feature depth to the nearest millimeter but then deciding that measurements to the nearest centimeter are more appropriate for our analysis). In such a case, we can easily collapse our old classes into new, larger classes. For example, classes of 1.1 and 1.2 cm could be collapsed into a single class of 1 cm, which can include everything from .5 cm to 1.5 cm. We can round numbers using the implied limits of our new classes; values that fall within the implied limits of each new class become members of that class. Thus analysts can chose the numbers of classes that they find useful and modify their visual representations accordingly while clearly indicating the new classes' resolution.

But what is to be done with a value at the edges of the implied limits between classes? For example, into which class would we place a value of 1.5 cm if we were rounding to the nearest centimeter? Does it belong to the 1 cm or 2 cm class, given that 1.5 is within the implied limits of both classes? When gathering data, the problem of such borderline cases causes people to become frustrated, ask for second opinions, or go to the bar for a drink. In reality such an intermediary value can be placed into either of the two classes because it is equally close to both of them, but that still doesn't help us tally class frequencies, given that the variate can only be represented in one of the two classes. We could arbitrarily choose one of the two possible classes, but this creates the problem of potential bias. Sure, values of 1.5 cm and 2.5 cm could be grouped in the class of 2 cm, but consistently assigning these border values to this class as opposed to 1 cm or 3 cm could artificially elevate the class's frequency.

Statisticians have given considerable thought to this problem and have devised the following rule of thumb (Sokal and Rohlf, 1996: 15). In cases where a digit is followed by a 5 that is either standing alone or followed by zeros, it is rounded to the nearest even number. For example, if we are rounding to the nearest centimeter, the value of 1.49 cm would be rounded to 1 cm, the values of 1.51 cm and 1.50 would both be rounded to 2 cm. Likewise, we would round 2.49 cm to 2 cm and 2.51 cm to 3 cm, but 2.50 cm would be rounded to 2 cm (just as 1.50 cm was), not 3 cm. This approach, often called the Banker's rule, is different than the "rounding up" rule commonly taught in math classes, in which a value of 5 is rounded to the next highest value. The Banker's rule is based on the rationale that about half the time the number will be rounded up, and half the time it will be rounded down. The number of values increasing is therefore roughly equal to the number of values decreasing, which will help prevent any distortion to statistical measures such as the average. Consistently rounding up in contrast will actually artificially elevate such measures.

The last digit of any variate should always be *significant*; that is, it should always indicate that the true range of the variate is halfway above and below the class unit. When measuring to the nearest centimeter, our values must be reported as 0 cm, 1 cm, 2 cm, etc. up to the largest variate, not as 0.0 cm, 1.0 cm, etc. Values such as 1 cm and 2 cm indicate that their implied limits are .5 cm larger and smaller than the classes (e.g., the class 5 cm has an implied limit of 4.5 cm to 5.5 cm). Values such as 1.0 cm and 2.0 cm in contrast indicate our data reflect measurements to the nearest millimeter with implied limits .05 cm above and below each class (i.e., a class of 1.0 cm has an implied limit of .95 cm to 1.05 cm). In regards to our data, 2 cm is different than 2.0 cm, which is different than 2.000000 cm, in that each value indicates that our data reflect measurements using progressively smaller units (centimeters vs. millimeters vs. micrometers). Adding excess numbers (including zeros) onto our data is confusing, and prevents the creation of meaningful frequency tables.

Returning to Figure 3.1, the tally marks provide a visual perspective on how the data are *distributed* across the variable classes. This perspective is called a *distribution*, and tally marks are one way of creating it. A visual inspection of the distribution created by the tally marks shows us that the observations (data) cluster toward the classes 9 through 13 cm with fewer variates toward the top and bottom of the table.

Figure 3.2 shows the data in a slightly different manner. Rather than employ a class interval of 1 cm, a class interval of 2 cm is used, halving the number of classes. The result is a more clearly clustered set of tally marks. The tendency evident in Figure 3.1 for most observations to fall toward the center of the distribution and fewer toward the extremes is accentuated in Figure 3.2. Because two values from Figure 3.1 are collapsed for each class in Figure 3.2 (e.g., the values of 3 and 4 now constitute only one class in Figure 3.2, where they were two classes in Figure 3.1),

Practical limit	Implied limit	Class mark	Tally mark	f
3–4	2.5–4.5	3.5	IIII	4
5–6	4.5–6.5	5.5	IIIII IIII	9
7–8	6.5–8.5	7.5	IIIII IIII	9
9–10	8.5–10.5	9.5	IIIII IIIII IIIII I	16
11–12	10.5–12.5	11.5	IIIII IIIII	10
13–14	12.5–14.5	13.5	IIIII IIIII	10
15–16	14.5–16.5	15.5	IIIII IIII	9
17–18	16.5–18.5	17.5	IIIII	5
19–20	18.5–20.5	19.5	IIIII IIII	9
21–22	20.5–22.5	21.5	IIIII	5
23–24	22.5–24.5	23.5	IIIII	5
				$\Sigma f = 91$

Figure 3.2 Frequency distribution resulting from grouping depths for Carrier Mills features into 11 classes: Interval = 2 cm

Practical limit	Implied limit	Class mark	Tally mark	f
3–7	2.5–7.5	5	IIIII IIIII IIIII II	17
8–12	7.5–12.5	10	IIIII IIIII IIIII IIIII IIIII IIIII I	31
13–17	12.5–17.5	15	IIIII IIIII IIIII IIIII II	22
18–22	17.5–22.5	20	IIIII IIIII IIIII	15
23–27	22.5–27.5	25	IIIII I	6
				$\Sigma f = 91$

Figure 3.3 Frequency distribution resulting from grouping depths for Carrier Mills features into five classes: Interval = 5 cm

it is no longer possible only to present these values under the column for the variable "*Y*" as in Figure 3.1. Instead, the *class mark* is used. The class mark represents the midpoint of the values combined into each of the 11 classes. The implied limits represent the new levels of accuracy, and the *practical limits* represent the limits of actual values that are encountered in Table 3.1. Once again, *f* represents the frequency of observations for each class, that is, a counting of the tally marks. Σf represents the sum of all frequencies in the figure. Figure 3.3 is similar to Figures 3.1 and 3.2 except that the class interval is 5 cm, and the class marks have been changed accordingly.

The differences in Figures 3.1, 3.2, and 3.3 show that the number of classes and the size of the intervals affect the shape of frequency distributions. Excessive numbers of classes tend to spread out the distribution and can obscure interesting patterns. Too few classes cause too many variates to be lumped together, which can also obscure meaningful patterns. Somewhere between these extremes is an adequate number of classes that will illuminate the characteristics of the distribution effectively.

Unfortunately, there is no hard and fast rule to determine the ideal number of classes. A general rule of thumb is that 10 to 20 classes provide a good representation of most distributions, although some data sets may be so large and varied that many more than 20 classes are needed. Researchers should try a number of class intervals to determine which interval best depicts the data. In our example, Figure 3.2 is the best. Figure 3.1 is quite spread out, and Figure 3.3 is so clustered that useful information is likely obscured. The investigator must consciously evaluate the number of classes and the class interval that are most useful, but oftentimes the decision is more art than science. While many computer programs use algorithms that make good decisions regarding class intervals using a variety of techniques, it's best not to slavishly follow their direction. Most good programs allow the analyst to manually specify the number of classes.

A common way of combining classes when making a frequency table is to reduce the number of significant digits. Thus, values such as 1.7 cm and 2.3 cm are collapsed into a single class of 2 cm, which in turn reduces the number of significant digits from two to one. Frequency distributions should always reflect the correct number of significant digits so the reader properly understands the exactness with which the data are grouped into classes, *even if the original data were measured using more exact units.*

Significant digits may seem self-evident, but they can be strangely confusing to people. How many significant digits does 200 have? The answer is, "It depends." In a data set measured to the nearest hundred (e.g., 100, 200, and 300), it has one significant digit, because the zeros trailing the 1 are simply placeholders that don't really reflect the exactness of the measurement (i.e., the value 200 reflects variates ranging from 150 to 250). In contrast, in a data set of 200, 202, and 199, it has three significant digits, because the zeroes do reflect the exactness of the measurement (i.e., the value 200 reflects variates ranging from 199.5 to 200.5). This and other "ambiguities" often cause confusion that leads to muddled analyses and thinking.

Why should we care about properly understanding significant digits and keeping them constant in a table, graph, or other quantitative context? The reason is that numbers with different significant digits are different, even when they superficially appear the same. For example, if you remember the discussion of implied limits, you know that 1 is not the same as 1.0; 1 has implied limits of .5 to 1.5 whereas 1.0 has implied limits of .95 to 1.05. This is a significant difference, and the careless addition of extra zeros changes the implied exactness of a variate. Likewise, needlessly eliminating significant digits (e.g., shortening 1.00 to 1) "discards" information that might be meaningful (to the analyst or the reader).

Here are some useful rules to keep in mind when dealing with significant digits:

- All non-zero digits are significant, as are zeros appearing between numbers (e.g., 123 and 103 both have three significant digits).

- All zeros and non-zero numbers appearing to the right of a decimal point are significant if there are non-zero numbers to the decimal's left (e.g., 103.04 and 103.00 both have five significant digits).
- Zeros to the right of a number without a decimal point that are not bracketed by non-zero values (typically) are not significant (e.g., 100 has only one significant digit, whereas 100.5 has four). We note as previously mentioned that values such as 100 could have three significant digits if they actually are measured to the nearest digit in the one spot.
- Zeros to the right of a decimal point but to the left of the first non-zero value that do not have non-zero values to the left of the decimal point are not significant. Thus, .0067 has only two significant digits (the zeros before the 6 aren't significant), but 1.0067 has five significant digits. The zeros in .0067 aren't significant because they are considered "placeholders" that reflect the absolute magnitude of a value, just like the zeros in 100 do.
- Avoid superfluous exactness. Dividing numbers can lead to long trails of decimals. For example, does a hunter included in an ethnoarchaeological study who captures 1100 calories in 1.75 hours of hunting really gain 628.571429 calories per hour? No, because the exactness of our measurements wouldn't allow the researcher to differentiate between 628.571428 and 628.571429 calories per hour. Instead, a measure of 630 calories per hour more accurately reflects the exactness of our measurements, and will consequently be more useful in the analysis and more easily communicated to other researchers. As a general rule, don't end up with more significant digits than the original measurement.

In order to demonstrate how frequency distributions are constructed, Figure 3.4 presents the framework for the construction of a frequency distribution of heights of individuals in your class (assuming you are using this text as a textbook). Should your instructor choose to do so, you may construct your own frequency distribution.

Histograms

Another common way of visualizing data is through the construction of a histogram. Histograms graphically represent the frequency of data in various classes by listing the class marks on the horizontal axis and class frequencies on the vertical axis. A histogram looks similar to tally marks from a frequency table laid on their side. Let us use an example of data gathered in the Gallina area of New Mexico to illustrate the histogram.

Table 3.2 presents 100 measurements of minimum sherd thickness measured to examine differences in ceramic technology across settlements in the project area. These measurements were taken using 0.5 mm increments. Thus, the value 4.5 mm reflects a sherd with a minimum thickness between 4.25 and 4.75 mm.

Implied limit	Y (inches*)	Tally mark	f
57.5–58.5	58		
58.5–59.5	59		
59.5–60.5	60		
60.5–61.5	61		
61.5–62.5	62		
62.5–63.5	63		
63.5–64.5	64		
64.5–65.5	65		
65.5–66.5	66		
66.5–67.5	67		
67.5–68.5	68		
68.5–69.5	69		
69.5–70.5	70		
70.5–71.5	71		
71.5–72.5	72		
72.5–73.5	73		
73.5–74.5	74		
74.5–75.5	75		
75.5–76.5	76		
76.5–77.5	77		
77.5–78.5	78		
78.5–79.5	79		
79.5–80.5	80		
			$\Sigma f =$

* 1 inch = 2.54 cm

Figure 3.4 Frequency distribution of heights in your class

Table 3.2 Minimum ceramic sherd thickness (mm) from the Gallina region of New Mexico

5.0	3.0	5.0	6.0	5.0	5.0	6.0	5.0	4.0	5.0
4.5	5.0	4.0	5.5	5.5	5.0	6.0	6.0	4.5	6.0
4.0	2.0	4.5	3.0	4.5	4.0	5.5	6.0	5.0	5.0
4.5	5.5	5.0	5.0	5.0	4.0	5.0	6.0	3.5	6.0
6.0	3.0	5.0	3.0	5.0	5.0	5.0	4.5	7.0	4.0
7.0	6.0	5.5	4.0	5.0	5.5	5.0	5.5	4.0	5.0
7.5	4.0	5.0	4.5	5.0	5.0	5.5	5.5	5.0	4.0
5.0	4.0	3.5	4.0	3.0	5.5	5.0	5.0	5.0	5.0
4.5	5.0	4.0	6.0	5.5	7.0	5.0	3.0	6.0	5.0
4.0	5.5	4.0	2.0	5.5	2.0	6.5	5.0	4.0	5.0

Again, patterns in the raw data are difficult to identify. To make sense of our data, we must first organize them. Figure 3.5 presents a frequency distribution of these data. Once a frequency distribution is constructed, it is quite easy to construct a histogram – a visual that is slightly more aesthetically pleasing (and publishable) than the tallies, lines, and numbers in Figure 3.5. Figure 3.6 is a histogram of the data presented in Table 3.2.

For Figure 3.6, the metric scale of measurement is depicted on the horizontal axis of the graph. The number of variates in each class is counted, and the counts of variates are depicted using vertical bars called *bins*. For example, there are 35 variates with a minimum sherd width of 5.0 mm. The bin associated with the class of 5.0 mm consequently rises to the level associated with a frequency of 35 variates. This histogram provides an excellent visual presentation of the shape, or

Implied limit	Y	Tally mark	f
1.75–2.25	2.0	lll	3
2.25–2.75	2.5		0
2.75–3.25	3.0	lllll l	6
3.25–3.75	3.5	ll	2
3.75–4.25	4.0	lllll lllll lllll l	16
4.25–4.75	4.5	lllll lll	8
4.75–5.25	5.0	lllll lllll lllll lllll lllll lllll lllll	35
5.25–5.75	5.5	lllll lllll lll	13
5.75–6.25	6.0	lllll lllll ll	12
6.25–6.75	6.5	l	1
6.75–7.25	7.0	lll	3
7.25–7.75	7.5	l	1
			$\Sigma f = 100$

Figure 3.5 Frequency distribution of the Gallina ceramic minimum sherd thicknesses (mm)

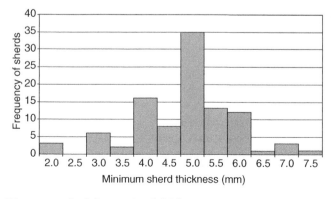

Figure 3.6 Histogram of minimum sherd thicknesses

distribution, of the data. We can see that the distribution is roughly symmetrical, and that values toward the center are more common than extreme values. By grouping the observations into categories of a consistent interval we have been able to gain information about the distribution of our variates that could not be obtained through a simple visual inspection of the raw data.

Visualize how an increase or decrease in the class interval would affect the shape of such a distribution. A larger class interval would result in fewer bins (bars), and they would tend to have more variates in each, causing higher peaks. A smaller class interval would leave many empty bins and result in a more "spread out" distribution with few members in each bin. In both cases, the distribution would remain roughly symmetrical.

Stem and Leaf Diagrams

Another useful way to view data is through the construction of a stem and leaf diagram. Figure 3.7 is a stem and leaf diagram of the Gallina sherd thickness data (Table 3.2). The diagram looks complex, but is easy to build and is very informative, once you know how it is constructed. The *stem* is the vertical column of numbers to the left of the vertical line. The *leaf* refers to those numbers to the right of the vertical line. Each leaf, along with the associated stem to its left, reflects an individual variate. In this particular example, the stem represents the number to the left of the decimal point in our data. The smallest value is 2.0, and the largest 7.5. Therefore, the limits of our stem are 2 (top) and 7 (bottom). The leaf reflects the value to the right of the decimal point in the data. In the Gallina data, the numbers to the right of the decimal points are all 0 or 5. We therefore know that each leaf will be 0 or 5. By looking at the stem and leaf diagram, we can both reconstruct our raw data and look at their distribution. For example, we can tell from the diagram that three variates are 2.0 mm, six are 3.0 mm, and two variates have a value of 3.5 mm.

The stem and leaf can be constructed by reading across your data from any direction, which results in the leaves being unordered with respect to magnitude, as in Figure 3.7. Reordering the leaves as depicted in Figure 3.8 often makes the

2	000
3	00500005
4	5055000050005050050000
5	00050500050500500005500050550050005500000000000
6	0000005000000
7	0500

Figure 3.7   Stem and leaf diagram of Gallina ceramic data

2	000
3	00000055
4	00000000000000055555555
5	000000000000000000000000000000000005555555555555
6	0000000000005
7	0005

Figure 3.8 Ordered stem and leaf diagram of Gallina ceramic data

2	000
2	
3	000000
3	55
4	0000000000000000
4	55555555
5	00000000000000000000000000000000000
5	5555555555555
6	000000000000
6	5
7	000
7	5

Figure 3.9 Stem and leaf diagram created using increments of .5 cm

diagram easier to use. Regardless of the reordering, the shape of the distribution is the same.

In the same way that it is useful to change the numbers of classes in a frequency distribution, it is sometimes desirable to change the numbers of classes in a stem and leaf diagram. Doing so is quite easy and can be accomplished by dividing the stems further or collapsing them together. For example, Figure 3.9 is a stem and leaf diagram of the Gallina data, but it differs from Figure 3.7 in that each stem is repeated twice (there are two 2s, two 3s, two 4s, etc.). The first stem of 2 includes all leafs between 2.0 and 2.4, whereas the second 2 reflects all leafs between 2.5 and 2.9. The same is true for all of the other number pairs. Leafs that are 0 are consequently grouped with the first in a pair of stems, whereas leafs that are 5 are grouped with the second. Using this same principle, we can divide the stem and leaf diagram into as many or as few classes as we wish. The only requirement is that each stem reflects the same increment of measurement.

Knecht (1991) provides an additional set of data with which to construct a stem and leaf diagram (Table 3.3). These data are maximum thickness measurements of a sample of losange-shaped (leaf-shaped) Early Upper Paleolithic projectile point bases. Use these data to place the leaves on the two stems provided in Figure 3.10.

4	
5	
6	
7	
8	
9	
10	
11	

4	
4	
5	
5	
6	
6	
7	
7	
8	
8	
9	
9	
10	
10	
11	
11	

Figure 3.10 Stems that can be used to create a stem and leaf diagram of data from Table 3.3

Table 3.3 Measurements of the maximum thickness of losange-shaped Early Upper Paleolithic projectile points

6.5	6.6	9.2	9.8	8.9
8.0	8.4	7.7	8.7	10.7
9.7	9.2	8.3	10.1	10.3
6.9	7.4	5.6	10.0	7.9
7.3	4.1	10.2	11.2	7.7
8.3	7.9	6.1	7.2	9.7
5.2	9.3	10.8	8.9	10.0
10.8	10.7	11.4	9.3	9.0
7.7	10.5	7.4	11.4	9.2
9.5	10.3	8.8	10.0	10.7
7.7	8.9	10.2	5.7	8.5
11.1	10.0	10.5	11.4	7.1
5.8	9.0	8.9	9.1	8.4
7.4	9.0	6.5	6.2	9.2
9.6	6.2	7.9	9.1	7.8
4.3	11.6	4.9	5.8	5.8
5.9	8.4	6.2		

Source: Knecht, Heidi Deborah (1991). *Technological Innovation and Design during the Early Upper Paleolithic: A Study of Organic Projectile Technologies*, pp. 628–30. Ph.D. Dissertation, Department of Anthropology, New York University.

Ogives (Cumulative Frequency Distributions)

Another kind of frequency distribution that is very useful is the cumulative frequency distribution or ogive. Figure 3.11 is a cumulative frequency distribution for
the Carrier Mills feature data (Table 3.1). It is similar to the previously illustrated
frequency distribution (e.g., Figure 3.2) except that the frequency corresponding to
a given class is created by summing the frequencies of all of the smaller classes with
that of the class itself. For example, for $Y = 4$, f (cumulative) is equal to the sum of
its frequency ($f = 2$) added to the frequency for the previous value of $Y = 3$ ($f = 2$).
Therefore, f (cumulative) for $Y = 4$ is equal to $2 + 2 = 4$. The f (cumulative) for the
last class, where $Y = 24$, is the sum of all of the class frequencies, or 91. Figure 3.12
provides a graphic presentation of Carrier Mills feature data (Table 3.1). The cumulative frequency distribution is often useful when we want to compare the shapes
of two distributions by contrasting their cumulative values visually.

Y	f	f (cumulative)
3	2	2
4	2	4
5	3	7
6	6	13
7	4	17
8	5	22
9	8	30
10	8	38
11	6	44
12	4	48
13	7	55
14	3	58
15	3	61
16	6	67
17	3	70
18	2	72
19	4	76
20	5	81
21	1	82
22	4	86
23	3	89
24	2	91
	$\Sigma f = 91$	

Figure 3.11 Cumulative frequency distribution of Carrier Mills feature depths (cm)

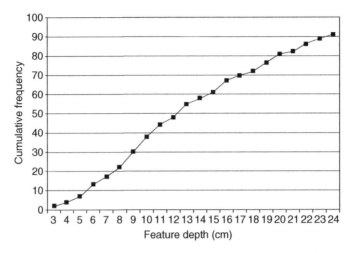

Figure 3.12   Plot of the cumulative frequency distribution of Carrier Mills feature depths

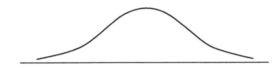

Figure 3.13   Plot of a symmetrical, or normal distribution

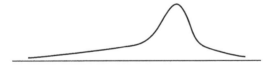

Figure 3.14   Plot of a left-skewed distribution

Describing a Distribution

There are a number of ways to characterize a distribution. A good place to start is by describing its degree of symmetry. Figure 3.13 presents a distribution that is perfectly symmetrical. This distribution may be more familiar to you as the *normal* or *bell-shaped distribution*, which we will discuss in greater detail later.

Non-symmetrical distributions are described as *skewed*. Figure 3.14 presents a distribution that is skewed to the left and Figure 3.15 is a distribution skewed to the right, a very common distribution in archaeological data. Many people find it difficult to remember the difference between distributions skewed left and distributions skewed right. Remember, a distribution is skewed in the direction with the longest tail.

We may also be concerned with the number of peaks that a distribution has. Figures 3.13 through 3.15 all have one peak and are therefore called *unimodal*

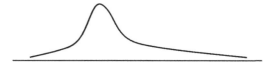

Figure 3.15 Plot of a right-skewed distribution

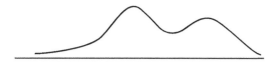

Figure 3.16 Plot of a bimodal distribution

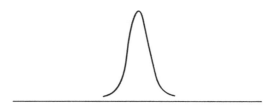

Figure 3.17 Plot of a leptokurtic distribution

distributions. Figure 3.16 presents a plot of a distribution that is *bimodal,* because it has two peaks. Distributions that have more than two peaks are typically called *multimodal.* Note that we are using the word "mode" in a different way here then that with which you may be familiar. A mode can also refer to a statistical measure of location, as measured by the class in a distribution with the highest frequency (see Chapter 4). Here we are using the word to refer to a general assessment of the number of peaks a distribution has. This may be confusing at first, but it is important to know both usages.

A characteristic that archaeologists often overlook when describing distributions is their relative heights. Often a distribution will superficially look like another distribution, but will in fact be more spread out or more narrowly distributed. Given similar numbers of variates, a narrow distribution will have a higher peak, whereas a more dispersed distribution will have a lower peak. For example, the two distributions look similar, but the distribution illustrated in Figure 3.17 has a higher peak than the distribution in Figure 3.18. Of course there are terms used to describe these differences. The high peak of the distribution in Figure 3.17 is called *leptokurtic,* and the flat peak of the distribution in Figure 3.18 is *platykurtic.* The classic bell-shaped curve depicted in Figure 3.19 is *mesokurtic.* If you use these words at parties, your friends will be impressed.

Figure 3.18   Plot of a platykurtic distribution

Figure 3.19   Plot of a mesokurtic distribution

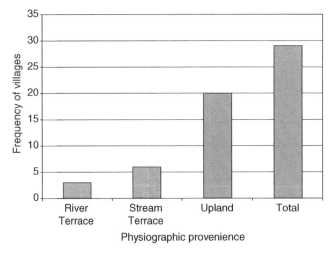

Figure 3.20   Bar chart of the physiographic provenience of villages during the Early Monongahela period in the Lower Monongahela and Lower Youghiogheny river basins

Bar Charts

All of the visual representations described to this point are useful for interval or ratio scale data, but not ordinal or nominal scale data. A very useful tool in archaeology for characterizing ordinal and nominal data is the *qualitative frequency distribution*, more commonly called a *bar chart*. Figure 3.20 provides a bar chart of the number of villages during the Early Monongahela period occupation of the Allegheny Plateau in several physiographic settings (Hasenstab and Johnson, 2001:6). Bar charts provide abundance information on variables measured at the nominal and ordinal scale. As with histograms, the bar chart allows the comparison of frequencies among classes. For example, it becomes obvious from Figure 3.20 that sites in the upland provenience are about four times more common than those on stream terraces, a trend that might not be so clearly evident from the raw data. Unlike histograms, though, bar charts cannot be described in terms of symmetry, peakedness, or modality, because the order that the data are entered from left to right is arbitrary. For example, river terrace sites listed on Figure 3.20 could just as easily be placed to the right of stream terrace sites as to their left. Even when using bar charts to illustrate the abundance of ordinal scale data, for which the order of classes is set, the symmetry, peakedness, or modality of a bar chart cannot be described, because the units do not reflect a consistent amount of variation. A bar chart reflecting a preponderance of "medium-sized" artifacts could indicate that the distribution of artifact size is symmetrical, with fewer large and small artifacts. Or it could reflect that the "medium-sized" category includes more variation, and therefore more variates, than the other size categories. It is important not to confuse bar charts with histograms, and one way to keep them distinct is to present gaps between the bars in the bar graph (e.g., Figure 3.20), and to close the bars together in the histogram (e.g., Figure 3.6).

Bar charts are particularly useful when we wish to compare differences within classes for multiple variables. For example, Figure 3.21 is a bar chart with the number of villages dating to the Early Monongahela, the Middle Monongahela, and the Late Monongahela/Protohistoric periods in different physiographic settings. By examining the relative heights of the bars, we might be able to identify changes in settlement patterns. For example, upland sites are common in all three time periods, but are more common during the Middle Monongahela period when compared to the Late Monongahela period, which has about an equal number of villages in the upland and stream terrace proveniences. Further, villages on river terraces are absent during the Late Monongahela period, which is not the case for the other

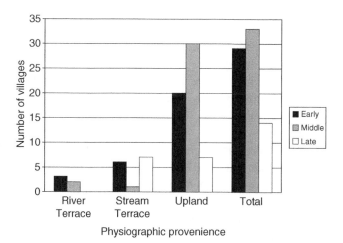

Figure 3.21 Bar chart of the frequencies of villages during the Early Monongahela, Middle Monongahela, and Late Monongahela periods in various physiographic proveniences

periods. The statistical significance of such differences can be evaluated using the various methods for nominal data analysis discussed later in Chapter 13.

Displaying Data like a Pro

Although this chapter has focused on graphs, we also presented tables in the form of frequency tables. Archaeologists rely on data tables because our data are too numerous and too complicated to present systematically in an understandable format without such tools. Interestingly, however, archaeologists rarely consider the attributes that make a table more or less useful. Requirements for making good tables (i.e., ones that effectively communicate the pertinent information) would seem to be commonsense, but even the most casual evaluation of the numerous tables in any given archaeological journal or technical report are likely to identify tables that: (i) fail to clearly present the trends/characteristics the author wants them to; (ii) undermine the author's argument in that they suggest relationships contrary to the author's description because of some error in the data reporting or presentation; and (iii) effectively misrepresent the data in such a way as to appear to support the author's argument as a result of some clever (or possibly unintentional) error.

Archaeology of course is not alone in facing the specter of bad tables. Because of their general importance in many disciplines, statisticians have developed several rules of thumb to help create good tables. Ehrenberg (1981) outlines six useful suggestions for table construction. These are:

1   Provide column/row totals or averages as appropriate. These will help the reader identify trends more easily than jumbles of numbers listed next to each other.
2   Order the rows/columns of a table according to size so that the reader can identify trends in the data. Keep the order the same when presenting multiple tables with the same (or similar) classes.
3   Place digits to be compared into columns, instead of rows.
4   Round to two significant digits.
5   Organize the table to guide the eye to the pertinent comparisons.
6   Give a brief verbal summary of the trends and exceptions that the reader ought to identify from the table.

Consider Table 3.4, which reports summary information for four sites. What trends do you see? We doubt that you don't see any. If you do, we suspect that you had to look at the table for a bit and mentally move some numbers around to see if some relationship that you suspect might be there actually is. If you have read very many archaeological reports or sat through presentations at professional archaeology conferences, you have likely seen a table that looks exactly like this. This table may accurately report data, but it isn't as effective as it could be if one were to follow Ehrenberg's (1981) "commonsense" suggestions.

Table 3.4 Summary feature information for four sites

	1	2	3	4
Average number of hearths in each house	2.45	3.34	1.79	3.01
Average number of storage pits per house	3.32	12.97	1.23	7.81
Average house size (sq. m.)	367.34	741.82	157.98	642.80

Table 3.5 The reorganized summary feature information

Site	Average house size (sq. m.)	Average number of hearths per house	Average number of storage pits per house
Justin Ruin	740	3.3	13.0
Agave Junction	640	3.0	7.8
Timber Ruin	370	2.5	3.3
Casita	160	1.8	1.2
Average	480	2.6	6.3

Reorganizing the table as Ehrenberg (1981) proposes creates Table 3.5. We suspect that the typical reader will find this table much easier to read, and will almost immediately notice that villages with larger than average house size also have more hearths and storage pits per house. Certainly the typical person will identify this relationship more quickly from Table 3.5 as opposed to Table 3.4. Several features make the second table more effective. First, the numbers to be compared are arranged in columns with the values arranged from the largest to the smallest. The columns also end with the average value from the four sites, which provides the reader with a "scale" to compare against the numbers presented above it. The variables have been rearranged to further help guide the reader's eye to the pertinent relationships, with the average house size being the first variable the reader encounters, and, as a result, the variable against which the trends in the other data are compared. Further the column and row labels have been clarified and the numbers rounded to two significant digits to make the table simpler to read. Mixed with the straightforward statement about the relationship between the average house size and the other variables, the trends in the data become instantly recognizable. The data in Tables 3.4 and 3.5 are the same, but Table 3.5 is much more effective at communicating them (see Lewis, 1986: 278–82 for a similar exercise with different data).

As useful as Ehrenberg's (1981) suggestions are, it is important to remember that they are rules of thumb and should only be used when they improve the table. At times, it might be beneficial to switch the order of data from table to table, if this helps to clarify the patterns that you are illustrating. In other contexts column and row averages/totals might not be useful, or even possible if you are dealing with individuals or nominal scale data. The construction of a table is as much a matter of art as it is quantitative rigor. The goal is to communicate information accurately. Whatever form does this best is, by definition, the best table.

The purpose of the table can help determine what form might be best in a given context. Lewis (1986: 277) defines three types of tables commonly used in archaeology: formal tables (which communicate research results to the reader); working tables (which help the analyst investigate a data set's structure); and reference tables (which are an archive of a data set). Formal and working tables are tools used by the archaeologist to explore and communicate information relevant to an argument. Reference tables in contrast are passive storehouses often presented as appendixes that store data for further analysis. Placing numbers to be compared in columns is an excellent suggestion for working and formal tables, but might be impractical for reference tables, given their size and the fact that the author is not trying to draw attention to particular comparisons. Instead of ordering classes according to their magnitude, it might be better to present data in reference and working tables using other criteria such as specimen number, room block, project number, stata, temporal period, or cultural affiliation. Rounding figures to two significant digits also may be inappropriate for working and (especially) reference tables, if it causes the loss of information that would interest those who wish to analyze the data from their own perspectives or using other methods. Ehrenberg's suggestions should consequently be disregarded if necessary, especially when compiling reference tables.

Archaeology and Exploratory Data Analysis

Frequency tables and the various graphing techniques presented above can be used to summarize and present data in order to clarify our thinking about the possible relationships among our data and to illustrate patterns we wish to communicate to others. These techniques are consequently an ideal starting place in statistical analysis precisely because they summarize data in a useful way that may in fact cause us to note possible relationships worth exploring. Inductively identified relationships may factor heavily into future analysis, and are in truth the source of some of the most significant anthropological insights archaeology has contributed. After all, it is exactly this sort of reasoning that led archaeologists such as William Flinders Petrie (1899) and James A. Ford (1938, 1962) to use graphing methods to develop frequency and occurrence seriations (O'Brien and Lyman, 1999).

Using graphs to identify otherwise "hidden" relationships that can become the subject of further analysis is an example of a more general statistical approach called exploratory data analysis (EDA). In general, quantitative methods can be used in two ways. The "typical" approach, which forms the core of this and most other quantitative texts, is called statistical hypothesis testing (or confirmatory data analysis). This tradition has certainly been the focus of archaeological analysis since the introduction of the hypothetico-deductive approach focused on deductively testing archaeological propositions, which was central to the New Archaeologists of the 1960s and 1970s (Watson *et al.*, 1971). We will discuss hypothesis testing in great detail in Chapters 6 and 7, but this framework employs statistical analysis to evaluate explicitly defined propositions that are presented prior to the application of statistical methods. Quantitative analysis then determines if the proposition is plausible or not at a given probability level.

In contrast, EDA uses quantitative approaches not to evaluate previously defined relationships, but to define hypotheses to be evaluated. This approach is perhaps most commonly associated with John Tukey (1977), but it actually has a long history in archaeology (e.g., the previously mentioned seriation techniques used by cultural historians, Albert Spaulding's (1953) proposal that statistical methods can be used to inductively define artifact types, and F. Fenenga's (1953) use of weight to differentiate between arrow and atlatl dart points based on the bimodality evident in graphs of projectile point weight). EDA has proven useful in archaeology to help maximize insights into complex data sets, determine which variables might be important in given cases, and to detect outliers and anomalies that both might complicate traditional hypothesis testing and that might reflect behaviorally meaningful differences in the archaeological record (e.g., Baxter, 1994).

We will discuss EDA in further detail at various places in this text but we do wish to note several of its characteristics here. First, EDA is not a series of unique statistical methods distinct from those used for hypothesis testing, but is instead a different application of the same techniques. Although some methods such as the graphs discussed in this chapter are particularly useful for EDA, virtually any statistical technique *could* be used if you are sufficiently clever.

Second, EDA is not so much an alternative to hypothesis testing, but is instead a means of complimenting it by providing a useful source for possible hypotheses. EDA isn't *the* way archaeologists should use quantitative analysis, because archaeologists wish to derive and test hypotheses. Likewise, archaeologists should not limit their application of quantitative methods only to hypothesis testing, because they are interested in discovering previously unknown relationships. Using EDA and traditional hypothesis testing in tandem can allow archaeologists to accomplish both tasks. Care must be taken to insure, though, that the process of generating and testing hypotheses does not become an exercise in circular reasoning. Evaluating a hypothesis with the same data from which it was derived using EDA is tautological, and may reify relationships that are accidental, as opposed to meaningful. EDA is a great place to start, but you must evaluate the significance of relationships identified using EDA with other lines of information and/or data. After all, even randomly generated numbers can occasionally create what appears to be a strong relationship. The presence of a relationship discovered using EDA does not constitute adequate proof that it is in fact meaningful.

Finally, EDA, especially when based on graphs, tends to include a subjective component resting on the analyst's interpretation of patterns. What one individual sees as "a clear pattern" can be another's "ambiguous mess of data." Many archaeologists might agree on the presence of some relationships, but it is entirely possible, in fact expected, that analysts will disagree at least some of the time when conducting EDA. In contrast, if we accept the applicability of a given method and analytic framework during hypothesis testing, the statistical outcome should be unambiguous and universally accepted in regards to the specific hypothesis a researcher is evaluating.

Now that you know how to characterize basic data visually, let us turn to procedures that allow archaeologists to make numerical comparison. That is the subject of Chapter 4.

Practice Exercises

1 Define and differentiate the following terms:

   (a) bar charts and histograms

   (b) practical limits and implied limits

   (c) class interval and class mark

   (d) stem and leaf (in the context of a stem and leaf diagram)

   (e) skewed and symmetrical distributions

   (f) unimodal and multimodal distributions

   (g) leptokurtic, platykurtic, and mesokurtic distributions

   (h) formal tables, working tables, and reference tables.

2 Would the class 2 cm include the value of 1.7 cm? Would the class of 2.0 cm include the value 1.7? What is the difference between the classes of 2 cm and 2.0 cm?

3 Maximum depths (cm) of archaeological features discovered at Charlestown Meadows in Westborough, Massachusetts, are listed below (Hoffman, 1993: 468).

20	7	30	10	5	9	13	26
15	21	12	7	16	19	37	15
5	20	15	18	22	23	7	15
5	11	21	17	11	26	15	11
20	30	9	5	13	43	10	10
40	28	15	8	11	15	34	5
13	28	8	11	8	29	25	31
29	9	4	20	11	13		

Using these data, complete the following tasks:

(a) Create a stem and leaf frequency distribution from these data. Make an ordered array from your stem and leaf distribution. Describe the distribution.

(b) Construct a frequency distribution. Be sure to include your implied limits, practical limits and class marks.

(c) Reduce the number of classes by half and reconstruct the frequency distribution.

(d) Double the number of classes and reconstruct the frequency distribution.

(e) Identify which of the three frequency tables best illustrates the shape of the distribution. Draw a histogram of the best frequency distribution.

4 Create an ogive from the data used in Question 3.

5 Following is a summary of the time periods reflected at various Paleolithic sites discussed by Runnels and Özdoğan (2001:72). Prepare a bar chart reporting the frequencies of remains from each time period within the four regions. Describe any differences you note in the frequencies of materials from the three time periods among the regions.

Region and site	Lower Paleolithic	Middle Paleolithic	Upper Paleolithic
Black Sea Coast (Asia)			
Kefken	–	X	X
Sarisu	–	–	X
Ağva	–	X	–
Domali	–	X	–
Black Sea Coast (Europe)			
Domuzdere	X	X	X
Gümüsdere	–	X	–
Ağaçli	X	X	X
Bosphorus			
Kemerburgaz	–	X	–
Göksu	X	X	–
Marmara			
Akçaburgaz	–	X	–
Eskice Sirti	X	X	–
Karababa	–	X	–
Davetpaşa	–	X	v

6   Find five tables from archaeological reports or published works. Based on the context in the author's discussion, determine whether each table is a formal table, a working table, or a reference table. Next evaluate whether the tables follow Ehrenberg's (1981) suggestions for creating effective tables. Could the tables be improved to be more effective for their stated purposes? If so, how?

4

Characterizing Data Numerically: Descriptive Statistics

While visual representations of data are very useful, they are only a starting point for gaining more information about using the data. Numerical, as opposed to visual, characterizations of the data allow a more formal means of both describing a distribution and comparing two or more distributions. Numerical description ultimately allows us to make the inferences we desire. These numerical characterizations are termed either *statistics* or *parameters*. *Statistics are descriptions of the characteristics of a statistical sample, whereas parameters refer to the characteristics of a statistical population.* A *statistical population* is similar to a population as presented in Chapter 2 in that it is defined in time and space, but differs in that it is composed of data (i.e., observations of things) as opposed to the objects themselves. Do not confuse statistical populations with the use of the term "population" in biology, physical anthropology, or even in Chapter 2. Statistical populations are not people, animals, rocks, or pots. They are data.

For example, the length of Folsom points manufactured during the Paleoindian occupation of the New World could be considered a statistical population, but this is distinct from the actual population of Folsom points. The weight of thimbles used in New York during the Historic period is another possible statistical population. As you can see, statistical populations can vary in their scope from data about all of a class of objects that ever existed to data for a more limited subset of objects. We could even define a statistical population at the scale of a site or component (e.g., the cutting dates of wooden beams used during the construction of a historic church).

A statistical sample is a subset of the data in a statistical population. An important point to remember from Chapter 2 is that the archaeologist defines populations

and samples. When collecting data, the archaeologist also defines statistical populations and statistical samples that correspond with the parent populations or samples
that interest him or her. Yes, it is confusing to use the terms "sample" and "population" to refer to both a group of objects and the data that are collected from them,
so we must take care to define how the terms are being used in each case. Even
when we define a population, say Mimbres pots from Dana Ana County in New
Mexico, we have not defined the variables we are going to measure, and hence do
not have a statistical population. Only by defining the variables of interest and then
collecting data about them can we create statistical populations. Our data cannot
and should not be equated with the empirical objects that we are actually studying.
Christopher Chippindale (2000: 605) provides an insightful discussion of this fact,
and argues (tongue in cheek?) that archaeologists should replace the term "data"
with "capta" to emphasize that data "are captured by the researcher, who seeks to
grasp from the material record the essentials of some complex and little-known
phenomenon."

Any particular group of data could be a statistical sample or population, depending on the question. For example, the grade point average of undergraduate anthropology majors at Eastern New Mexico University could be considered a population,
if the group of interest is anthropology majors from that school, or a sample, if we
are interested in all undergraduates from the institution.

Despite the importance of maintaining the distinction between populations/
samples and statistical populations/samples, we, along with most researchers, find
the repeated use of the terms "statistical population" and "statistical samples" irritating. As a result, we will simply shorten these down to population and sample,
and specify that we are talking about statistical populations and samples from here
on out, unless otherwise noted.

We can describe a population's parameters (e.g., the average value, the maximum
value, the minimum value) without error, assuming that the data are derived using
a method that is adequately precise and accurate. These values cannot change for
the simple reason that the population contains all of the data that is to be considered; no matter how many times a person calculates the average height of the 2008
US Olympic basketball team, the outcome will always be the same. Only the units
of measurement might change (e.g., meters vs. feet). Assuming that the statistical
population does in fact satisfactorily reflect the population of objects being studied
(i.e., the data are properly derived and recorded using appropriate tools), the
parameter is a perfect (in the sense of not having error) description of some
attribute of the objects.

In an ideal world, archaeologists could always measure their populations of
interest to derive statistical populations. Typically though, archaeologists can only
access a sample of their population. This in turn necessitates that they use statistical
samples to derive statistics that estimate the parameters of the statistical populations
(which in turn correspond with the populations of interest). For example, we might
measure the cranial capacity of a sample of Neandertal skulls using one or more

statistics, and then use these statistics to estimate population parameters and draw inferences regarding the population of Neandertal skulls of interest. In contrast to the immutable, unchanging parameters of a population, though, statistics will vary as the members in the sample vary (e.g., samples composed of different Neandertal skulls will likely provide a slightly different estimate of the population parameters). Statistics derived through the analysis of a sample are thus estimates, which can be very good or quite poor, of the actual population parameters. Using various quantitative methods, we can determine how close these statistics likely are to the population parameters that we are trying to measure.

Numerical descriptions of distributions can either be parameters or statistics depending on whether they describe a population or a sample. By convention, population parameters are denoted with Greek letters, and sample statistics by Roman letters. Numerical descriptions can also be categorized into two types: *measures of central tendency* and *measures of dispersion*. Measures of central tendency, as the name suggests, provide a numerical account of the center, or midpoint of a distribution. Measures of dispersion provide information about the spread of data around a distribution's midpoint.

Measures of Central Tendency

Measures of central tendency are also called *measures of location*, which may actually be the better term in that these measures provide a numerical account of the location of the center of a distribution. Three measures of central tendency are commonly used: the *mean*; the *median*; and the *mode*.

Mean

Everyone is familiar with this measure of central tendency, which is also called the arithmetic mean, or the average. The mean is calculated by summing all of the values in a data set, and then dividing this sum by the number of observations.

To illustrate the calculation of the mean, consider the data in Table 4.1 that Jim O'Connell (1987: 85) collected in an ethnoarchaeological study on the number of residents in Alyawara camps. The mean is calculated by following the instructions signified by the symbols in Equation (4.1).

Table 4.1 Number of residents in Alyawara camps

21	18	12	31	23	31	44	7

The mean

$$\mu = \frac{\sum_{i=1}^{i=n} Y_i}{n} \text{ or, more simply, } \frac{\sum Y_i}{n} \text{ for populations}$$

(4.1)

$$\bar{Y} = \frac{\sum Y_i}{n} \text{ for samples}$$

The formulas look menacing but they are not, once you understand how to read them. μ (vocalized as *mu*) represents the population parameter of the mean of the variable Y. $\bar{Y}$ (vocalized as *Y-bar*) is the mean of a sample, and is an estimate of μ. In this case, $\bar{Y}$ refers to the average number of residents of Alyawara camps recorded by O'Connell from Gurlanda A in August, 1971, and from Bendaijerum Ridge in May, 1974, which is a sample of modern Alyawara camps. The Greek symbol Σ is the friendly symbol for summation introduced in Chapter 3. What might not be familiar is the notation above and below Σ, and the subscript to the right of Y.

$\sum_{i=1}^{i=n} Y_i$ represents the following set of instructions: for the variable Y, beginning with the first value of Y (symbolized by $i = 1$), sum all values continuing to the last value of Y (symbolized by $i = n$). Y_i (vocalized as *Y sub i*) simply refers to all values of Y, each of which can be numbered individually using the data from Table 4.1, as follows: $Y_1 = 21$, $Y_2 = 18$, $Y_3 = 12$, ..., $Y_8 = 7$. For the set of instructions $\sum_{i=2}^{i=4} Y_i$ we begin summation with $Y_2 = 18$, and continue to sum through $Y_4 = 31$, or $18 + 12 + 31 = 61$.

The left term of the formula for μ in Equation (4.1) shows the full symbolism for the calculation of the mean. The term on the right denotes the same thing, but implies all of symbolism associated with the term to the left. Outside of introductory texts such as this one, people rarely express the formula's full symbolism. Instead, people tend to use the abbreviated version for ease. So, when you encounter the symbol ΣY_i, it can be assumed that the instructions are to sum all values of Y, beginning with the first, and continuing to the last.

The calculation of the mean in our Alyawara example is as follows.

$$\bar{Y} = \frac{21+18+12+31+23+31+44+7}{8}$$

$$\bar{Y} = \frac{187}{8}$$

$$\bar{Y} = 23.375$$

The mean, or average, number of people in Alyawara camps is 23 people.

Median

The median of a data set is the variate with the same number of observations of greater and lesser value. Consider the following data: 14, 15, 16, 19, 23. The median value is 16, because it is bounded on both sides by two variates. Consider an additional set of variates that consist of an even number: 14, 15, 16, 19. Unlike the example above, no single number has an equal number of variates numerically greater and lesser than it. Is the median 15 or 16, or does this set of numbers simply have no median? No, in this case the median is determined by averaging the two values in the middle of the distribution. Here, the median is the average of 15 and 16, which is 15.5.

To determine the median of the Alyawara data, we first make an ordered array that presents the data in Table 4.1 from the smallest to the largest variate: 7, 12, 18, 21, 23, 31, 31, 44. The median value for these data is the average of 21 and 23, which is 22 people.

Mode

The mode is the most popular, or most abundant, value in a data set (i.e., the value with the highest frequency). An inspection of Figure 3.7 shows that the most popular value in the Gallina ceramic data is 5.0. In the Alyawara data in Table 4.1, the mode is 31. It is possible to have two or more modes, if the most popular classes have the exact same number of variates.

Which measure of location is best?

For the Alyawara data, the average number of people in the camps is 23, the median is 22, and the mode is 31. Depending on the situation, all three measures may be useful, but oftentimes, one will be preferred over the others. In the Alyawara example, the mean and the median provide values that are reasonably close to one another, and both constitute good measures of central tendency. The mode in this instance, however, is not representative of the distribution's central tendency.

When a distribution is perfectly symmetrical, the mode, mean, and median are identical. In general, when dealing with a symmetrical distribution, the mean is the most useful measure of central tendency, followed by the median, and then the mode. The mean's preeminence is because of its utility in the analysis of variance and regression analysis, subjects of later chapters. The mean, however, can be inordinately influenced by extreme values, typically called *outliers*. As a result, its utility for describing heavily skewed distributions is questionable, causing many to prefer the median in such cases.

For example consider again the following values for the Alyawara data: 21, 18, 12, 31, 23, 31, 44, 7. Let us replace one of the variates of 31 with 100 so that the data are: 21, 18, 12, 31, 23, 100, 44, 7. In comparison with $\overline{Y} = 23$ for the original set of data, the mean of the revised data set is $\overline{Y} = 32$, a value that is larger than six of the eight variates. The mean no longer is "in the center" of the distribution. The median, however, is unaffected by the extreme value, and remains the same. It is in such cases where extreme outliers are present that the median becomes a better indicator of central tendency than the mean.

Measures of Dispersion

While measures of central tendency provide important information about a distribution's location along some measurement scale, they offer no information about the shape of that distribution. Two distributions might have the same location, but not resemble each other at all in terms of shape, or dispersion (e.g., Figure 4.1).

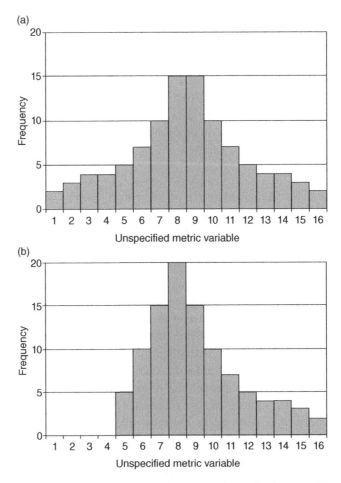

Figure 4.1   Two distributions with identical means and sample sizes but different shapes

Measures of dispersion will help us identify these differences. Three measures of dispersion are commonly used: the range; the variance; and the standard deviation.

Range

The range is the difference between the largest and smallest values in a set of data. For the Alyawara example:

Largest value	44
Smallest value	(−)7
Range	37

Although the range grants insight into the dispersion of the distribution of a sample, the sample range almost always underestimates the population parameter. It is unlikely, after all, that both the largest and smallest variates in a population will be selected in a sample because samples contain just a portion (and sometimes only a small portion) of a population. In archaeological contexts, the range is likely to be especially problematic given that imperfect preservation and recovery makes the likelihood of recovering the largest and smallest (or oldest and youngest, or heaviest and lightest, etc.) artifact or feature quite small. Still, a large sample comprising most of a population is more likely to give an accurate estimate of the population's range. Regardless of sample size, the range is greatly affected by outliers because it only takes into account two variates in the data set, the largest and smallest.

Despite its simplicity, the range is often a misunderstood statistic because of the different use of the verb and noun forms of the word *range*. With the Alyawara data, it is appropriate to state that the data range (verb form) from 7 to 44. However, *the* range (noun form) is 37, not 7 to 44.

Interquartile range

Interquartile range is a measure of variation that is closely related to the range, except that it attempts to measure the variation in variates towards the center of a distribution. It is calculated by subtracting the variate demarcating the lower 25% of a distribution from the variate demarcating the upper 25% of the distribution. This value in turn reflects the range of the middle 50% or the "body" of a distribution. To prevent confusion, the demarcation of the lower 25% of the distribution is called the *25th percentile* whereas the demarcation of the upper 25% is called the *75th percentile.*

Using the Alyawara data, 25% of the variates are equal to or less than 18 and 25% are equal to or greater than 31. Consequently the 25th percentile is 18 and the

75th percentile is 31. The interquartile range is 31 − 18 = 13. Thus, the variates that comprise the middle 50% of the Alyawara data differ by no more than 13 people.

Variance and standard deviation

The variance and the standard deviation are related statistics used to describe, and ultimately compare, the shapes of distributions. The range and interquartile range are useful, but they reflect the spread between only two variates. Given that a distribution is created by the values of all of the variates, it would be useful to have some measure of dispersion that reflects each of them. But how can such a measure be created?

One way to do this would be to measure the distance, or variation, of each variate from a fixed point and somehow express this variation in a meaningful way. Broadly dispersed distributions would have a large measure of variation whereas tightly distributed distributions would have low values. We could even compare the spread of the variates in one distribution with the variation present within other distributions. But what fixed point should be used?

A good choice would be one of the measures of location we just discussed. Comparing this central value to the variates would allow us to determine if the variates are spread widely or tightly clustered together. This could be very useful information for describing the shape of a distribution and comparing different distributions with each other. Thus, a useful way to measure the distribution of variates is to characterize their distance from the mean, which is a robust (but not the only) measure of central location.

Again using the Alyawara example, the distance or deviation from the mean for $Y_1 = 21$ is:

$$y = Y_1 - \overline{Y}$$

$$y = 21 - 23$$

$$y = -2$$

Note that the lower case y is used as the symbol of this deviation, and by convention, the mean is subtracted from the variate in order to provide the measure of distance.

Knowing the difference between Y_1 and $\overline{Y}$ is potentially useful, but we are concerned with the shape of the complete distribution. It comes to mind that perhaps if we sum all of the deviations of all variates from the mean, and divide this sum by the number of variates, we could create a kind of "average" deviation. Large values would indicate a broadly spread distribution, and small values a narrowly spread distribution. Unfortunately, this is not the case, as the sum of all deviations from the mean is equal to zero; the amount of deviation is equal on both sides of the mean, because of the way it is calculated, causing the deviation of values greater

Table 4.2 The sum of *y*, where $\bar{Y} = 23$ people

Size of Alyawara settlement (people)	y $(Y_i - \bar{Y})$
21	−2
18	−5
12	−11
31	8
23	0
31	8
44	21
7	−16
	$\Sigma y = 0$

and lesser than the mean to cancel each other out when summed. This fact is illustrated in Table 4.2. As you can see, the sum of *y* is 0.

The problem then, is not with the magnitude of the deviation, but with its sign. All of the plusses and minuses cancel each other out. One way to circumvent this problem is to square each deviation, which results in only positive numbers. This is precisely the solution used to calculate the *variance*, which is determined using Equation (4.2):

The variance

$$\sigma^2 = \frac{\Sigma y^2}{n} \text{ and } s^2 = \frac{\Sigma y^2}{n-1} \tag{4.2}$$

σ^2 is the symbol representing the population variance and s^2 represents the sample variance.

The variance is useful for many purposes, but remember that it is transformed by squaring the deviates before they were summed. Thus, it is not in the original units used to measure the distribution and often doesn't make much intuitive sense. For example, calculating the variance for the numbers of rooms in prehistoric settlements would result in a unit of "rooms^2", which isn't a meaningful unit. Likewise, when one calculates the variance of the Alyawara data in Table 4.1, its units are "people^2". We know what people are, but people^2 is not a generally acknowledged, meaningful unit for anthropological inquiry.

By taking the square root of the variance, we can return those squared values to their original units, providing another measure of dispersion that makes more intuitive sense – the standard deviation (σ for population parameter and *s* for sample statistic). The standard deviation is calculated using Equation (4.3).

The standard deviation

$$\sigma = \sqrt{\sigma^2} = \sqrt{\frac{\sum y^2}{n}} \text{ and } s = \sqrt{s^2} = \sqrt{\frac{\sum y^2}{n-1}} \tag{4.3}$$

Table 4.3 presents in table form the squared deviates used to calculate the sample variance and standard deviation for the Alyawara example, where $\bar{Y} = 23.375$. Column 1 of Table 4.3 presents each Y. Column 2 presents their frequency *f*. This column is necessary because the frequency of occurrence of a given Y may vary (e.g., the variate 31 appears twice in these data). If we counted each Y with multiple occurrences only once, we would incorrectly estimate the true variance. Column 3 presents *y*, which is $Y - \bar{Y}$, the deviation of the variate from the mean. Column 4 provides a solution to the problem with signs by squaring the deviations created in Column 3 as symbolized by y^2. Column 5 is the frequency of the occurrence of Y as presented in Column 2 multiplied by the squared deviations calculated in Column 4. This column is necessary in order to take into account the values of Y with more than one observation. The sum of Column 5 ($\Sigma[fy^2] = 973.9$) is also called the *sum of squares*. Using this value to solve for Equations (4.2) and (4.3) produces:

$$s^2 = \frac{\sum y^2}{n-1} = \frac{973.9}{7} = 139.1$$

$$s = \sqrt{\frac{\sum y^2}{n-1}} = \sqrt{139.1} = 11.8$$

We treated the data from Table 4.1 as a sample of the population of Alyawara camps occupied during the time when O'Connell conducted his research. You have probably noticed that the sample statistics are calculated by dividing the sum of squares

Table 4.3 Computations of the sample variance and standard deviation for the number of residents in Alyawara camps

(1)	(2)	(3)	(4)	(5)
Y	f	$y = Y - \bar{Y}$	y^2	fy^2
21	1	-2.375	5.641	5.641
18	1	-5.375	28.891	28.891
12	1	-11.375	129.391	129.391
31	2	7.625	58.141	116.281
23	1	-.375	0.141	0.141
44	1	20.625	425.391	425.391
7	1	-16.375	268.141	268.141
	$\Sigma = 8$	0		$\Sigma = 973.877$

by $n - 1$, not n as is the case when calculating population parameters. Through experimentation, it has been determined that dividing by n in a sample tends to underestimate the true variance and standard deviation, but that dividing by $n - 1$ provides a better estimate. Therefore, when calculating the population parameters of σ or σ^2, divide the sum of squares by n, but when calculating s or s^2, divide the sum of squares by $n - 1$.

The procedure illustrated in Table 4.3 is a terrific but cumbersome means of calculating the variance and standard deviation. One of the most pleasant characteristics of quantitative methods is that there are often simple ways to calculate otherwise complex calculations. Equation (4.4) offers an equivalent method of calculating the sum of squares that is less computationally intensive and time consuming than Table 4.3.

Calculation formula for the sum of squares

$$\sum y^2 = \sum Y^2 - \frac{\left(\sum Y\right)^2}{n} \tag{4.4}$$

For the Alyawara example:

$$\sum y^2 = 5345 - \frac{(187)^2}{8} = 5345 - \frac{34969}{8} = 5345 - 4371.1 = 973.9$$

This is the same result we calculated using Table 4.3.

We have chosen to present all arithmetic operations above because of potential confusion in reading Equation (4.4). The left-hand term of the equation $- \sum Y^2 -$ is an instruction to square each Y and then sum the resulting values, which in this case sum to 5345. The numerator of the right-hand term $- (\sum Y)^2 -$ is an instruction to sum all Ys, then square the resulting value, which equals 187 squared, or 34,969. The difference in notation is subtle, but extremely important.

The standard deviation we just calculated ($s = 11.8$ people) is an extremely useful measure of dispersion and will form the basis for describing the variation within the distribution and for comparing this variation to the variation in other distributions as will be discussed at length in other chapters. In addition, principles behind the calculation of the standard deviation provide extremely important conceptual tools for understanding many quantitative methods we will be learning through the remainder of this book.

Calculating Estimates of the Mean and Standard Deviation

Sometimes an archaeologist may want to quickly gain information about the location and spread of a distribution (e.g., at a professional presentation where one wants to explore some relationship in data being presented or at the start of analysis

Table 4.4 Denominators for deriving an estimate of the standard deviation

Sample size	Divide range by
5–29	3
30–99	4
100–499	5
500–999	6
1000+	6.5

where one intuitively thinks there might be a difference/similarity worth exploring further). A thorough and wise researcher may also wish to quickly check the mean and standard deviation calculations for possible errors caused by inevitable blunders created during the measurement or data entry process. Using the following procedures, we can calculate a quick estimate of both the mean and the standard deviation.

To estimate the mean, we can compute the *midrange*. The midrange is similar to the range except that the largest and smallest values are averaged rather than subtracted. In the Alyawara example, the midrange is $(44 + 7)/2 = 25.5$ people. This value is fairly close to the computed mean of 23.4 people and may serve as a reasonable estimate in a pinch. If the midrange is very different, you may want to recheck the mean's computation.

While estimates of the mean are easy to come by, a good estimate of the standard deviation is a bit more difficult. However, the standard deviation can be estimated by dividing the range, calculated as described previously, with the appropriate value from Table 4.4. Applying this method to the Alyawara data produces an estimate of $(37/3) = 12.3$ people, which is not too far off from the computed value of $s = 11.8$ people.

Coefficients of Variation

Oftentimes archaeologists calculate standard deviations to compare the spread of two or more distributions. In such cases distributions with larger variances and standard deviations relative to other distributions are thought to reflect populations with greater variation. This assumption seems intuitively obvious but is in fact problematic because the variance and standard deviation are strongly influenced by the size of the objects being measured. For example, Clovis projectile points used to hunt mammoth during the Paleoindian occupation of the New World tend to be much longer than the proto-historic "bird points" used for warfare and small game hunting by the Pueblo Indians in the American Southwest. Standard deviations are calculated using the sums of deviations of each variate from the group mean. The standard deviation for the proto-historic points consequently will always be smaller than the corresponding standard deviation for Clovis points because the proto-historic points are themselves smaller; each proto-historic point cannot differ at the same magnitude from its mean as can a Clovis point. As a result the standard deviation and variance are inappropriate for comparing the relative amount of variation within groups with significantly different means, because they will tend to overestimate the amount of variation in large variables and underestimate the amount of variation in smaller variables relative to each other.

This problem is resolved by calculating the *coefficient of variation*. The coefficient of variation (*CV*) is an expression of the standard deviation as a percentage of the mean from the parent distribution. It standardizes the standard deviation so that the size of the variable being measured is controlled. Instead of reflecting the absolute size of the variation from the mean, *CV*s reflect the *proportion* of variation from the mean. Thus, a Clovis point and a proto-historic projectile point that are two-thirds of their respective mean lengths will demonstrate the same proportional variation as reflected by the *CV*. The *CV* therefore allows the variation within distributions with significantly different means to be compared. The coefficient of variation is computed using Equation (4.5).

The coefficient of variation

$$CV = \frac{s \times 100}{\overline{Y}} \text{ for samples, and } CV = \frac{\sigma \times 100}{\mu} \text{ for populations} \qquad (4.5)$$

For the Alyawara example:

$$CV = \frac{11.8 \times 100}{23.4} = 50.5\%$$

The coefficient of variation has meaning only in a comparative sense to other coefficients of variation. For example, we might be interested in comparing the variation in camp sizes of the Alyawara with another group with $\overline{Y} = 30$ people, $s = 12$ people, and $n = 8$ camps. For this second group:

$$CV = \frac{12 \times 100}{30} = 40.0\%$$

50.5 > 40.0, indicating that the variation in the Alyawara residential size is greater than that of the second group. The significance of this finding is of course contingent on your theoretical perspective, methodological and research design, and research questions.

Statisticians have determined that the *CV* is a biased estimate of the population parameter in small samples, but that this bias can be easily corrected. The corrected coefficient of variation (corrected *CV* or *CV**) is computed using Equation (4.6). Obviously, the impact of the correction is minimal for large sample sizes (e.g., for samples larger than 25, the corrected *CV* will be less than .01% different than the original *CV* value), but it can be significant for extremely small samples (*CV* values for samples of five will increase 5%). Of course in archaeological contexts, we will frequently wonder if the underlying distribution is accurately reflected in such small samples.

Correction formula for the coefficient of variation

$$CV^* = \left(1 + \frac{1}{4n}\right)CV \qquad\qquad (4.6)$$

For the Alyawara example:

$$CV^* = \left(1 + \frac{1}{4(8)}\right)50.5$$

$$CV^* = 52.0\%$$

For the second group:

$$CV^* = \left(1 + \frac{1}{4(8)}\right)40$$

$$CV^* = 41.25\%$$

Our conclusions regarding the variation in our Alyawara example are the same, since 52.0% > 41.3%, yet we now likely have better estimates of the parametric values.

Coefficients of variation are commonly used for comparing variation in populations or samples with significantly different means. For example, coefficients of variation are necessary to compare variation in the skeletal morphology of different primate groups that differ in their size. Another common application that is unique to archaeology is the use of *CV** to evaluate the organization of craft production. The "standardization hypothesis" holds that small coefficients of variation suggest greater standardization of products, which in turn is indicative of specialized production, whereas larger coefficients of variation more likely reflect generalized production at the household level (e.g., Arnold and Nieves, 1992; Crown, 1995; Longacre *et al.*, 1988; Mills, 1995; Sprehn, 2003). Crown notes that with respect to the manufacture of ceramics, known specialist groups rarely produce ceramics with coefficients of variation above 10%, causing her to suggest that coefficients of variation smaller than 10% likely reflect specialized production. Based on this measure, she concludes that the large *CV**s associated with morphological attributes of

Salado polychromes in the American southwest reflect little, if any, standardization and subsequently little, if any, specialized production.

Box Plots

Now that we have an understanding of the median and the range we can return to the topic of visual representations of data discussed in the previous chapter and introduce an additional means of visually characterizing data – box plots. Box plots are extremely useful for characterizing multiple distributions at the same time, although they can be used to characterize a single distribution, because they provide information about the variation and central tendencies of data in a very condensed manner (e.g., Figure 4.2).

Box plots reflect the median, the interquartile, and the range. They are created in three steps. First, calculate the distribution's median and the quartiles. Second, plot the location of the median and the two quartiles. Draw a box using the quartiles as limits. This box reflects the distribution of the middle 50% of the data – the "body" of the data. Third, draw a line from the lower quartile to the value of the smallest variate and from the upper quartile to the value of the largest variate.

The utility of box plots can be demonstrated in Figure 4.2. This figure is the length of four sets of 12 flakes made as part of an experiment studying the flaking characteristics of lithic raw materials (Table 4.5). The box plots in Figure 4.2 provide a quick means of describing both the structure of each distribution and the differences among them. By simply glancing at the boxes, we can identify differences in the central locations and dispersion of the distributions for each raw material type.

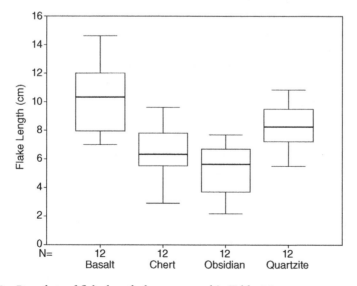

Figure 4.2 Box plots of flake length data presented in Table 4.5

Table 4.5   Flake length (cm) by raw material

Basalt	Chert	Obsidian	Quartzite
7.0	2.9	2.2	5.5
7.0	4.8	2.4	5.5
7.7	5.3	3.1	7.0
8.2	5.8	4.3	7.4
10.3	5.8	5.0	7.7
10.3	6.2	5.5	7.9
10.3	6.5	5.8	8.6
10.8	7.7	6.0	8.9
11.0	7.7	6.2	9.4
13.0	7.9	7.2	9.6
13.9	8.9	7.4	10.6
14.6	9.6	7.7	10.8

Characterizing Nominal and Ordinal Scale Data

The measures of dispersion and central tendency discussed above are applicable to ratio and interval level data. As people learn how to use them, most are amazed at how much information these measures communicate about a distribution. However, the measures aren't suitable for ordinal or nominal data. Consider the oddity of averaging a distribution of pots equally divided between "extremely large" and "miniature" containers to determine that the "average" pot is "medium-sized." Far from helping us understand the distribution of pot sizes, this would clutter our thinking at best.

Unfortunately, anthropologists routinely make this mistake. For example, ethnographic surveys frequently ask respondents to rank their reaction to some question or event on a scale of 1 to 5 or something similar, with 1 reflecting strong approval/desire, 3 reflecting neutral feelings, and 5 reflecting disapproval/loathing. The data derived using the questionnaire are ordinal, in that the difference between a response of 4 (mild loathing) and 5 (extreme loathing) is not quantifiably comparable in the same manner as is the difference between 20°C and 40°C. Further, how can the researcher guarantee that different people "mean" the same thing when marking a particular response (i.e., how does one objectively quantify "loathing")? Unfortunately, ethnologists and other anthropologists using these types of data frequently run headlong into the numerical descriptors previously presented, calculating and comparing averages and standard deviations as if they were meaningful. They shouldn't do this. More specifically, *you* shouldn't do this. Any analysis based on averaging ordinal or nominal data is inherently flawed; we cannot add, subtract, multiply, or divide ordinal or nominal data, even when numbers are used as class labels (e.g., miniature vessels = 1, small vessels = 2, and so forth).

So are there ways to describe the distribution of nominal and ordinal data? Well, yes. We can, for example identify the most popular classes, which is sort of like, but not the same as, talking about a distribution's mode. When using ordinal data, we can also identify the largest and smallest classes in a distribution, which is sort of analogous to maximum and minimum values in ratio and interval scale data. However, these are merely "commonsense" descriptions when compared to the quantitative methods used for interval and ratio data. More useful measures are the index of dispersion and the index of qualitative variation.

Index of dispersion for nominal data and the index of qualitative variation

The index of dispersion ($\hat{D}$, pronounced D-hat) provides a measure of dispersion within data classified using asymmetrical classes whether they are nominal or ordinal scale data. It does so by measuring the likelihood that a pair of independently selected variates will come from different classes. Consider for example the distribution of ceramic types in Table 4.6. What is the probability that a pair of randomly selecting vessels will reflect different classes? Many will intuitively realize that it is 3 out of 4, or 75%. If it isn't intuitive, think about it this way: the probability of selecting a vessel from any of these classes is 1 in 4. By extension, the probability that the first vessel selected will come from the *same* class as the second vessel selected is 1 in 4, or 25%. Thus, the likelihood that the first vessel will come from a *different* distribution than the second is 3 in 4, or 75%. Although intuitive in Table 4.6, the probability of selecting members of different classes can be formally calculated using Equation (4.7), where n_i is the count in each class, and n is the total assemblage size.

The index of dispersion for nominal data

$$\hat{D} = 1 - \sum (n_i/n)^2 \qquad (4.7)$$

For Table 4.6, $\hat{D}$ is calculated as follows:

$$\hat{D} = 1 - \sum \left[(25/100)^2 + (25/100)^2 + (25/100)^2 + (25/100)^2 \right]$$

Table 4.6 Frequency of ceramic vessels classified using cultural historical types

Ceramic type	Frequency
Carretas polychrome	25
Ramos polychrome	25
Villa Ahumada polychrome	25
Madera black-on-red	25

$$\hat{D} = 1 - \sum (.0625 + .0625 + .0625 + .0625)$$

$$\hat{D} = 1 - .25 = .75$$

If we convert .75 into a percentage by multiplying it by 100, we find that the probability of selecting variates from different classes is 75%. (We will discuss the relationship between probabilities and percentages in detail in Chapter 5.) Why would this bit of information be useful?

Compare $\hat{D}$ for Table 4.6 with $\hat{D}$ for Table 4.7, which is calculated as:

$$\hat{D} = 1 - \sum \left[(25/28)^2 + (1/28)^2 + (1/28)^2 + (1/28)^2 \right] = .20$$

The likelihood of selecting a pair from different classes in Table 4.7 is only 20%. Compared with the $\hat{D}$ value corresponding with Table 4.6, this indicates that Table 4.7 is less diverse in that some classes are far more common than others (i.e., the assemblage is more homogeneous). Looking at Table 4.7 this is certainly the case; Carretas polychromes are far more common than the other pottery types. As illustrated by this example, the $\hat{D}$ value allows us to describe the amount of variation within a sample of nominal or ordinal data, and even compare the diversity reflected in different samples.

There is a complicating factor when comparing $\hat{D}$ values, however. The amount of *potential* variation increases as the number of classes increases. Classifying an assemblage using six as opposed to three classes actually increases the amount of potential variation because the additional categories reduces the average number in each of the categories (e.g., 100 vessels divided into 10 categories averages fewer members in each category than 100 vessels divided into two categories). The probability of selecting a pair of variates from the same category will decrease as the frequencies within each category decrease. Thus, $\hat{D}$ values for assemblages classified using many classes are expected to be lower than those using few classes for no other reason than differences in the number of classes.

The correspondence between the number of classes and $\hat{D}$ values is potentially significant to archaeologists, who might very much like to compare the dispersion in nominal or ordinal data. Consider for example a study of mortuary goods in which an archaeologist wishes to determine if there are changes in the diversity of grave goods within burial populations through time using published data. If some of the data are reported using more categories than others (e.g., sometimes culture historical types are used to classify pottery as opposed to a less detailed division between plainware, bichromes, and polychromes), then some graves might appear to be more diverse (i.e., have higher $\hat{D}$ values) for no reason other than the classifications used by the investigator. This would complicate the mortuary goods analysis, and potentially cause the analyst to reach erroneous conclusions about grave good diversity.

An effective means of solving this problem is calculating the index of qualitative variation (*IQV*), which standardizes $\hat{D}$ by the total number of categories. It is calculated using Equation (4.8), where *I* is the total number of classes present within an assemblage. *IQV* values reflect the variation within nominal or ordinal data as a percentage of the *maximum potential* variation that would be possible given the number of classes. It thus ranges from 1 (which is equivalent to 100% and indicates the maximum potential variation) to 0 (which is possible only when all of the variates are classified into a single class, causing there to be no variation at all).

The index of qualitative variation

$$IQV = (I/[I-1])\hat{D} \tag{4.8}$$

For Table 4.6, *IQV* is calculated as $IQV = (4 / [4 - 1]) \times .75 = 1$. For Table 4.7, *IQV* is calculated as $IQV = (4 / [4 - 1]) \times .20 = .27$. The *IQV* values thus indicate that the data in Table 4.6 is as diverse as it possibly can be given the number of classes used in its classification. The *IQV* value of Table 4.7 is .27, indicating that these data reflect 27% of the possible variation given the number of classes. *IQV* values thus allow the variation in nominal and ordinal data to be directly compared, even if the classification used is very different. Ultimately, $\hat{D}$ and *IQV* are potentially very useful to archaeologists, who frequently make pronouncements about "increased" or "decreased" variation in attributes of the archaeological record. Using these measures, the variation in flaked stone raw materials, pottery morphology and designs, ground stone artifact morphology, faunal assemblages, macrobotanical assemblages, settlement types, grave goods, architectural forms, settlement locations, and any of the other hundreds of ordinal and nominal scale data that archaeologists collect can be quantified.

Identifying differences in variation in qualitative data is in fact a great place to start exploratory data analysis. Consider a case in which an archaeologist compares the diversity in pottery assemblages from various sites in a region using *IQV*, and finds that one appears to be substantially more diverse (i.e., has a large *IQV*) than the others. Perhaps this site is an important trading center and the high *IQV* reflects

Table 4.7   A second series of ceramic vessels classified using cultural historical types

Ceramic type	Count
Carretas polychrome	25
Ramos polychrome	1
Villa Ahumada polychrome	1
Madera black-on-red	1

trade wares from throughout the region. Perhaps some group of immigrants joined a previously existing community, resulting in greater ceramic diversity than is typical. Or, perhaps the site was a "pilgrimage" settlement where people from throughout the region met for feasting and other ritual activity. The archaeologist might not know which of these or hundreds of other possible scenarios might be present, but he or she can start to explore these hypotheses. Thus, being able to identify differences in the diversity of ordinal and nominal data will suggest interesting and productive research avenues, in addition to allowing us to present straightforward descriptions of the variation in assemblages.

We now know how to characterize data visually and numerically. The next step is to learn about the role of probability in statistical reasoning. Knowledge of probability underlies hypothesis testing and the various extremely useful applications that fill the rest of this book. As a result, we now turn to this topic in the next chapter.

Practice Exercises

1 Define "measures of central tendency". What are the common measures? What information do they provide when characterizing a distribution? Which measure is best?

2 Define "measures of dispersion". What are the common measures? What information do they provide when characterizing a distribution? Which measure is best?

3 Following are data for the height of ceramic vessels recovered from a Bronze Age site in England: 14.10 cm, 13.86 cm, 14.86 cm, 14.52 cm, 14.07 cm, 12.01 cm, 13.09 cm, 12.57 cm, 14.00 cm, 13.22 cm.

   (a) Determine the mean, median, and mode of the sample.

   (b) Determine the sample's range.

   (c) Using both the calculation formula presented in Chapter 4 and the long method presented in Table 4.3, determine the sum of squares, standard deviation, and variance of the sample.

4 If the data presented in Question 3 were considered a population instead of a sample, which of the measures of central tendency and dispersal would change? What are the new values?

5 The maximum length (cm) of flakes made using three different reduction methods are presented in the following table.

Generalized hard hammer reduction	Generalized soft hammer reduction	Bifacial reduction
7.15	4.10	0.91
4.89	3.86	2.15
4.52	4.86	0.89
7.06	4.52	1.44
6.94	4.07	1.27
5.32	2.01	1.76
4.85	3.09	1.48
5.86	2.57	1.99
6.01	4.00	2.30
5.73	3.22	0.98

(a) Determine the mean, range, variance, and standard deviation for each sample.

(b) Compare the standard deviations between groups and rank the samples in order of the size of their respective standard deviations. Do the same for the range.

(c) Compute the coefficient of variation and the corrected coefficient of variation for each sample.

(d) Rank the samples according to the size of their respective corrected coefficients of variation.

(e) Which reduction method produced the most variation in maximum flake length? Which produced the least? How do the rankings based on the range, standard deviation, and the corrected CV compare to one another? Which measure is the best for evaluating the differences in variation in the samples? Write a brief (paragraph or two) analysis of any differences present between the rankings based on the three measures of dispersion.

6 The length in millimeters of 120 flakes excavated from a Neolithic site in
   Egypt are presented in the following table.

58.9	79.6	23.2	37.6
83.4	68.3	58.8	25.6
53.2	15.4	35.8	25.9
32.1	20.2	17.8	35.4
25.0	19.6	13.3	20.6
25.5	70.2	45.4	34.2
23.6	26.7	50.8	30.1
43.0	60.4	73.4	29.2
30.0	38.9	15.1	42.5
26.1	52.2	15.8	25.4
15.2	23.3	15.5	24.5
38.1	26.4	23.6	18.5
23.0	27.7	16.8	20.0
22.0	17.9	13.3	19.9
10.1	20.0	40.6	14.5
12.1	23.8	41.9	17.3
45.3	21.4	11.0	21.3
18.9	15.5	30.3	44.1
27.1	25.4	36.7	16.9
11.7	43.4	11.3	33.0
18.9	5.4	31.6	18.8
13.9	20.2	27.3	55.6
28.9	23.7	21.1	55.1
17.9	33.5	17.5	43.6
12.2	23.8	14.0	76.0
11.8	31.4	14.8	24.8
25.1	14.4	20.7	14.2
13.4	17.1	11.4	18.3
18.2	19.3	14.5	18.7
23.8	19.0	27.8	14.8

(a) Determine the sample mean and standard deviation for these data.

(b) Estimate μ and σ for the data using the midrange and range. How
    similar are the two estimates of μ and σ?

7 Create a frequency table and histogram describing the distribution of the
   flake length data presented in Question 6. Next, calculate the median, the
   mode, the range, the maximum value, and the minimum value of the data.
   Is the median, the mean (determined in Question 6), or the mode the
   better measure of central tendency for this distribution in your opinion?
   Why?

8 Create a box plot of the following data, which are samples of the thickness (mm) of differently shaped projectile points from La Ferras, an Upper Paleolithic site (Knecht, 1991: 628–31):

Spindle-shaped simple based points	Losange-shaped simple based points
8.1	8.9
7.7	9.3
7.9	11.4
5.4	10.0
8.1	11.2
6.7	10.7
4.8	10.3
5.0	10.1
5.7	11.4

How do the distributions compare to one another based on your graph? Compute the corrected coefficients of variation for the distributions. How do these compare?

9 Following are data derived from Crown and Fish (1996:810) reflecting the grave lot values for graves of males and females from the Hohokam culture of the American southwest. Using $\hat{D}$ and IQV, describe the variation in the grave lot values of females and males. Does there seem to be a difference in their homogeneity? If so, which sex appears to reflect more grave lot diversity?

Period and body treatment	Females	Males
Pre-Classic	10	90
Classic cremations	160	95
Classic inhumations	250	520

An Introduction to Probability

Probability is a very powerful tool. It starts with the assumption that the world is consistent, that knowledge we have gained by observing some portion of the world can be used to understand the portion we have not directly observed. Probability is the means by which we make inferences about populations from samples. It also allows us to make predictions about individual variates based on a sample or a population. While probability is very important in making archaeological inferences, its abstract nature is better first illustrated through hypothetical examples using coins, dice, and cards. The observant reader will notice that we are reneging on our promise not to rely heavily on such "Vegas-style" discussions. We promise that this is merely a brief interlude. Archaeological illustrations of the application of probability will be presented once the fundamental principles have been outlined.

Theoretical Determinations of Probability

As you probably have figured out by now, quantitative analysis relies on standardized symbolism. This symbolism looks arcane to the uninitiated, which is part of the reason why so many people are scared or at least uncomfortable with statistical analysis. People look at a statistical procedure/outcome and simply don't understand what it means. It might as well be a foreign language. As a result, they skim the statistical discussions, trusting that the analyst knows what he or she is doing, and go on to read the parts of the paper that are written in a language they understand. There might be those who like others' confusion and use statistics to obscure their reasoning while giving themselves an air of authority and competency, but obscuring one's analysis is not the purpose of the symbols used in quantitative analysis. Quite to the contrary, the consistent use of symbolism is intended to remove any ambiguity about what is being done and how data are being connected to specific analytic results. Probability in particular is frequently confusing when trying to communicate it using standard written language. Any ambiguity in wording makes it impossible to accurately communicate probabilities between researchers. Unfortunately, colloquialisms, poor wording choices, and simple typos ensure at least occasional mistakes. As a result, statisticians and others such as archaeologists who deal with probability use a very rigid, formalized notation system for explicitly communicating probabilities. You must learn this notational system to effectively use probability.

Consider the simple flip of an unbiased coin. We know that we have two possible outcomes: a head or a tail. In quantitative contexts, the specific outcome of a coin flip is called an *event*. As heads and tails are our only possible outcomes, they constitute a *finite probability space* (i.e., a limited number of possible outcomes). We can symbolize the probability of heads as $P(H)$, and the probability of tails as $P(T)$. The finite probability space is $\{H,T\}$; the result of a coin flip must be one or the other. Being part of a world of coin flippers, we all know that we have a one out of two chance of a coin flip resulting in the event "heads" or the event "tails." So, the probability of heads is 1/2, or expressed as a proportion, .50.

Let us express these probabilities with the appropriate symbols: $P(H) = 1/2 = .50$ and $P(T) = 1/2 = .50$. (Probabilities are always expressed as fractions or proportions.) Notice that probabilities in the finite sample space can vary between 0 and 1, and must sum to 1 (i.e., $P(H) + P(T) = .5 + .5 = 1$). The probabilities of H and T were derived theoretically without a single flip of a coin, but we intuitively accept that they can be used to predict outcomes in the real world. Given our experience of flipping coins, we realize that this theoretical expectation regarding probabilities will be realized empirically within an expected range of error so that we will have roughly equal numbers of heads and tails after flipping the coin several times. The standardized notation makes it possible to communicate this expected empirical pattern clearly.

Now that we have the basic notation down, let's consider an example where the finite probability space is more complex than $\{H,T\}$ – the case of rolling two dice. Each face of each die is designated with one through six dots. There are a total of 36 unique combinations between the two dice (if the first die is 1, the second die can be 1 to 6; if the first die is 2, the second die can be 1 through 6, and so on). There is only one way of rolling a sum of 2: 1 on both dice. The probability of rolling a sum of 2 is therefore 1/36 or .03, which is derived by dividing the number of ways that an event can occur (in this case 1) by the total number of possible outcomes (in this case 36). We can roll a 3 two ways. The first die may be 1 and the second 2, or vice versa. And so on for 4, 5, etc. These probabilities are listed in Table 5.1 and can be graphed as we do so in Figure 5.1. Note the perfect symmetry of the distribution of probabilities. Given these probabilities, we can determine the

Table 5.1   Probability of rolling a given sum using a pair of dice

Y	P(Y) as a ratio	P(Y) as a probability
2	1/36	.03
3	2/36	.05
4	3/36	.08
5	4/36	.11
6	5/36	.14
7	6/36	.17
8	5/36	.14
9	4/36	.11
10	3/36	.08
11	2/36	.05
12	1/36	.03
Σ	36/36	1.00

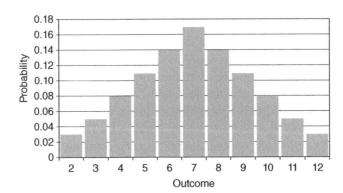

Figure 5.1   The distribution of the probabilities of rolling a sum with a pair of dice

likelihood of any given outcome of a roll of dice, a point to which we will return presently.

Empirical Determinations of Probability

People like using theoretically derived probabilities, because they are often intuitively meaningful as in the cases above. Sometimes deriving probabilities theoretically isn't possible, because of the lack of *a priori* knowledge of the pertinent factors. This is frequently the case in archaeological contexts. What is the probability that large sites are located on floodplains instead of highland locations? We can't determine this theoretically the same way we can when determining the probability of getting a head when flipping a coin. We might even suspect that the probability changes through time and across space. We can, however, determine probabilities empirically in these cases. When doing so, the probability for an event can be determined using Equation (5.1).

Calculation of an empirically derived probability

$$P(\text{event}) = \text{frequency of the event}/\text{number of trials} \qquad (5.1)$$

Consider the following example. The University of New Mexico (UNM) has a severe parking problem. While a graduate student at UNM, the senior author noticed that it was quite easy to find a parking place early in the morning, but that it became increasingly difficult as the day progressed until it was nearly impossible as noon approached. By mid-afternoon parking spaces became easier to find and by late afternoon they were abundant until students and faculty began to arrive for evening classes and open spaces again became infrequent.

Assume that we had the patience, the free time, and the desire to conduct the following experiment. For each of the following hours of the day, one of the authors tried on 10 different days to find a parking space. These results are presented in Table 5.2. Column 1 depicts the time of day, Column 2 the number of successful attempts at obtaining a parking place, Column 3 the number of failures, Column 4 the number of successes divided by the number of trials (which is the outcome of Equation (5.1)), Column 5 the number of failures divided by the number of trials, and Column 6 the sums of Column 4 and Column 5. Column 4 is the probability of successfully finding a parking place at a given time, whereas Column 5 is the probability of being unsuccessful (e.g., a .30 probability or 30% chance of successfully finding a parking place between 11 and 12, as P(parking space 11 to 12) = 3/10, and a 70% chance of having to park elsewhere). The distribution of the

Table 5.2 Probabilities of success or failure of finding a parking space at UNM during different times of the day

1	2	3	4	5	6
Time of day	S	F	S/n trials, where n trials = 10	F/n trials, where n trials = 10	∑S/n trials + F/n trials
7am–8am	10	0	1.0	0.0	1
8am–9am	8	2	0.8	0.2	1
9am–10am	7	3	0.7	0.3	1
10am–11am	2	8	0.2	0.8	1
11am–12pm	3	7	0.3	0.7	1
12pm–1pm	0	10	0.0	1.0	1
1pm–2pm	1	9	0.1	0.9	1
2pm–3pm	2	8	0.2	0.8	1
3pm–4pm	4	6	0.4	0.6	1
4pm–5pm	7	3	0.7	0.3	1

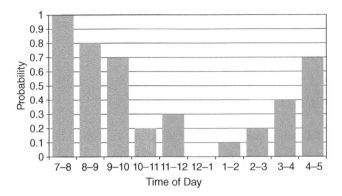

Figure 5.2 Distribution of the probability of finding a parking spot in UNM's overcrowded parking lot

probability of a success is depicted in Figure 5.2. With empirical cases, larger numbers of trials increase the accuracy of the prediction. One hundred trials would certainly yield a more reliable predictor than 10 trials.

Empirically derived probabilities are very useful, but the assumption of continuity (the idea that the world will behave consistently) is sometimes problematic when dealing with them. In the preceding case, the probability of rolling a sum of 2 on two dice will hold so long as the dice are unbiased. Assuming that the dice aren't "fixed", we can calculate the probability of rolling any sum from 2 through 12, and these probabilities are expected to hold in the past, right now, and in the future, a fact upon which the gambling houses in Las Vegas rely. (They wouldn't be able to stay in business if the probability of rolling a 7 while playing craps changed every day.) We do not need to roll the dice 100 times and then compute the actual probability of rolling the sum of 2. Instead, we need only ensure that someone has not tampered with the dice.

In contrast, the use of empirically derived probabilities requires that we create an argument justifying the application of that probability to future and (as is relevant in most archaeological cases) past events. For example, we expect that the probability of finding a parking spot between 11 and 12 during the spring semester is 0.30, but will this probability hold during spring break? Probably not, given that many of UNM's students escape Albuquerque and travel to somewhere that has water for the week. Many of UNM's staff members also take vacations during spring break. As a result, we expect that the probability of finding a parking spot between 11 and 12 is higher during spring break than is typical of the rest of the semester. Likewise, the probability of large settlements being built on floodplains may have changed as ecological and social conditions shifted in the past.

Our empirically derived probability is consequently based on underlying factors that affect the variable of interest. Ideally we could determine what these factors are and adjust our probabilities to take them into account. We could for example further refine our probability estimates of successfully finding parking spots by establishing probabilities for different days of the week, the time of year, the presence of special events on campus such as high school band competitions, and student enrollment, among other variables. The manifestation of each of these variables could then be factored into our probabilities.

In general, theoretical determinations are preferred, as they are not subject to sampling vagaries and historical (empirical) contingencies, but empirical determinations are common and useful in archaeological contexts. When using probabilities determined empirically, you should be very clear on the sample size and conditions under which the observations were taken. In general, large numbers of observations increase our accuracy of estimating the probability of an event (we will return to this issue in future chapters), and our research design must justify the use of the probability in the context in which it is used.

Complex Events

The examples presented above determined probabilities for simple events. At times, we seek probabilities for more than one event. Such situations are called complex events. To illustrate the difference between complex and simple events, let us consider a deck of playing cards.

We know that each deck contains 52 unique cards (discard the jokers) divided into four suits: hearts, spades, clubs, and diamonds. Each suit contains 13 cards: ace, king, queen, jack, 10, 9, 8, 7, 6, 5, 4, 3, and 2. We know that the probability of obtaining any unique card is 1/52. For example, P(ace of spades) = 1/52, P(three of diamonds) = 1/52, and so on. These events are simple events. We also know that the probability of drawing a card of a specific suit is 13/52, as there are 13 cards in each suit. So, P(heart) = 13/52, P(spade) = 13/52, and so forth. These too are simple events.

For a particular game, we might be interested in the probability of drawing a heart or a diamond. This constitutes a complex event. The probability of such a complex event is P(heart) + P(diamond), or 13/52 + 13/52 = 26/52 = .50. This complex event is appropriately symbolized by P(heart $\cup$ diamond) = .50, which is verbalized as "the probability of heart union diamond is equal to .50." This union is simply the sum of independent probabilities. Note that the categories "hearts" and "diamonds" are mutually exclusive. A card cannot be both a heart and a diamond. This is illustrated with the Venn diagram in Figure 5.3.

What if our probability spaces overlap? We can also create a Venn diagram of the possibilities of the intersection of events. Figure 5.4 illustrates one such intersection, that of obtaining a heart or an ace. The states of being a heart and being an ace are not independent in that the ace of hearts is both. As a result we cannot simply add the probability of hearts (P(heart) = 13/52) with the probability of aces (P(aces) = 4/52) because doing so would count the ace of hearts twice (once as an ace and once as a heart). We must therefore eliminate the probability of the co-occurrence of events. This set of non-mutually exclusive events is symbolized as:

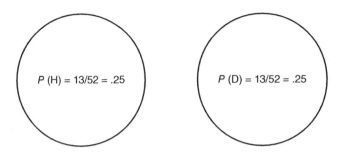

Figure 5.3   Venn diagram of probabilities of hearts and diamonds

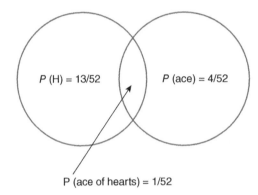

Figure 5.4   Intersection of probabilities of obtaining a heart or an ace

$P(\text{heart} \cup \text{ace}) = P(\text{heart}) + P(\text{ace}) - P(\text{heart} \cap \text{ace}) = (13/52) + (4/52) - (1/52)$ = 16/52. Here we have controlled for the intersection of the two events (heart and ace) represented by the ace of hearts as symbolized by the notation $P(\text{heart} \cap \text{ace})$, which is read as "the probability of heart intersection ace". Without controlling for this intersection, our calculated probability would be in error.

Venn diagrams are incredibly useful, and the calculation of $P(\text{heart} \cup \text{ace})$ becomes self-evident from looking at the Venn diagram, yet many archaeologists don't wish to be bothered drawing "silly circles" on a scrap of paper. We empathize with both students and professional archaeologists who like cutting "unnecessary" steps out of their work, but we strongly recommend that everyone draws Venn diagrams whenever they are dealing with complex events. Venn diagrams are a quick and effective means of clarifying the relationships between various probabilities and preventing confusion and error that can cost time and prestige as one is forced to correct "silly mistakes".

Using Probability to Determine Likelihood

You probably intuitively already realize it, but we can use the probabilities we have discussed to determine the likelihood that a particular event will occur. For example, as we mentioned at the end of the section on theoretically derived probabilities, we can determine the likelihood of an event by understanding the probability of its occurrence. When rolling two dice, $P(7) = .17$, so we expect to roll a seven 17 times out of 100 rolls (see Table 5.1). Likewise when choosing a random card from a complete deck $P(\text{heart}) = .25$, so we expect to draw a heart 25 times out of 100 tries. In both cases, the specific events would be likely, in the sense that they would be common relative to the other possible outcomes.

Some events (like winning the lottery) might be unlikely. Consider for example the probability of finding a parking spot between 1pm and 2pm at UNM. The probability of doing so in our contrived example is only .1, which corresponds with 10 times out of 100 tries. While finding a parking spot between 1pm and 2pm isn't as unlikely as winning the lottery, we are always pleasantly surprised when we are able to do so. If someone finds a spot between 1 and 2, then that person is lucky, but what about someone who is able to find a spot on two consecutive days? We know intuitively that this is even less likely than just simply saying that we ought to be able to find a spot twice out of 20 tries, which doesn't imply anything about the order of the events. Finding a spot on three consecutive days is even more unlikely, as is finding a spot four times in a row, and so on.

We can even mathematically determine the likelihood of such consecutive "successes" by multiplying the probabilities associated with each event. Consider that we know that on Day 1, the probability of finding a spot is .10, or 1 out of 10 tries. After "being lucky" enough to find a spot, the probability of finding a spot on the second day remains .10. These probabilities are independent of each other, such that the probability of a success in one trial (a parking attempt) doesn't influence the probability of success during another trial. Because the probability of parking success is independent for each day, we anticipate that only once out of every 10 times we successfully park will we find a parking spot the next day. The other nine times we won't. Thus, only once out of every 100 tries are we expected to find spots on two consecutive days. Using this same logic, only once out of 1000 days is it likely to find a spot on three consecutive days, only once out of 10,000 days to find a spot on four consecutive days, and once out of 100,000 days to find a parking spot five days in a row. This relationship can be mathematically expressed as illustrated in Equation (5.2).

Calculation of the probability of repeated events

$$P(n \text{ events}) = (\text{frequency of an event} / \text{number of trials})^n \qquad (5.2)$$

For the example above, P(successfully finding a parking spot between 1 and 2 on three consecutive days) $= p$(successfully finding a parking spot between 1 and 2)$^3 = (1/10)^3 = .001$ or once out of 1,000 trials.

We have already discussed how we can add and subtract probabilities in the section on complex events, so the idea of using other mathematical processes to manipulate probabilities probably isn't that surprising to you. A handy trick to remember is that we add and subtract probabilities when determining the likelihood of an outcome of a single event (e.g., P(heart or a spade) $= .25 + .25 = .50$) and we multiply probabilities when dealing with the outcome of multiple events (e.g., P(drawing a heart twice in a row) $= .25 \times .25 = .063$).

Being able to determine the likelihood of a series of events is a very handy skill to have when dealing with the world, because it allows us to differentiate between co-occurring events that are likely and those that aren't. For example, at some point we will become incredulous about someone who is able to consistently find a parking spot between 1pm and 2pm, so much so that we will conclude that they have the competition for a parking spot rigged in their favor. Perhaps the overly successful "parker" has a reserved parking spot as many administrators do. Perhaps he or she has a friend who consistently leaves between 1 and 2 and is willing to wait until the successful individual can claim the friend's newly vacant spot. Regardless of the mechanism, at some point we come to the conclusion that the individual's parking success is so unlikely that his or her probability of succeeding is not the same as everyone else's. Archaeologists use this exact same reasoning when talking about "burials with a disproportionate number of artifacts," "excessively large settlements," "faunal assemblages with an overabundance of jackrabbits," or any of the other myriad differences we note in the archaeological record. What we are really saying is that the preponderance (or lack) of some attribute is so improbable, it seems likely that there is some factor underlying this difference that is not applicable to other members of the same class as reflected elsewhere.

But at what probability should we reach such a conclusion? Returning to the parking spot example, no outcome of finding a spot is impossible. Even finding a spot 10 times in a row can happen, it just isn't likely. (Using Equation (5.2), we can determine that the probability is $(.1)^{10} = 0.0000000001$ or 1 in 10,000,000,000 tries.) Despite the fact that it is possible to be this lucky, we might reasonably conclude that it is more than just blind luck going on, and as a result begin looking for alternative explanations. As will be discussed in more detail in the next chapter, archaeologists and others often use a probability of .05 as a means of demarcating between likely and unlikely events; events that are likely to occur fewer than 5 times out of 100 tries are considered unlikely enough to prompt us to conclude that they reflect meaningful differences that aren't the product of blind luck. Using likelihood in this way forms the core of hypothesis testing, and will be the focus of most of this book, but before we explore this further, let's consider the binomial distribution and its utility for archaeological research.

The Binomial Distribution

The binomial distribution is a means of formalizing the previously discussed structure of probability. It is a means of determining the likelihood of the outcome of multiple events in which there are two and only two possible outcomes (hence the name "binomial"). One of these possible outcomes is designated a "success" (symbolized as p) and the other a "failure" (symbolized as q). Don't read into these labels; they are arbitrary, and don't reflect anything about a desired or preferred outcome; unpleasant things like suffering from anemia or dying a violent death might be considered successes for the purposes of a binomial analysis.

The binomial distribution builds on the principles outlined previously. Our introduction to it represents one of the most mathematically intensive discussions in this text, so please bear with us. Remember, we are only asking you throughout this text to add, subtract, multiply, and divide. If you can do that, then you can understand and perform all of the steps we discuss here.

Given that there are only two possible outcomes when using the binomial distribution, we know $P(p + q) = 1$ and that $P(q) = 1 - p$. If we define a success as drawing a heart from a complete deck of cards, then $P(p) = .25$ and $P(q) = 1 - .25 = .75$. In one event (i.e., one card draw), the probability of a success is $P(p) = .25$ and of a failure is $P(q) = .75$. But what is the probability of having two successes (drawing a heart) in two tries? We can answer this question by multiplying the probabilities to determine the likelihood of all possible outcomes: two successes; one success and one failure; and two failures.

There is only one way to get two successes – draw two hearts. Likewise there is only one way to get two failures – draw two cards of other suits. But there are two ways to get one success and one failure – draw a heart and then another suit, or draw another suit and then a heart. The probability of getting two successes (hearts) in consecutive trials is $P(p \times p) = .25 \times .25 = .0625$. The probability of getting two failures (non-hearts) is $P(q \times q) = .75 \times .75 = .5625$. Given that there are two ways to get a success and a failure, the probability of $P(p \times q$ and $q \times p) = 2(pq) = 2(.25 \times .75) = .375$. (Because $[p \times q]$ and $[q \times p]$ are mathematically equivalent, we can replace them with $2pq$.) Note that $P(p \times p) + P(p \times q)$ and $q \times p) + P(q \times q) = .0625 + .375 + .5625 = 1.0$. These probabilities reflect all possible outcomes of drawing two cards. This series of probabilities could be generalized for all binomials with two trials as follows: $P([p + q] \times [p + q]) = P(p + q)^2 = p^2 + 2pq + q^2$. The probabilities of each possible outcome (two successes, one success and one failure, and two failures) will change as the probability of a success changes, but the formula of $p^2 + 2pq + q^2$ will always remain the same.

Similar binomial formulas are presented for any number of trials to determine the probability of any possible outcome, so long as each trial has the same probability of a success or failure. For a single event, the probability of all possible outcomes can be determined as $P(p + q)$. For two events, the probability of all possible outcomes can be determined as $P(p + q)^2$. For three events, $P(p + q)^3$. For four events, $P(p + q)^4$. And so forth, with the power changing according to the number of events. This is identical to the structure of Equation (5.2), except that we are considering the probabilities of all possible events, not just one possible outcome. Thus, the binomial distribution is derived from Equation (5.3) in which p is the probability of success, q is the probability of failure, and n is the number of trials.

The binomial formula

$$P(p+q)^n \tag{5.3}$$

Those who haven't blocked out algebra can solve the formula for any given number of trials. Then, we can simply insert the probabilities for p and q to determine the likelihood of a given outcome of events. For example, what is the likelihood of being able to find a parking spot between 1 and 2 once out of three days? The binomial equation for this is $P(p + q)^3 = p^3 + 3p^2q + 3pq^2 + q^3$. Reading from the left to the right, these values reflect that there is only one way to get three successes (p^3), there are three ways to get two successes and a failure ($3p^2q$), there are three ways to get one success and two failures ($3pq^2$), and there is only one way to get three failures (q^3) out of three tries. Since we are interested in P(one success and two failures), $3pq^2 = 3(.10 \times .90^2) = .243$. Using the binomial equation, we can determine the probabilities of each possible outcome, and even the likelihood of a range of outcomes. For example, P(finding a parking spot one or more times) $= p^3 + 3p^2q + 3pq^2 = .1^3 + 3(.1^2 \times .9) + 3(.1 \times .9^2) = .001 + .027 + .243 = .271$.

It is important to note that the binomial distribution does not make any assumptions about the order of events. P(two successes and one failure) doesn't specify whether the failure is on the first, second, or third trial. It thus is not the same as specifying the probability of finding parking spots on two consecutive days and then failing to find a parking spot, which species an order of events. P(two consecutive successes followed by a failure) $= p \times p \times q = .1 \times .1 \times .9 = .009$. P(two successes and one failure) $= 3p^2q = .027$, which is considerably larger than .009. Why the difference? There is only one way to have two successes followed by a failure (p then p then q), but there are three ways to have two successes and a failure (p then p then q AND p then q then p AND q then p then p). As a result, the probability of two successes *and* a failure is three times larger than the probability of two successes *then* a failure.

Unfortunately researchers (and humans in general) frequently get the distinction between such ordered and unordered events confused, and in our experience this confusion underlies most errors when dealing with probability. A real-world example of the foolishness of people, which we call the psychic's trick, can help illustrate this point more forcefully.

The psychic's trick

One of the most effective parlor tricks for both professional and amateur psychics/magicians is "sensing" that two people in a room full of people share the same birthday. This trick plays to our perception that the likelihood of someone sharing a birthday with someone else is very unlikely, which is true. The probability associated with one person having the same birthday as another is only 1/365 (excluding leap years for simplicity's sake). If someone had the ability to identify people who shared the same birthday without any foreknowledge, then that person truly would be psychic. However, that is not the psychic's claim, which is instead that two (or more) people in the room share the same birthdays. In a room with two people, the probability of two people sharing a birthday is indeed 1/365, but that is not the case when a third person enters the room. Assuming the first two don't share a birthday, the third person could share a birthday with the first person or the second person, which corresponds with P(sharing a birthday)= 1/365 + 1/365 = 2/365. Assuming none of the previous people share birthdays, P(sharing a birthday) for a fourth person would be 3/365, for a fifth person = 4/365, and so forth. Although we intuitively realize that the odds corresponding with two or more people sharing a birthday are considerably larger in a crowded room than they were initially, the probability of a match when the twentieth person enters the room still seems unlikely (P(sharing a birthday) = 19/365).

If a psychic could consistently beat odds of 19/365, then he or she truly does have a sixth sense. However, the binomial distribution demonstrates that the probability is not 19/365 as one intuitively thinks, and that the reasoning that led to the determination of this probability is wrong despite its intuitive appeal. Instead, in a room of 20 people, the first person could share a birthday with the second, third, fourth, … twentieth person; the second person could share a birthday with the third, fourth, … twentieth person; the third person could share a birthday with the fourth, fifth, … twentieth person; and so on. The probability of at least two people sharing a birthday is thus much larger than simply 19/365, despite the fact this is counterintuitive to many. Think about it this way; observing that P(the twentieth person entering a room sharing a birthday with someone else) = 19/365 is true, but ignores the probability that two or more of the people already in the room share a birthday. The psychic is counting on you not to realize this.

Approaching the birthday trick with a room of 21 people from the perspective of the binomial, we realize that the probability that any given person shares the same birthday with someone else is 20/365. With this knowledge, we can solve for one or more successes where $p = 345/365$, $q = 20/365$, and $n = 20$ trials. Here, we have chosen to call the absence of a birthday match a success to: (i) emphasize the point that "success" and "failure" are arbitrary; and (ii) to make the computation of the binomial a bit easier for reasons that will be discussed presently. The binomial formula $(p + q)^{20}$ will provide the equation that can be used to determine the probability of each outcome. This is a heck of an algebra problem, and would take some time to compute by hand. Given that we have selected the absence of the same birthday to be a success, we can fortunately forgo most of the computation for the formula; there is only one way for all 21 people to not share a birthday – 20 successes. All other possible outcomes include at least two people sharing a birthday (i.e., at least one failure). The probability of 20 successes is consequently $p^{20} = (345/365)^{20} = .32$. Given that the sum of all possible outcomes must be 1, we know that P(at least one failure) = 1 − .32. Put another way, the probability that there is at least one failure (a match) is 68%. This is pretty good betting odds, and is certainly far more likely than 20/365 = 5%.

Increasing the number of people in the room increases the likelihood of a match by increasing the "number of trials" for the binomial while decreasing the likelihood that no one shares a birthday. Increasing the number in the room to 31 decreases the probability of a success (again, defined as no one sharing a birthday) to 335/365, while increasing n (the number of trials) to 30. $P(p^{30})$ in now only 8%, which means that there is a 92% chance that at least two people in the room share a birthday. Those are exceptional betting odds. With 41 people in the room, p is reduced to 325/365, and $P(p^{40})$ is only 1%. In a room full of 41 people, it is a near certainty that at least two people share a birthday, not 40/365 = 11% as it intuitively seems to be. Magicians, fraudulent psychics, and, on occasion, even journalists, economists, lawyers, and politicians hope that you don't realize that probability works this way. The only way to guard against such sophistry (and errors in your own thinking) is to be sure that you recognize probabilities reflecting multiple trials and treat them as such, as opposed to treating them as simple events.

Simplifying the binomial

We didn't have to expand the binomial expression $(p + q)^{40}$ because of the way we established the analysis of the psychic's trick; there is only one way to get 40 successes so the only portion of the expression that interested us was p^{40}. Sometimes we won't be able to rig our consideration of binomials to be so easy. Imagine that we wanted to determine the probability of three or fewer people sharing a birthday. We would have to expand $(p + q)^{40}$ to determine this probability. While $(p + q)^2$ is easy to expand (if you haven't blocked out a painful college algebra class), computing $(p + q)^5$ is onerous enough that we suspect most archaeologists would mildly balk at the idea. Computing $(p + q)^{40}$ would probably be regarded by most as madness, or at least a sign of too much time and too little motivation to do something important. Is there an easier way to determine a binomial equation?

The short answer is yes. The long answer requires us to introduce two useful techniques. The binomial is composed of three components: the powers to which p and q are raised and the coefficient against which they are multiplied. The powers for p and q are easy to determine, and simply refer to the specified number of successes and failures. For example, if we are determining the probability of five successes and three failures out of eight trials, then p will be raised to the power of five and q will be raised to the power of three. This process can be generalized as illustrated in Equation (5.4), where k = number of trials and Y = the number of successes.

Calculation formula for the binomial power terms

$$p^Y q^{k-Y} \tag{5.4}$$

Determining the coefficient that goes in front of the p and q terms is the more difficult step when computing the binomial. As should be obvious at this point, the coefficient reflects the number of ways we can obtain a given number of successes and failures. We know that there is only one way to get all successes or all failures given that this requires every event to have the same outcome, so the coefficient for the first and last terms in the binomial will always be one. However, how many ways are there to get two successes and two failures out of four events? We could go through and figure out all of the possibilities ($ppqq$, $pqpq$, $pqqp$, $qqpp$, $qpqp$, $qppq$) to determine that there are six ways, but frankly this will be impractical when dealing with five successes and 35 failures. We would be better off just doing the algebra associated with $(p + q)^{40}$.

An easier method is to use a table created by Blaise Pascal that is called Pascal's triangle, which provides the coefficients for the binomial distribution. Pascal's triangle, illustrated in Figure 5.5, starts with row "0" and can be enlarged to whatever size we might want. It works as follows:

- Trace down the left-hand column until the row number equals k, that is, the number of trials.
- The contents of the row starting at the left and moving to the right represent the integers that are placed as coefficients starting with all successes, then $k - 1$ successes and 1 failure, then $k - 2$ successes and 2 failures, and so forth until we have no successes and all failures.

That's it. To illustrate this, consider the computation of the binomial equation related to the results of four trials: $P(p + q)^4$. Here we trace down to row 4 and find that Pascal's triangle gives us the values of 1, 4, 6, 4, and 1. These will be the coefficients, so what are the p and q terms? By convention, we start with $Y = $ number of trials, so, using Equation (5.4), our first term is $1(p^4q^{4-4})$. $4 - 4 = 0$, of course, so the q term becomes 1 (any non-zero number taken to the power of 0 equals 1). The binomial is consequently $1(p^4 \times 1)$, which can be shortened simply to p^4. The second term, which corresponds with three successes ($Y = 3$), will be $4(p^3q^1)$.

Number of trials					Binomial coefficients					
0						1				
1					1		1			
2				1		2		1		
3			1		3		3		1	
4		1		4		6		4		1
5	1		5		10		10		5		1
6	1		6		15		20		15		6		1
7	1		7		21		35		35		21		7		1

Figure 5.5 Pascal's triangle

The third term will use the coefficient of 6 and $Y = 2$, and will be $6(p^2q^2)$. Note that this is the same coefficient we determined when counting the number of possible combinations to get two successes and two failures previously. The term for one success is $4(pq^3)$ and for no successes is $1(p^0q^4)$, or, more simply, q^4. Using Pascal's triangle, we have determined that $P(p + q)^4 = p^4 + 4(p^3q^1) + 6(p^2q^2) + 4(pq^3) + q^4$. Using the exact same steps, we can quickly determine that $P(p + q)^7 = p^7 + 7(p^6q) + 21(p^5q^2) + 35(p^4q^3) + 35(p^3q^4) + 21(p^2q^5) + 7(pq^6) + q^7$ without having to bother with all of that "boring algebra stuff".

The version of Pascal's triangle presented here is limited to seven rows, but it can be expanded to an infinite number of rows if need be. This is accomplished easily by added the cells immediately above each new cell to derive its integer. Remember that the first and last cells in each row are always 1.

Even though expanding Pascal's triangle is easier than performing the algebra, there is an even easier way to determine a coefficient when only a small portion of the entire binomial distribution is needed. Returning to our example of wishing to determine the probability of three or fewer people sharing a birthday in a room with 41 people, we could expand Pascal's triangle to $k = 40$. This is easier than expanding the binomial by hand, but that is a lot of work, when all we need is three coefficients (those associated with 0, 1, and 2 shared birthdays), not all 40 of them. An easier way is to use Equation (5.5) to determine them.

Calculation formula for the binomial coefficient

$$C(k, Y) = \frac{k!}{Y!(k-Y)!} \tag{5.5}$$

Here, k is the number of trials, Y is the number of successes (which must be equal to or fewer than the number of trials), and $C(k,Y)$ is the binomial coefficient corresponding with k and Y. You will notice the exclamation marks in the formula. Their presence doesn't reflect some sort of jubilance on our part about this formula, but instead represents the factorial mathematic process. When calculating the factorial of a number, take that number and multiply it by all integers between itself and 1 (e.g., $4! = 4 \times 3 \times 2 \times 1$).

To continue our example, we wish to determine $C(40,40)$, $C(40,39)$, and $C(40,38)$. We already know that $C(40,40)$ is 1, because there is only one way to get all successes. Equation (5.5) confirms this as $C(40, 40) = \dfrac{40!}{40!(40-40)!}$. $0!$ equals 1 by definition, so $C(40,40)$ is $\dfrac{40!}{40!} = 1$. $C(40, 39) = \dfrac{40!}{39!(40-39)!}$. Given that $40! = 40 \times 39!$, the $39!$ in the denominator will cancel out the $39!$ in the numerator, necessitating that the coefficient for $C(40, 39) = \dfrac{40 \times 39!}{39!} = 40$. Using the same process $C(40, 38) = \dfrac{40!}{38!(40-38)!} = \dfrac{40 \times 39 \times 38!}{38! \times 2} = \dfrac{40 \times 39}{2} = 780$. We now know that there are 780 ways to get 38 successes and two failures in a sample of 40 trials and 40 ways to get 39 successes and one failure. Using these coefficients with Equation (5.4), we determine that the relevant binomial terms are $P(38$ or more successes$) = (325 / 365)^{40} + 40(325 / 365)^{39}(40 / 365) + 780(325 / 365)^{38}(40 / 365)^2 = .17$. Calculating this probability took far less effort than either expanding $(p + q)^{40}$ or Pascal's triangle would. The inclusion of Equation (5.5) also allows us to present a unified formula for each binomial term in Equation (5.6), where $C(k,Y)$ is the product of Equation (5.5), Y is the number of successes, and k is the number of trials.

Unified formula for specific binomial terms

$$C(k, Y) p^Y q^{k-Y} \tag{5.6}$$

Probability in Archaeological Contexts

Now that you are familiar with the basic structure of probability and its manipulation, we can abandon our dice, card, and parking examples and return to more relevant (and interesting) aspects of using probability to study the archaeological record. As good a tool as the human mind is, it tends to make mistakes when dealing with probability. An archaeologist must be on guard against these lapses. The importance of probability to archaeological research is not often mentioned, but it forms a central tenant to most research. Given that archaeologists generally deal with samples, whether these are at the level of sites in a region, artifacts from a site, or collections of artifacts from a single structure or feature, we are constantly faced with the necessity to determine if some perceived or documented pattern is similar to, or different than, patterns observed from other samples or some theoretically expected outcome. The only means of systematically doing this (as opposed to expressing an intuitive, subjective opinion on it) is through probability.

The binomial is a tremendously useful tool for archaeological analysis, so long as we can limit our outcomes to two states. Given that archaeologists often don't organize their data into two outcomes (a success and a failure), using the binomial might require a bit of creativity on the analyst's part. An easy way to do this might be to reclassify variables using a new nominal or ordinal scale taxonomy. For example, a faunal collection of many different species might be reduced to deer/not deer for use with binomial analysis, a multiplicity of subsistence resources could be reduced to high ranked/low ranked resources, and stone artifacts could be classified as formal tools/lithic debitage. The usefulness of such classes is contingent on the theoretical and analytic structure the researcher is using, but the binomial distribution can be an extremely easy and effective means of determining if some apparent association is likely given random chance or unlikely (and therefore likely reflecting some archaeologically significant relationship).

Consider the following example. In an analysis of decorative icons on Medio period Casas Grandes effigies from northwestern Mexico, Christine VanPool noted that bird icons were far more common on women then on men; seven women effigies had bird icons whereas only two male effigies had bird icons, despite the fact that male effigies were more common (VanPool and VanPool, 2006). This proportion seems intuitively unlikely if in fact males and females are equally liable to be associated with bird images. In fact, the preponderance of bird images on females could indicate that women were more closely associated with these images than men, which in turn could provide insight into Medio period gender distinctions and other social factors. Yet all of this is based on an "intuitive" feeling, and, as the psychic's trick demonstrates, using our intuition can lead to mistakes when dealing with probabilities. Is the ratio of 7 to 2 really improbable, assuming that effigies of males and females are equally likely to be associated with bird icons? The binomial distribution will help us answer this question.

Here we use a theoretically derived probability of P(bird icon on woman effigy) = .5 and P(bird icon on male effigy) = .5. Our question is, "what is the probability that birds would be associated with two or fewer males out of a sample of nine effigies with bird icons?" For simplicity's sake, we define a success as the association of a bird with a female effigy. Thus, we seek to determine P(7 or more females) using the binomial $(p + q)^9$. Using Equation (5.6), we know that:

$$P(\text{7 or more females}) = p^9 + \frac{9!}{8!(9-8)!}p^8q + \frac{9!}{7!(9-7)!}p^7q^2$$

$$= p^9 + 9p^8q + 36p^7q^2 = .002 + .018 + .070 = .090$$

There is a 9% likelihood that in a sample of nine effigies with bird icons, seven or more of them would be on females, assuming that males and females were equally likely to have bird images. While .09 is unlikely, this may not be as improbable as some intuitively would suspect, and it may not be sufficiently improbable to warrant an argument about some socially important association between women and birds (although some might make the argument anyway). Importantly, though, archaeologists have a yardstick by which to measure the probability associated with this pattern, as opposed to being forced to argue about the issue from a strictly intuitive, commonsensical framework. The same technique could be used to determine whether some portion of a graveyard has more members of one sex than would be expected by chance, whether some portion of the site has more storage pits than expected by chance, or any of a multitude of comparisons that archaeologists might find useful. You are limited only by your imagination and ingenuity. Understanding probability truly opens up a new world for framing and making comparisons using all of the wonderful data archaeologists have generated for over a century.

However, as illustrated by the various examples above, understanding probability does not allow us to determine the certainty of some association. Improbable events happen in the real world every day. Even something that has only a one in a million chance of happening is nearly certain to have happened after a million trials. Understanding probability does not mean that we can interpret the implications of that probability with certainty. Does the 9% probability associated with the bird icons on Medio period effigies determined above reflect a meaningful association of birds and women? We cannot know from the probability itself. Instead, this probability must be understood within appropriate theoretical and empirical frameworks. Perhaps other lines of evidence support this association, thereby helping the argument for an association. Perhaps instead other evidence undermines the apparent association, thereby strengthening the position that it is simply a product of random chance. Or perhaps the researcher decides to collect more data to help determine if some vagary of the sample is either creating or obscuring the apparent association. There is no way of determining with certainty which associations *are* archaeologically meaningful, and which simply reflect random chance in some way.

What would be useful to archaeologists (and all other scientists for that matter) is some means of clearly assigning significance to various probabilities while explicitly identifying the likelihood of making erroneous conclusions. Then we could both specify the likelihood that a relationship is meaningful while also specifying the likelihood that it isn't. Fortunately there are means of doing so. This process is called hypothesis testing, and is the core of the discussion contained in Chapter 7 and throughout much of the remainder of this book. But before we discuss hypothesis testing, we wish to further explore how we can determine probabilities focusing on a specific distribution, the normal distribution.

Practice Exercises

1   Assume you have a normal deck of playing cards:

(a)   What is the probability of selecting a heart or a diamond? Draw a Venn diagram reflecting this probability.

(b)   What is the probability of selecting a diamond or a queen? Draw a Venn diagram reflecting this probability.

(c)   What is the probability of selecting a diamond, a heart, or a queen? Draw a Venn diagram reflecting this probability.

(d)   What is the probability of selecting a heart, a diamond, a queen, or a jack? Draw a Venn diagram reflecting this probability.

(e)   What is the probability of drawing two hearts in a row, assuming that the card from the first draw is reshuffled into the deck before the second draw (i.e., that the probability of a success doesn't change)?

(f)   What is the probability of drawing at least one heart OR jack OR diamond in two draws, assuming that the card from the first draw is reshuffled into the deck before the second draw?

2   The following data reflect the manufacturers of glass bottles recovered during an excavation of a historic homestead.

(a)   Determine the probability of bottles from each manufacturer at the site.

(b)   In a random sample of 20 bottles, how many bottles from each manufacturer would one expect to recover?

(c)   Determine the probability of recovering a bottle from the Nehi Bottling Company or the Royal Crown Cola Company. Draw a Venn diagram illustrating these probabilities.

Bottle company	Number of bottles
Nehi Bottling Company	36
Royal Crown Cola	49
Ritters Bottling	12
Phoenic Bottling	32
Seven-up Bottling Company	7

3   A researcher who excavated a homestead near the historic homestead mentioned in Question 2 wishes to determine if there are significant differences in the frequencies of bottles from the various manufacturers to help identify possible differences in status and consumption patterns. Using the probabilities from Question 2, answer the following questions.

(a)   Would the probability of recovering only one bottle from the Nehi Bottling Company in a sample of seven bottles from the nearby homestead be less than .05?

(b)   What would the probability of recovering three or fewer bottles from the Nehi Bottling Company be in the sample of seven bottles?

4   Five skulls excavated from a large Pleistocene cave in mainland China had the following cranial capacities: 1120 cc, 1140 cc, 1050 cc, 1230 cc, and 1020 cc. Given this sample, what is the probability of a skull smaller than 1200 cc? In a sample of three skulls, what is the probability of each possible outcome defining a success as a skull smaller than 1200 cc (e.g., three successes, no failures; two successes, one failure, etc.)?

5   In a historic cemetery, grave stones are made of a locally available lime-stone and an imported marble, which was known from historic records to be more expensive. Across the cemetery, a researcher determines that the probability of limestone grave stones is .83 compared to a probability of marble gravestones of .17. During the years 1880 and 1890, there seems to be an increase in the frequency of marble gravestone ($n = 5$) compared to limestone gravestones ($n = 3$), which might reflect increased wealth and/or investment in mortuary ritual within the community during that time period.

(a)   Determine the statistical likelihood of the apparent increase in the frequency of marble gravestones given the general frequencies of marble and limestone gravestones. (Do this by determining the like-lihood of three or fewer limestone gravestones in a sample of eight gravestones.)

(b)   Based on the resulting probability, does it seem intuitively likely to you that the choice in gravestone material is consistent with the general pattern reflected throughout the entire cemetery, or that marble gravestones were in fact more commonly used from 1880 and 1890 than typical during other periods reflected at the cemetery? Justify your conclusion to the best of your ability.

6

Putting Statistics to Work: The Normal Distribution

Statistical comparisons are based on probabilities. These probabilities can be used to evaluate the likelihood of specific propositions (hypothesis testing) or discovering unknown associations that are suggestive of some significant relationship (EDA). Prior to the binomial, we discussed various measures of dispersion and central tendency that characterize continuous distributions of data measured at an interval or ratio level of measurement (see Chapter 4). These measures are not useful with binomial distributions, but they are central to using probability to evaluate relationships reflected in many archaeological data. Knowledge of the mean and standard deviation might let us identify "unusual" (i.e., improbable) members of a distribution, and thereby identify meaningful differences in artifacts, sites, features, bones, or whatever we happen to be evaluating at the moment; help us determine similarities and/or cultural continuity where we might not expect it; or simply evaluate similarities or differences when we wish to know more about the structure of our data.

For example, consider the means and other summary statistics of the length of unbroken adze rejects from different production locations on the island of Hawaii (Table 6.1). The production and distribution of adzes has been argued to reflect chiefly control and specialized production (Bayman and Nakamura, 2001). Are there significant differences in the adze reject lengths among the production locales? Would an interested researcher likely be able to differentiate adzes made at each location based on length? Do the adze rejects reflect a level of standardization indicative of specialized production and economic centralization as some authors have suggested? With which production locale would an adze 73 mm long be associated? How about an adze that is 140 mm long? Does the use of only two adze rejects from Pōhakuloa impact our ability to characterize the underlying distribution?

Table 6.1 Lengths (mm) of unbroken adze rejects from three production locales on Hawaii

Locale	Number	Mean	Minimum	Maximum	Range
Pōhakuloa	2	82.5	63	102	39
Pololū	19	123.7	90	170	80
Mauna Kea	63	176.2	74	349	275

Source: Bayman, James M. and Jadelyn J. Moniz Nakamura (2001). Craft specialization and adze production on Hawaii Island. *Journal of Field Archaeology*, **28**: 247.

These are all questions that are, at their heart, issues of probability; measuring probability will help us determine whether the adzes have similar or different lengths, whether adzes from each source can be differentiated according to their lengths, and whether the distributions reflect little or great variation. Probability will also help us determine when we can make a conclusion with great certainty (i.e., when the probability that our conclusion is correct is high) and when we can't (i.e., when the probability is comparatively low).

Given that the probabilities associated with any given distribution of archaeological material differ as reflected by differences in their means and standard deviations (and other descriptive statistics such as those reflected in Table 6.1), the probabilities associated with each distribution must presumably be unique. A projectile point 5 cm long might be quite likely in an assemblage of atlatl dart points with a mean length of 4.7 cm but improbable in an assemblage of arrow points with a mean of 2.1 cm. Given that we may have an infinite number of means as well as an infinite number of standard deviations, ranges, etc. that describe potential distributions, it seems that measuring probabilities using continuous distributions is a daunting task that requires one to determine the characteristics of each individual distribution and assign probabilities based on these. With the exception of being "daunting", this is indeed the case.

There are two ways probabilities can be determined for continuous distributions. First, we can use (re)sampling from a larger distribution to empirically determine probabilities. This is called "Monte Carlo simulations", and is somewhat common in archaeology. A second, more traditional (and easier) approach is to cluster the unique distributions into different "types" based on their similarities in important characteristics. These groups of similar distributions can be further characterized by an ideal (i.e., theoretical) distribution that typifies the distributions' important characteristics. Statistics for measuring probabilities can then be developed based upon our knowledge of the ideal distribution and applied to the real distributions by extension.

Archaeologists are of course familiar with the concept of types, and, if it helps, we can conceive of these groups of distributions as analogous to artifact types, with each type possessing its own defining characteristics shared by its members. These similarities allow the use of the same techniques to measure probabilities for each group of distributions, just as the methods described in Chapter 5 can be used for

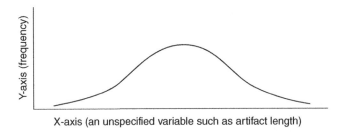

X-axis (an unspecified variable such as artifact length)

Figure 6.1   The normal distribution

all binomial distributions. Perhaps the most important ideal distribution is the "normal" distribution (Figure 6.1). Given its importance in many scientific contexts and its general familiarity to most social scientists, we will use it to introduce the general principles associated with measuring probability using continuous distributions.

Most people are familiar with the normal distribution, often described as a "bell-shaped curve." You may have even encountered it as a scale for grading. The bell-shaped curve is nothing but a special case of the normal distribution; the words "bell-shaped" describe the general shape of the distribution (it looks sort of like a bell), and the word "curve" is used as a synonym for distribution. While we generally refer to *the* normal distribution, there are really an infinite number of normal distributions; there are as many different normal distributions as there are possible means and standard deviations, both theoretically and in the real world. However, all of these normal distributions share five characteristics:

1 *Symmetry.* If divided into left and right halves, each half is a mirror image of the other.
2 *The maximum height of the distribution is at the mean.* A consequence of this and the previous characteristic is that the mean, the mode and the median are the same value.
3 *The area under a normal distribution sums to 1.* This proposition can be said in a more obtuse, but more precise, way by saying that the area under the curve must sum to unity. This requirement probably doesn't make intuitive sense, but it isn't really hard to understand. Consider that the distribution itself reflects a finite probability space that corresponds with the frequencies of particular variates. The mean is the most common variate and as a result is the highest point of the distribution. This in turn means that it is the value with the highest probability of occurring. In contrast, outliers far from the mean are less common than variates close to the mean, and, as a result, their frequencies are closer to zero as reflected by the closeness of the distribution's line to the *X*-axis. A consequence of this relationship between the shape of the distribution and the likelihood of individual variates is that the distribution can be used to determine the probability associated with a variate or range of variates. Given that the distribution contains all possible outcomes, then the likelihood of each possible

variate can be calculated and the sum of all possible outcomes must be 100% (which of course corresponds with a probability of 1). Further, because the distribution is symmetrical, half of the possible outcomes and their associated probabilities are on either side of the mean.

4 *Normal distributions are theoretically asymptotic at both ends, or tails, of the distribution.* Asymptotic is a fancy way of saying that the tails of the distribution never actually touch the X-axis, but instead become incrementally ever closer to zero without ever quite reaching it. This aspect of the normal distribution follows from Characteristic 3 above and is necessary because the distribution *must* include all possible outcomes; we need to consider every possible variate to infinity. Put another way, every single possible variate can be assigned *some* probability of occurring, even if it is infinitesimally small. In practicality, there may be limits to the size ranges we observe. For example, a historic beer bottle won't have a volume as small as 1 cc, and we don't expect to find the skeletal remains of a 9-foot tall Iron-age woman. However, we can use the normal distribution to assign a probability to any possible outcome, even an outcome that is a practical impossibility. In other words, we can quantify exactly *how unlikely* such finds might be.

5 *The distribution of means of multiple samples from a normal distribution will be normally distributed.* Considering this commonality among normal distributions requires thinking about means somewhat differently. As you know, means characterize groups of variates. In this special context we need to consider calculating individual means on repeated samples from the same population, and *plotting these means as variates* that collectively create a new distribution. This new distribution will be normally distributed.

All normal distributions share these five commonalities by definition, but they can otherwise be quite different from one another. As Figures 3.17, 3.18 and 3.19 illustrate, normal distributions may appear leptokurtic, platykurtic, or mesokurtic relative to each other. Additionally, any combination of means and standard deviations is possible, and there is no necessary relationship between the mean and the standard deviation for any given distribution. Normal distributions may have different means and the same standard deviation (Figure 6.2) or the same means and different standard deviations (Figure 6.3).

If σ is large, variates are generally far from the mean. If σ is small, most variates are comparatively close to the mean. Despite such differences, the underlying shared characteristics of normal distributions cause them to share additional attributes reflected in probabilities. Regardless of the standard deviation, variates near the mean in a normal distribution are more common (and therefore more probable) than variates in the distribution's tails. In fact, regardless of the value of σ or μ:

$\mu \pm 1\sigma$ contains 68.26% of all variates
$\mu \pm 2\sigma$ contains 95.44% of all variates
$\mu \pm 3\sigma$ contains 99.73% of all variates

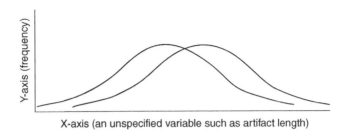

Figure 6.2 Two normal distributions with different means and the same standard deviation

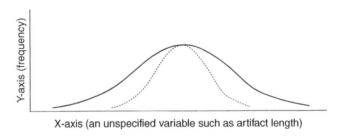

Figure 6.3 Two normal distributions with the same mean and different standard deviations

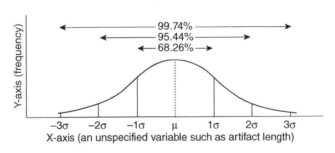

Figure 6.4 Percentages of variates within one, two, and three standard deviations from μ

This is illustrated in Figure 6.4, and is true (and is only true) for all normal distributions. It is also possible to express this relationship in terms of more commonly used percentages. For example:

50% of all variates fall within $\mu \pm .674\sigma$
95% of all variates fall within $\mu \pm 1.96\sigma$
99% of all variates fall within $\mu \pm 2.58\sigma$

If $\mu \pm 1\sigma$ contains 68.26% of all variates, $\mu \pm 2\sigma$ contains 95.44% of all variates, and $\mu \pm 3\sigma$ contains 99.74% of all variates (Figure 6.4), we know that values beyond

$\mu \pm 2\sigma$ are rare events, expected fewer than five times in 100, and $\mu \pm 3\sigma$ is even more rare, expected fewer than one time out of 100.

The consistency of normal distributions allows us to consider the probability of individual variates occurring within some portion of the distribution. We know that the percentages mentioned above may be converted to probabilities such that the probability that a variate is within $\mu \pm 1\sigma$ is .6826, within $\mu \pm 2\sigma$ is .9544, and within $\mu \pm 3\sigma$ is .9974 (Figure 6.5). These probabilities are unchanging for all normal distributions regardless of their means or standard deviations. Furthermore, probabilities may be calculated for any area under the curve, not just around the mean. Assuming a normal distribution, we could calculate the probability of an artifact being longer than 10 cm, a feature being between 1 and 3 cm deep, or a building being twice as long as one that is of average length. These "areas under the curve" vary depending on the location and the shape of the distribution as described by the mean and the standard deviation, but by understanding the characteristics of normal distributions, we can calculate the various probabilities that interest us *using the same principles.*

Calculating probabilities using the normal distribution really isn't that much different, then, than the binomial distribution we previously discussed, in that the underlying similarities among all possible binomial distributions make it possible to use the same principles to effectively determine probabilities for each individual binomial distribution. Still, given that each normal distribution differs according to its mean and standard deviation, how can we determine the probabilities associated with all of them? The probability that a variate is larger than 10 cm will be drastically different for different distributions. It would be easier for us if we were dealing with distributions that all had the same means and standard deviations. Then we could calculate only one set of probabilities that could be applied to all normal distributions. This would seem impossible given that normal distributions do in fact differ from one another, but statisticians have created a clever way to create a *standardized normal distribution,* which makes all normal distributions identical. This is done using Equation (6.1) to standardize the distribution using the relative sizes of the mean and the standard deviation. The result is a normal distribution where $\mu = 0$ and $\sigma = 1$ regardless of the original values of μ and σ.

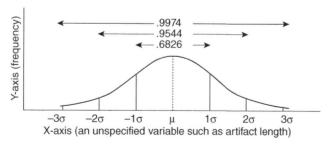

Figure 6.5 Areas under the normal distribution corresponding with the standard deviation

Calculation formula for a Z-score reflecting a standardized normal distribution

$$Z = \frac{Y_i - \mu}{\sigma}$$ (6.1)

The calculation of a Z-score, the outcome of Equation (6.1), establishes the difference between any variate and the mean $(Y_i - \mu)$, and expresses that difference in standard deviation units (by dividing by σ). In other words, the product of the equation, called a Z-score, is how many standard deviations Y_i is from μ. This in turn creates a new distribution in which the value of each variate is expressed in the number of standard deviation units it is from the mean (e.g., a variate that is one standard deviation larger than the mean will have a value of 1, a variate that is two standard deviation units less than the mean will have a value of −2). A consequence of this is that a single set of probabilities can be easily applied to every normal distribution using the exact same regularities that allow Figure 6.4 to reflect that 68.26% of the variates lie within 1σ of μ. Appendix A lists probabilities for the standard normal distribution as areas under the curve from the mean to a given Z-score. The probabilities can be used to determine the exact probabilities under the curve between any points we might chose. To illustrate this let us consider the following example.

Donald K. Grayson, in his analysis of the microfauna from Hidden Cave, Nevada, notes that only one species of pocket gopher, *Thomomys bottae*, occurs in the area today, although it is possible that other species were present in the past. Grayson (1984:144) presents the following descriptive statistics on mandibular alveolar lengths for a population of modern *Thomomys bottae*: $\mu = 5.7$ mm and $\sigma = .48$ mm. Archaeological specimen number HC-215 has a value $Y_1 = 6.4$ mm. What is the probability of obtaining a value between the mean, $\mu = 5.7$ mm, and $Y_1 = 6.4$ mm (Figure 6.6)? We can calculate the Z-score as follows:

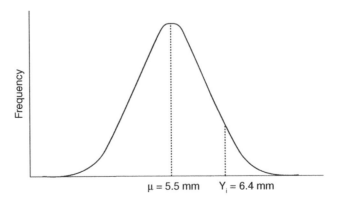

μ = 5.5 mm   Y₁ = 6.4 mm

Figure 6.6   The relationship between the mean and $Y_i = 6.4$ mm

$$Z = \frac{Y_i - \mu}{\sigma} = \frac{6.40 - 5.70}{.48}$$

$$Z = 1.46$$

The Z-score tells us that $Y_i = 6.4$ mm is 1.46 standard deviations from the mean $\mu = 5.7$ mm. Is this a common or rare event? We know in general it is common, as the value lies between one and two standard deviations from the mean. Appendix A, though, allows us to determine the exact probability. To find the probability for 1.46 standard deviation units, look down the left side of the table until the value 1.4 is located. Follow this row until it intersects with the column value for .06. At that intersection is the value .4279, which represents the probability of a variate falling between the mean and $Z = 1.46$. A value in that interval is therefore a common event; we expect it to occur about 43% of the time. There is no reason to stop with this probability, though.

We can also find the probability of having a value greater than and less than $Y_1 = 6.4$ mm ($Z = 1.46$). Since we know that the total probability represented in the curve is equal to 1.0, and that .50 lies on each side of the mean, we can determine that the probability of a value less than $Z = 1.46$ is $.5 + .4279 = .9279$. The probability of a value greater than $Z = 1.46$ is $1 - .9279 = .0721$. We can then conclude that roughly 93% of the variates should be less than $Y_1 = 6.4$ mm, but that larger values are relatively rare events in that we would expect them only seven times out of 100 trials.

The above example illustrates finding probabilities based on areas under the normal curve. It should be noted that we cannot determine exact values that represent individual variates (e.g., the probability of having the exact value of $Y_1 = 6.4$ mm), because this is a point, instead of an area. Unlike the binomial distribution that provided the probability of specific outcomes, the standard normal distribution can only be used to find the probability associated with some range of possible outcomes (e.g., the probability of a value between the mean and a variate). As a result, the way to determine the probability associated with a specific value is to find the area under the curve that corresponds with the value's implied limits. Thus, if we wish to find the probability corresponding with $Y_1 = 6.4$ mm, we should look for the area under the curve corresponding to Z-scores for 6.35 and 6.45 cm, which are the implied limits for $Y_i = 6.4$ cm in this case.

Note that Appendix A only presents values for areas where Y_1 is greater than the mean. What happens if Y_1 is less than the mean? Let us illustrate this using Grayson's gopher data again. What is the probability of an alveolar length between 5.3 mm and 6.8 mm (Figure 6.7)? We can illustrate this probability in the following way:

$$P\{5.3 < Y_i < 6.8\}$$

$$P\left\{\frac{Y_1 - \mu}{\sigma} < Z < \frac{Y_2 - \mu}{\sigma}\right\}$$

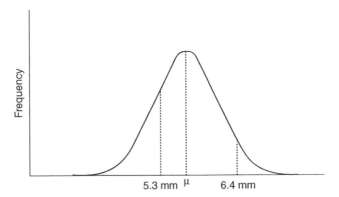

5.3 mm μ   6.4 mm

Figure 6.7 The area under the standardized normal distribution between $Y_1 = 5.3$ mm and $Y_2 = 6.8$ mm

$$P\left\{\frac{5.3-5.7}{.48} < Z < \frac{6.8-5.7}{.48}\right\}$$

$$P\{-.83 < Z < 2.29\}$$

Since the normal curve is symmetrical, it is possible to ignore the negative sign for −.83 to use Appendix A to find the area between this value and the mean. The tabled value for $Z = .83$ is .2967. The tabled value for $Z = 2.29$ is .4890. Since we are interested in the area under the curve between −.83 and 2.29, we can sum the two individual probabilities to determine that the

$$P\{-.83 < Z < 2.29\} = .2967 + .4890 = .7857.$$

In this case, we added the areas under the curve. However, we might need to subtract areas under the curve to find some probabilities. For example, calculating $P(.83 < Z < 2.29)$ requires us to subtract the probability associated with the distance from μ to $Z = .83$ mm from the distance between μ and $Z = 2.29$ mm. We previously talked about adding and subtracting probabilities in Chapter 5, so mathematically manipulating them should be straightforward. However, we can easily make mistakes by confusing when we should add or subtract specific probabilities. While it may seem tedious, it is important that you explicitly write and draw a sketch of your probability space. It is a useful and easy way to keep track of the probability space you are after while ensuring you do not make a simple mistake.

It has probably already occurred to you that being able to measure probability using quantitative methods such as the binomial distribution or the standard normal distribution can be incredibly useful. For example, we may now conclude that specimen HC-215, a gopher mandible from Hidden Cave, does not differ in a significant manner from the sample of modern *Thomomys bottae*. If it did, it may have led us to suggest that another species of *Thomomys* was present at Hidden

Cave in the past – a conclusion of significant paleoenvironmental and archaeological significance. Here, our conclusion is largely commonsensical based on our determination that roughly 7% of the *Thomomys bottae* mandibles should be larger than HC-215, meaning that the mandible is not unusually large for a gopher of this species. These sorts of commonsense conclusions can be useful, but they open up a lot of room for subjective interpretations, especially given that our "commonsense" varies so greatly from person to person and through time. The process of using probabilities to evaluate propositions about the world, which are typically called hypotheses, can be formalized through statistical hypothesis testing, which is the topic of the next chapter.

Practice Exercises

1 What are the five characteristics shared by normal distributions? Define the following terms: distribution, symmetrical, sum to unity, and asymptotic.

2 How does *a* normal distribution compare to *the standardized* normal distribution? How can a normal distribution be transformed into a standardized normal distribution?

3 Given a population of metates with an average grinding surface width of 21.2 cm ($\mu = 21.2$ cm) and a standard deviation of 4.7 cm ($\sigma = 4.7$ cm), answer the following questions:

   (a) What is the probability that a metate has a grinding surface of less than or equal to 10.4 cm in width? Draw a diagram of a normal distribution illustrating this area.

   (b) What is the probability that a metate has a grinding surface of greater than or equal to 20 cm in width? Draw a diagram of a normal distribution illustrating this area.

   (c) What is the probability that a metate's grinding surface is between 15.4 and 25.3 cm wide? Draw a diagram of a normal distribution illustrating this area.

4 A project director has determined over the years that the average crew member of an archaeological excavation team can excavate a 10 cm level in a 1 meter square test unit in about 1.5 hours, with a standard deviation of 15 minutes. What is the probability that a randomly selected crew member will take longer than 1 hour to excavate a 10 cm level? What is

the probability that the crew member will take less than 2 hours? Draw diagrams of a normal distribution illustrating both areas.

5   A normally distributed population of masonry stone lengths has $\mu = 50\,cm$ and $\sigma = 12\,cm$. There are 250 variates between 45 cm and 69 cm. How large is the population?

6   McKenzie (1970: 357) has demonstrated that fluted projectile points from the state of Ohio average 27.01 mm in width, with a standard deviation of 4.28 mm.

(a)   What is the probability that a randomly selected point will have a width less than 23 mm?

(b)   If there are a total of 772 fluted points from Ohio, about how many would you expect to be wider than 28 mm?

7

Hypothesis Testing I: An Introduction

Hypothesis testing is the heart of science. In its broadest sense, it is the procedure used to evaluate our competing ideas about how our world is structured. Statistical hypothesis testing is merely one formulation of the more general procedure operative in science. In archaeology, the ideas subject to evaluation are our ideas about how (and why) the archaeological record is structured as it is.

It is common for archaeologists, as well as scientists in general, to be interested in statistical procedures for hypothesis testing that involve comparisons of one kind or another. For example, we are frequently interested in comparing a variate to a population mean, a sample mean to a population mean, or two sample means to each other. Often the relationships we wish to evaluate are conceptually difficult and/or linguistically cumbersome to express in common language, yet it is crucial to state what we wish to evaluate unambiguously so that others can understand our research and so that we do not become confused and make preventable mistakes. As mentioned in Chapter 5, scientists and statisticians have consequently developed a uniform method of statistical notation to help simplify the process. A major portion of this notational language focuses on stating hypotheses so that they can be properly understood within and between disciplines. The procedure is ritualized to ensure consistency and relies on very specific notation and conceptual structures, which you must master to use or understand quantitative methods in your own research or the work of others. Interestingly the cornerstone of hypothesis testing is not a statistical issue at all, but is instead the analytically central "hypothesis of interest" that frames our analysis. We will consequently begin by discussing it.

Quantitative Analysis in Archaeology, Todd L. VanPool and Robert D. Leonard
© 2011 Todd L. VanPool and Robert D. Leonard

Hypotheses of Interest

Statistical hypothesis testing is a mathematical means of calculating the probability that some relationship we posit is correct. This relationship reflects our hypothesis of interest. A hypothesis of interest can be any empirical statement. Perhaps a researcher thinks that copper bells made in West Mexico are smaller than those made elsewhere in Mesoamerica because of some difference in manufacturing techniques. This hypothesis of interest provides a clear statement of the suspected relationship between classes of objects that can be evaluated with data collected using appropriate theoretical, empirical, and methodological tools. Hypotheses of interest can posit specific expectations regarding possible differences such as "West Mexican bells are smaller", but they don't have to. A common hypothesis of interest for archaeologists is whether or not two sites are contemporaneous, a statement that might not include any presupposition about which site is older or the duration of any temporal difference that might be present. The point is that the hypothesis of interest states the relationship that an archaeologist wishes to consider about the world.

Most archaeologists, especially those who argue for "hypothesis testing" as a key to scientific archaeology, hold that it is better to form your hypothesis of interest before collecting data (Binford, 1964; Black and Jolly, 2002). There are several reasons why this is true:

- Having a previously existing hypothesis of interest allows us to more reliably collect the right information to generate meaningful data for our analysis. There is nothing more irritating than returning from the field and realizing that there was some important piece of information necessary to analyze a significant hypothesis of interest that was not properly collected. Anyone who has tried to use excavated collections created by someone else realizes the significance of the relationships between what we seek to study and the collection and reporting of information and artifacts. Otherwise interesting hypotheses of interest might have to be discarded from the analysis for no other reason than the data necessary to evaluate them don't exist and can't be generated. Clearly stating your hypothesis of interest *before data collection* can help minimize this problem.
- Granting and permitting agencies/clients often take the research design into account when deciding if archaeological research will be supported. Clearly defined and meaningful hypotheses of interest may consequently be a prerequisite to conducting archaeological research, especially in cultural resource management where the hypotheses of interest address relationships central to the Section 106 compliance process or similar legal frameworks.
- Clear hypotheses of interest can save time and money, and promote the archaeological record's conservation by allowing us to concentrate on collecting data from the applicable portions of the archaeological record. There is no reason to excavate the post-Classic Maya site if your interests are focused on the nearby

early pre-Classic village, or to record the volume of pots when your interest is their paint composition. While this may seem obvious, we have been awestruck several times by the amount of extraneous data some archaeologists generate with no apparent connection to their research questions. Thinking about your hypotheses of interest before conducting the research can save months, if not years, of work while improving the ultimate quality of the research.

· A clearly stated hypothesis of interest clarifies the appropriate theoretical and methodological frameworks an archaeologist should use at a given point of time. This will in turn lead to a more coherent integration between our empirical and conceptual tools and will help prevent problems arising from incorrectly gathered data or incompatible methods and interests.

Still, there will be times when a hypothesis of interest may arise only after data are collected. This can be tremendously significant in archaeological contexts as we identify associations or differences that haven't previously occurred to us. EDA with its rich history in archeological analysis is in fact a formalized method of finding these associations. There is nothing analytically or conceptually wrong with EDA as a means of identifying hypotheses of interest, but relying on it exclusively increases the likelihood of wasted effort collecting unnecessary data, incompatibility between data and theoretical/methodological structures, and investing resources investigating chance associations that initially appear meaningful but aren't. Having at least some idea of your hypotheses of interest is a good place to start, even when using EDA techniques.

Another key element to a good statistical analysis is realizing that a single analysis can have implications for several integrated hypotheses of interest. Although there are exceptions, archaeologists rarely have a single hypothesis of interest when conducting research. More commonly they wish to evaluate a series of propositions that may overlap considerably. Testing multiple competing hypotheses is a straightforward example, but certainly not the only case in which hypotheses of interest can inform on one another. The utility of quantitative analyses can be greatly improved if the researcher takes the articulation among hypotheses of interest into account.

Shifting the consideration of hypotheses of interest down from an abstract level to an empirical application, our discussion of the normal distribution in the previous chapter employed a hypothesis of interest focused on evaluating the similarity of a variate to a population mean in that we used the Z-score to determine the likelihood that a prehistoric gopher mandible was from a member of the species *Thomomys bottae* based on its length. Another common archaeological hypothesis of interest is the function of ceramic vessels. Imagine that researchers in a particular culture area have argued that performance requirements cause the maximum wall thickness of water storage pots and cooking pots to be different. An archaeologist investigating this relationship might have two related hypotheses of interest:

1   to determine if the average maximum wall thickness of vessels identified as cooking pots (μ_1) based on archaeological context, residue analysis, sooting, etc.

is meaningfully different than the average maximum wall thickness of vessels identified as water storage pots (μ_2) based on archaeological context, porosity, and other lines of evidence; and if so,

2   to determine if newly excavated pots are likely water storage or cooking vessels based on their maximum wall thickness (Y_1, Y_2, ... , Y_n).

Let us imagine that we do find a difference between μ_1 and μ_2 (using techniques that will be introduced in Chapter 8), and then use a Z-score to determine that two variates, Y_1 and Y_2, are 0.73σ and 3.5σ from μ_1 but 3.1 and $-.62$ from μ_2. We know that an individual variate does not differ very much from a population mean if it is less than 1σ away from μ. In fact, variates that are within $\pm 1\sigma$ of a mean are very common in a population, but those more than $\pm 3\sigma$ away from μ are unlikely (more than 99.84% of the variates in a normal distribution are closer than $\pm 3\sigma$ to the mean). We therefore recognize that it is extremely unlikely that Y_1 was drawn from population μ_2, but is quite consistent with population μ_1; and Y_2 likely isn't a member of μ_1, but could be a member of μ_2. We may intuitively conclude that Y_1 is probably a cooking pot and that Y_2 is more likely a water storage pot based on the proximity of each variate to the means, but this is just a commonsense conclusion at this point. Placing our commonsense argument within a statistical framework using formal hypothesis tests of these hypotheses of interest will clarify our reasoning and make their evaluation and communication to others easier.

Formal Hypothesis Testing and the Null Hypothesis

While we might be satisfied with intuitive interpretations when the probabilities associated with the variates are as extreme as the previous example, we will likely find it dissatisfying in most circumstances. A subjective, intuitive approach leaves us with questions such as: "What values differentiate variates that probably aren't from a population from those that might be?" and "What relationship are we really addressing?" To answer such questions, statisticians have developed a system of formal hypothesis testing that is designed to ensure the statistical hypotheses under consideration are clearly defined, the terms of rejecting them are understood, and the outcome of our hypothesis testing is unambiguously presented. This system works by taking a hypothesis of interest, no matter how it is phrased, and transforming it into a statement that assumes there is no difference between the things being compared. The hypothesis comparing the size of a gopher mandibular alveolar length to *Thomomys bottae* mandibles can be more formally stated as $H_0: Y_i = \mu$. H_0 is by convention the symbol for the *null hypothesis*; the hypothesis of no difference. As you know, Y_i represents our variate and μ represents the population mean.

In plain English, the null hypothesis can be summarized as the statistical proposition that there is no difference between a variate Y_i and the population parameter μ. With respect to hypothesis testing, the presentation of the null hypothesis is the basic statement formalizing our intent to compare, which can produce two possible outcomes. The first is of course support for H_0 – that there is no difference between the two things being compared. The second possible outcome is that there is indeed a difference. This outcome is symbolized by the alternative hypothesis, H_a. After all, if the relationship specified in the null hypothesis is unlikely, then some other relationship must be true (i.e., Y_i must be from a different population). Therefore, once a null hypothesis is defined, an alternate hypothesis has been defined too. In this example, the alternate hypothesis is simply the inverse of the null hypothesis – $H_a : Y_i \neq \mu$ (i.e., the alveolar length of the prehistoric gopher mandible is significantly different than the average length of *Thomomys bottae* mandibles). So, the general terminology is as follows:

- $H_0 : Y_i = \mu$ constitutes the null hypothesis specifying no difference; and
- $H_a : Y_i \neq \mu$ constitutes the alternative hypothesis where there is indeed a difference.

An astute reader will notice that the null hypothesis (there is no difference between the alveolar length) states exactly the opposite of our hypothesis of interest (that there is a difference, which could indicate the presence of gopher species other than *Thomomys bottae* in the past) and that the alternate hypothesis agrees with the hypothesis of interest. That is true. Null hypotheses are statistical propositions that specifically state the statistical relationship we will evaluate, not necessarily the specific relationship that interests us. This is part of the reason why we must take care to clarify exactly what relationships are being evaluated at each stage of the analysis. *By convention, the null hypothesis states that there is no difference, and it is this relationship that the statistical test directly evaluates. We must then tie the results of this analysis to our hypotheses of interest in order to give our statistical analysis analytic meaning.*

We now have a null hypothesis that we can evaluate, in this case with a Z-score. Although we use the Z-score in this example, the relationship between the null and alternate hypotheses described above and illustrated below characterizes all hypothesis testing. The actual statistical test used to evaluate competing hypotheses is dictated by the data and the nature of the hypotheses themselves, but the philosophy of hypothesis testing remains the same.

As you learned in the previous chapter, the Z-score (like a great many statistical tests) is designed to determine the probability of the occurrence of a set of observations (variates) as defined by an area under a specific distribution (in this case the normal distribution). How do we connect these probabilities to a null hypothesis so that it can be evaluated? We do so by defining a specific probability at which we hold it is more likely that the alternate hypothesis is true as opposed to the null hypothesis. In this case, we will select a probability at which we conclude it is unlikely that a variate came from a member of *Thomomys bottae*, the population specified in the null hypothesis. If it is unlikely to have a value as far or farther from the mean as a particular Y_i, then we can conclude that it is unlikely that the null hypothesis is correct. Let us clarify how this works further.

In Chapter 6 we considered the likelihood of specimen number HC-215 ($Y_1 = 6.4$ mm) being drawn from the population of where $\mu = 5.7$ mm and $\sigma = .48$ mm. We can formalize this test as an evaluation of the null hypothesis $H_0 : Y_1 = \mu$, and an alternative hypothesis $H_a : Y_1 \neq \mu$. Consider our null hypothesis, $H_0 : Y_1 = \mu$. We of course know that 6.4 mm does not exactly equal $\mu = 5.7$ mm. What we are really assessing is whether or not 6.4 mm is *close enough* to μ to likely be from the population that the mean describes.

We concluded in the previous chapter that it was very likely to get a variate that is smaller than 6.4 mm ($p = .93$) and somewhat unlikely to get one larger ($p = .07$), but this doesn't constitute a test of the null hypothesis. Instead, we merely concluded that it is perhaps somewhat unlikely to observe a *Thomomys bottae* alveolar length of 6.4 mm given that only 7% of the variates will be longer. Based on the probabilities we calculated, which hypothesis – the null or the alternate hypothesis – is likely correct? There is no cut and dry answer to this question. Instead, the researcher must specify the probability at which he or she will reject the null hypothesis of no difference in a given situation based on his or her research design and knowledge of the relationships under consideration. This rejection value is always arbitrary, in that some other value *could* be chosen. For whatever reason, archaeologists and many other scientists generally use a probability of .05 as the cutoff mark. Probabilities greater than .05 are taken as supporting a null hypothesis whereas probabilities less than .05 are interpreted as supporting the alternate hypothesis. We will discuss why we shouldn't blindly use the .05 demarcation value presently, but we will use it now for the sake of simplicity.

Using .05 as a cutoff mark, we will *reject H_0* in favor of the alternate hypothesis if the probability of $Z < .05$ (i.e., we will conclude that H_0 is unlikely to be true if the probability for obtaining a value of Z or greater is smaller than .05). By statistical convention, this cutoff value is now called our *alpha value*, and is symbolized as α. In this case, $\alpha = .05$, and is illustrated in Figure 7.1. Note that the "area of rejection" in Figure 7.1 is split between the portions of the distribution greater than and less

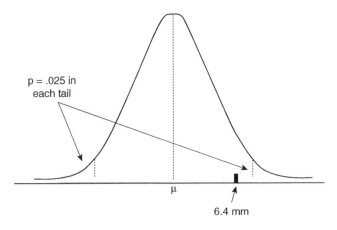

p = .025 in each tail

μ

6.4 mm

Figure 7.1 Areas of rejection associated with $\alpha = .05$

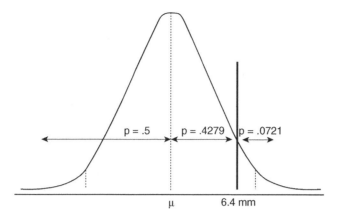

Figure 7.2 The areas under the normal distribution associated with $Z = 1.46$

than the mean. The reason for this is that our null hypothesis states that Y_1 equals μ. There are two ways this could not be the case: Y_1 can be greater than μ or less than μ. As a result, the probability of rejecting the null hypothesis must be equally divided between both sides of the distribution. Given that our alpha value is .05, the area of rejection must be .025 (half of .05) in both tails of the distribution.

After specifying the alpha value, the hypothesis testing continues by calculating Z:

$$Z = \frac{Y_1 - \mu}{\sigma}$$

$$Z = \frac{6.40 - 5.70}{.48}$$

$$Z = 1.46$$

When Z was calculated in the previous chapter, we went to Appendix A to find the probability of a variate falling between μ and Z. That probability is .4279, which is larger than .05. Does this mean that we do not reject the null hypothesis? Well, no, because the value of .4279 reflects the area between the mean and the variate. To test the null hypothesis, we are really interested in the probability of an event as large as Z or greater, not the probability of a variate between the mean and Z. The probability that interests us can be calculated using the tabled value, though. It is equal to the area under the curve to the right of $Z = 1.46$, which is calculated as $1 - (.5 + .4279) = .0721$. Figure 7.2 graphically illustrates these areas. The 1 reflects the entire area under the curve (i.e., unity). .5 reflects the area to the left of the mean, which includes half of the total area under the curve. .4279 is the area between the mean and $Z = 1.46$, which we determined from the table. Adding .5 to .4279 defines the probability of a variate smaller than $Z = 1.46$, which is .9279. Given that we seek to determine the probability of a variate larger than Z, we can subtract this value from 1 to determine that the probability of a variate larger than $Y_1 = 6.4$ mm is .0721. This is a relatively rare event (again, about seven times out of 100), but is not smaller than our level of rejection. Because our area of rejection is split between the two halves of the distribution, we cannot reject H_0 unless the probability is less than .025. Therefore, we must conclude that $Y_1 = 6.4$ mm is not significantly different than $\mu = 5.7$ mm. Our analysis indicates that the null hypothesis is plausible and that Y_1 could be from the *Thomomys bottae* population.

While using the standardized normal distribution to calculate the specific probability of a given test is useful, it is not always necessary. An alternative strategy is to simply determine the Z-score associated with a particular alpha value. You will recall from Chapter 6 that the $\mu \pm 2\sigma$ roughly corresponds with 95% of the variates in a normal distribution; the exact value associated with 95% is actually 1.96σ. As a result, evaluating the hypothesis $H_0 : Y_i = \mu$ can be streamlined by using the Z-score of 1.96 as a *critical value*. Critical values are the values that mark the cutoff line that describes the point demarcating the areas of rejection in which we reject the null hypothesis. We know that Z-scores greater than 1.96 or less than −1.96 represent variates that are outside of the area containing 95% of the distribution and Z-scores reflecting variates between 1.96 and −1.96 represent variates that are common in the distribution. In this example $Z = 1.46$ is between 1.96 and −1.96, indicating that the probability associated with the Z-score must be greater than .05. We can therefore conclude, without bothering to calculate the specific probability associated with it, that the variate $Y_1 = 6.4$ mm is common enough in this distribution that we cannot reject the null hypothesis.

You will notice, of course, that we used the clumsy phrase *cannot reject the null hypothesis* throughout the previous discussion instead of the more elegant phrase *accept the null hypothesis*. This is a result of the nature of hypothesis testing. If a value is close (say within 1σ) to the mean, it has a high probability of occurring within the distribution associated with μ. However, can we conclude with certainty that it is a member of the population represented by μ as opposed to some other population? No. It could be a member of population μ_2, μ_3, or another population, yet still be close to μ, depending on the differences between the populations. This point can be illustrated using the Alyawara settlement size data presented in Chapter 4. Based on our sample of settlement sizes, we determined that the average Alyawara settlement has a mean of 23 individuals with a standard deviation of 11.8 people. If we found a village with 23 individuals can we conclude the village is an Alyawara settlement? Of course not. The village could be anywhere in the world with residents of any number of ethnic groups. While the number of residents in our hypothetical village is not significantly different from the average number of residents in an Alyawara camp, we cannot conclude that it *is* an Alyawara settlement.

In contrast, we are able to use the phrase *reject the null hypothesis* because a variate that is sufficiently far from the mean is unlikely to be part of the parent distribution. A village with 1,230 residents likely *is* not an Alyawara camp, assuming our sample of Alyawara settlements is representative of the population of Alyawara settlements. Using hypothesis testing, then, we can often conclude a null hypothesis is false (i.e., unlikely), but we cannot conclude it is true. When we do fail to reject the null hypothesis, we can merely conclude that the null hypothesis may be true, and consequently remains plausible. Establishing that the relationship actually is present requires us to provide logical arguments, not statistical analyses, to justify its validity.

The discussion above should have driven one point home; hypothesis testing in a statistical framework is not a magical gateway to "the truth." This disappoints some people, who hope (or have been told) that statistics is a way to prove whether a relationship is true or not. It isn't. Statistics merely provides a framework to assess the probability that a relationship is or is not present. One might even say statistics merely reveals to us which of our educated guesses are plausible and which are not. Even then, we may be wrong. This leads us to our next topic of discussion: errors in hypothesis testing.

Errors in Hypothesis Testing

Because hypothesis testing is based on the assessment of probabilities, we are never certain that any specific conclusion derived from our statistical analyses is necessarily correct. There are two correct and two incorrect decisions that can be made during hypothesis testing, as illustrated in Table 7.1. The two correct decisions are that we may fail to reject the null hypothesis when it is in fact true, and reject the null hypothesis when it is in fact false. Likewise, two kinds of errors may be made. We may reject the null hypothesis when it is indeed true (thereby making a Type I or alpha (α) error), or we may accept (fail to reject) the null hypothesis when it is indeed false (a Type II or beta (β) error).

In general, Type I errors are easier to understand. An analyst sets the probability of making such errors each time he or she sets α. This alpha level specifies the probability of rejecting the null hypothesis when it is in fact true, and is properly determined before we begin the quantitative analysis. Setting α at .05 means that 5% of the time, we will *incorrectly* conclude that a member of a population is more likely a member of an alternate population. Thus, we expect to commit a Type I error once out of every 20 statistical analyses. When evaluating 20 or more statistical propositions, it is not really a question of *if* we are going to commit a Type I error, given that we are almost certain to do so. Instead, the question is determining which

Table 7.1 Possible outcomes of hypothesis testing

Null hypothesis is:	*Results of statistical analysis cause null hypothesis to be:*	
	Accepted	*Rejected*
True	Correct decision	Type I error (α)
False	Type II error (β)	Correct decision

differences are spurious, reflecting Type I errors, and which are meaningful reflections of real differences. Likewise, setting α at 0.10 necessitates that we will commit a Type I error 10% of the time whereas an α of .01 reduces the probability to one out of every 100 tests. When expressed as a percentage, the alpha value is known as a *significance level*.

The importance of the alpha level and Type I error is illustrated by Figure 7.3, which presents the distribution of the *Thomomys bottae* alveolar lengths previously discussed. It is composed of a distribution of individual alveolar lengths normally distributed around μ. In this figure, the areas of rejection for $\alpha = 0.50$ and $\alpha = 0.05$ are illustrated. When evaluating the null hypothesis $H_0 : Y_i = \mu$ using an α of 0.50, we will expect to incorrectly reject the null hypothesis (i.e., conclude the variate is not representative of a *Thomomys bottae* when in fact it is) 50% of the time. Likewise, when evaluating the null hypothesis with $\alpha = 0.05$ we will expect to incorrectly reject the null hypothesis 5% of the time. As illustrated by this figure, the larger α is, the more often we expect to commit a Type I error.

As mentioned previously, many researchers use $\alpha = .05$ by convention. The general selection of $\alpha = .05$ is more or less done out of laziness, and partly from the fact that a probability of .05 generally seems intuitively small to us. There is neither any particular reason for choosing this value as opposed to any other, nor a rule of thumb dictating what α is the most useful value for any given situation. The investigator simply needs to decide how frequently he or she is comfortable with rejecting the null hypothesis when it is indeed true after fully considering the consequence of this decision.

The natural inclination when people realize that α corresponds to the likelihood of committing a Type I error is to make α as small as possible. After all, if an α of 0.05 suggests we will incorrectly reject a null hypothesis when it is true 5% of the time and an α of 0.001 suggests we will reject a null hypothesis when it is true only

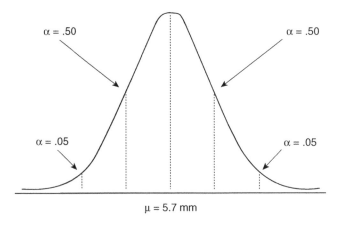

Figure 7.3   Areas of a normal distribution of *Thomomys bottae* alveolar lengths associated with $\alpha = .05$ and $\alpha = .01$

0.1% of the time (i.e., one time out of 1000), why not minimize our error and make the α level 0.00001 or even 0.0000000001?

The answer is that doing so would make it more likely that we will incorrectly fail to reject a null hypothesis when it is in fact false and another hypothesis is true (i.e., commit a Type II error). By reducing α we are making it more likely that we will not identify cases when the null hypothesis *is not* true. That means in this case that we will not readily differentiate when variates are drawn from another population of alveolar lengths. After all, the only way to be sure of never committing an alpha error is to never reject the null hypothesis. Yet if we do this, we are guaranteed to incorrectly conclude that the variates from all other possible distributions are possibly members of the distribution associated with μ_1. Coming to this conclusion guarantees that we will commit β errors when dealing with variates from alternate distributions. Therefore, when setting our α level, we must consider how willing we are to commit a β error.

Oddly enough, archaeologists seldom determine β. It seems to us that this is the result of three factors. First, archaeologists seldom compare hypotheses that are mutually exclusive. As a result, the potential for β (or Type II) errors is extreme and potentially damaging to any number of otherwise useful arguments. Archaeologists therefore find it convenient to ignore the possibility of β errors instead of trying to justify their conclusions further. Second, the probability of committing a β error must actually be calculated instead of being arbitrarily set as is the case with α. Calculating β frequently is a labor-intensive process for individuals with little statistical or mathematical background. Finally, there has been a common misperception that β errors cannot be calculated at all. Regardless of the reasons, the failure of archaeologists to consider β is unfortunate, as it is only with the determination of β that we are able to determine the *power* of a hypothesis test. We will return to this issue and demonstrate how the probability of committing a β error can be calculated in Chapter 9. First, though, we need to understand the role of *confidence limits* and their *critical regions* in evaluating hypotheses. These are the subjects of Chapter 8.

Practice Exercises

1 Define Type I errors, Type II errors, and α. What are the correct decisions that can be made when evaluating a null hypothesis?

2 Differentiate between a hypothesis of interest and a null hypothesis. Define three hypotheses of interest that might interest you.

(a) Define null hypotheses that correspond with them.

(b) To the best of your ability, consider the implications of making a Type I and a Type II error while evaluating the null hypotheses you defined. How would either of these impact the evaluation of the hypotheses of interest?

3 Evaluate the null hypothesis that a fluted projectile point 18.45 mm wide is a member of the population of Ohio fluted projectile points introduced in Question 5 from Chapter 6 ($\mu = 27.01$ mm and $\sigma = 4.28$ mm) using $\alpha = .05$. Retest the null hypothesis using $\alpha = .01$. Did changing the alpha level alter your conclusion concerning the null hypothesis? How did changing the alpha level affect the likelihood of committing a Type I error while evaluating the null hypothesis? How did changing the alpha level affect the likelihood of committing a Type II error while evaluating the null hypothesis?

4 As part of an ethnoarchaeological analysis, Shott and Sillitoe (2004: 346) report the steel axes used by the Wola people of highland New Guinea have a use life of 12.47 years with a standard deviation of 8.35 years.

(a) What is the probability that a steel axe will have a use life between 10 and 14 years?

(b) Using $\alpha = .10$, test the null hypothesis that the use lives of each of the following five axes are consistent with the population described by Shott and Sillitoe: 2.1 years, 37.3 years, 7.5 years, 16.9 years, and 20.3 years.

(c) Which of your conclusions regarding the hypotheses tested in Part (b) could reflect a Type I error? Which could reflect a Type II error?

8

Hypothesis Testing II:
Confidence Limits, the t-Distribution,
and One-Tailed Tests

As illustrated in Chapter 7, we can evaluate a null hypothesis by calculating a specific probability and comparing it to an alpha value, or by comparing a Z-score (or any of the other statistical measures discussed later in the book) to a critical value. These two methods produce identical results when evaluating the plausibility of a null hypothesis, but they can be cumbersome when testing a number of closely related null hypotheses, say, $H_0 : Y_i = \mu$ for 200 variates. If we were comparing the alveolar lengths of 200 gopher mandibles with the average alveolar lengths of *Thomomys bottae* to determine which mandibles are likely members of that species, we could calculate probabilities for each variate using a standardized normal distribution as illustrated in Chapter 7. Doing so would be time consuming, though, given that probabilities would have to be independently calculated for each variate, which would require us to calculate 200 Z-scores and then determine the corresponding probabilities using Appendix A. Comparing the Z-scores to a critical value (say ±1.96 for $\alpha = .05$) could save time given that it would eliminate the need to calculate individual probabilities for all 200 null hypotheses, but would still require us to calculate all 200 Z-scores using Equation (6.1). This seems like a lot of work for archaeologists who didn't choose their field out of a love of mathematics. There is a third option, confidence limits, which is widely used in archaeology and other statistics-using disciplines. This method produces the same result as previous methods described, but has the advantage of giving a result that can be directly compared to our raw data as opposed to transforming the data using a formula. Consequently we would need to perform only a single set of calculations instead of calculating 200 Z-scores, which can save a lot of time and computational effort.

Quantitative Analysis in Archaeology, Todd L. VanPool and Robert D. Leonard
© 2011 Todd L. VanPool and Robert D. Leonard

In a nut shell, confidence intervals work by identifying the real-world values that correspond with the critical values. These values define the upper and lower boundaries of the region of acceptance (i.e., the values that differentiate between when a null hypothesis will and when it won't be rejected). Using the *Thomomys bottae* alveolar length example again, we previously evaluated the null hypothesis $H_0 : Y_i = \mu$ using a critical value of ±1.96 associated with $\alpha = 0.05$. If the Z-value is equal to or between −1.96 and 1.96, we failed to reject the null hypothesis. If Z is less than −1.96 or greater than 1.96, we rejected the null hypothesis and concluded that the specimen probably was not from a *Thomomys bottae*. Underlying this test is the recognition that 95% of the variates are contained within $\mu \pm 1.96\sigma$. Using this same relationship, we can determine which variates aren't within the region of acceptance by simply identifying what values correspond with $\mu \pm 1.96\sigma$ for the distribution of *Thomomys bottae*. Here, $\mu = 5.7$ mm, $\sigma = 0.48$ mm, and $\alpha = .05$ (which corresponds with $Z = \pm1.96$). The confidence limits are:

$L_1 = \mu - 1.96\sigma$
$L_1 = 5.7 - (1.96 \times .48)$
$L_1 = 4.76$ mm
$L_2 = \mu + 1.96\sigma$
$L_2 = 5.7 + (1.96 \times .48)$
$L_2 = 6.64$ mm

The *confidence interval* around the mean, defined as the area between L_1 and L_2, is thus 4.76 mm to 6.64 mm, with L_1 representing the *lower confidence limit* and L_2 the *upper confidence limit* of the mean. Using our knowledge of probability, we know that 95% of the variates in the distribution represented by $\mu = 5.7$ mm should be between 4.76 mm and 6.64 mm (Figure 8.1). We will consequently fail to reject the null hypothesis for any variate within the confidence interval and will reject the null hypothesis for all variates outside of it. For $Y_1 = 5.7$ mm (the variate specified when we first introduced the example in Chapter 6) we would fail to reject the null hypothesis, but we would reject the null hypothesis for $Y_2 = 7.7$ mm, which exceeds

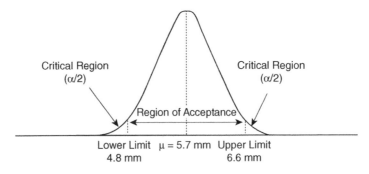

Figure 8.1   Illustration of the confidence limits and critical region of *Thomomys bottae* alveolar length distribution

the upper confidence limit. The exact same evaluation could be completed for as many variates as we might have, all without needing to calculate individual Z-scores or probabilities for each variate.

You now know three different ways to evaluate the same null hypothesis $H_0 : Y_i = \mu$ using the standardized normal distribution: determining a specific probability; using a critical value; or establishing a confidence interval. Any of these can be used in any given situation, and the results of each of them are identical. The choice of technique is up to you. Some find using probabilities most useful, because it allows the reader to directly compare the test's result to a specified alpha value. (Probabilities are the most common output for quantitative software.) Others find determining probabilities too cumbersome and prefer to use either critical values or confidence limits, because of their comparative simple calculations. These individuals are often especially fond of confidence intervals because they define the region of rejection using intuitively meaningful real-world values as opposed to values on the standardized normal distribution. You may of course select whichever style you like best, but you do need to be comfortable with all three approaches because they are all used in archaeological contexts.

The three approaches to hypothesis testing can be applied to testing a variety of other hypotheses. Archaeologists do frequently compare variates to a population mean $(H_0 : Y_i = \mu)$, but we also evaluate a host of other hypotheses too. These include $H_0 : Y_i = \bar{Y}$ (a variate to a sample mean), $H_0 : \bar{Y} = \mu$ (a sample mean to a population mean), and $H_0 : \bar{Y}_i = \bar{Y}_2$ (two sample means to each other). All three of these hypotheses deal with sample means, which are estimates of the population parameter μ. A fundamental tool in understanding the relationship between these sample means and their corresponding population mean is a statistical proposition called the *central limit theorem*, which stipulates that:

1 regardless of sample size n, the means of samples from a non-normally distributed population are normally distributed; and
2 as sample size gets larger, the means of samples drawn from a population of any distribution will more closely approximate the population parameter.

The central limit theorem is quite straightforward but it does require a subtle shift in our thinking. As mentioned in Chapter 6, we are accustomed to considering means and standard deviations as summary information about the distribution of variates. The central limit theorem calls for us to think about means themselves as variates. If we calculate means from multiple samples of a population, then we could plot them as variates, thereby building a distribution of means. This distribution will have its own mean (called a *grand mean*) as well as its own variance and standard deviation.

The central limit theorem states that regardless of the shape of the original distribution, a new distribution built out of sample means from the same population will be normally distributed. If you are like us when we first started studying quantitative methods, this doesn't make intuitive sense; it seems as if the shape of the distribution of sample means should be the same shape as the original distribution from which they are drawn. The reason why sample means from non-normal distributions will be normally distributed is actually quite simple, however. Statistically valid sample means are calculated from each sample of a population using randomly selected variates. The population mean by definition is the center point for the probabilities on each side of the distribution, such that 50% of the area under the distribution's curve is on each side of the mean regardless of the distribution's shape. As a result, *each half* of the distribution is expected to be reflected equally in a sample, even though the number of variates on each side of the mean may not be the same in heavily skewed distributions. Variation between sample means is expected, though, as different variates are selected. Sample means will consequently cluster around μ differing from it only as a result of the randomness of each sample's composition, with the mean, median and mode of the sample means equaling μ. Further, the frequency of sample means will decrease as we move further away from μ regardless of the shape of the parent distribution, because of the unlikelihood that one-half of the distribution will be disproportionately reflected in any given sample. The result of this is a normal distribution, regardless of the original distribution's shape. The larger the sample size of each sample, the less likely it is that one side of the distribution will be underrepresented in a sample. As a result, the accuracy of the sample means as an estimate of μ will increase with sample size, and the distribution of the sample means will become more tightly constrained around μ.

Let us illustrate the central limit theorem with the maximum lengths of 166 flakes recorded during the survey of the Medio period Casas Grandes site of Timberlake Ruin in southern New Mexico (Figure 8.2). The distribution is unimodal as is the normal distribution, but it is skewed to the right. Treating this distribution as a population, the mean (μ) equals 3.1 cm and the standard deviation (σ) is 1.5 cm.

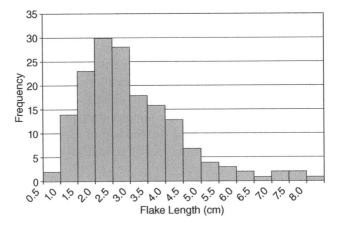

Figure 8.2   The distribution of 166 maximum flakes lengths

We took 15 samples of three variates each and calculated their means, which are graphed on Figure 8.3. Notice how the means cluster around $\mu = 3.1$ cm. The distribution only loosely resembles a normal distribution, but if we were to take many more samples, the distribution would eventually be perfectly normally distributed. Now let's sample the population again, only this time increasing the size of our samples to $n = 10$ for 20 samples (Figure 8.4). This reflects an increase in both the number of variates in each sample and an increase in the number of samples. The distribution more closely matches the shape of a normal distribution and is even more tightly constrained around μ than the previous distribution. Finally, let's increase our sample size to 30 samples of $n = 20$ flakes (Figure 8.5). Here, the distribution is indeed normally distributed, and reflects even less variation from $\mu = 3.1$ cm than Figure 8.4. These three figures perfectly demonstrate the central limit theorem. As the size of the samples increased, the means clustered more closely around μ. Further, as the number of samples increased, their distribution more closely resembled a normal distribution, despite the underlying distribution's shape.

The central limit theorem is important because it allows us to use the normal distribution to make statistical inferences about means even if the original distributions are not normally distributed. In other words, the normal distribution is *always*

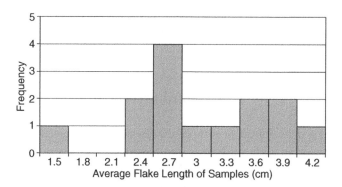

Figure 8.3   Distribution of the mean maximum flake lengths for 15 samples of three flakes

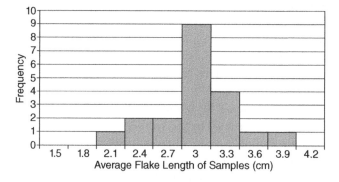

Figure 8.4   Distribution of the mean maximum flake lengths for 20 samples of 10 flakes

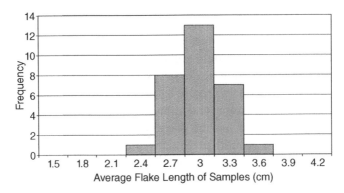

Figure 8.5   Distribution of the mean maximum flake lengths for 30 samples of 20 flakes

applicable to the distribution of sample means, assuming that they are from randomly selected samples. This fact greatly simplifies the evaluation of hypotheses such as $H_0: \bar{Y} = \mu$ and $H_0: \bar{Y}_1 = \bar{Y}_2$, because it allows them to be evaluated using a consistent set of tools appropriate for normal distributions. These tools share substantial similarities to the Z-score previously introduced, but there are a few differences. Primary among these is the difference between the standard deviation (which characterizes the variation of variates around a mean) and the standard error (which characterizes the variation of sample means around μ), the topic to which we now turn.

Standard Error

In considering a distribution of sample means, we can calculate a variance and a standard deviation in addition to the grand mean (the average of the means). As illustrated by the differences between Figures 8.3, 8.4, and 8.5, the measures of dispersion will decrease as the size of our samples increases. Samples of 50 variates will provide more precise and accurate estimates of μ than samples of five. The distribution of sample means will consequently be much more tightly constrained around μ for the larger samples. As n increasingly gets closer to N (the population parameter for the size of a population), the standard deviation of the sample means approaches zero (less variation). In addition to this, the spread of sample means will be a product of the variation in the parent populations. Samples of a widely spread population will tend to have greater variation among them than samples from a tightly clustered population.

For clarity's sake, the standard deviation of a distribution of means is called the "standard error" and is represented by the symbol $\sigma_{\bar{Y}}$. Standard deviations describe the spread of variates around a mean whereas standard errors describe the spread of sample means around the population parameter μ. They are otherwise identical, but using different terms helps ensure that the analyst and his or her audience understand what is being measured.

When comparing sample means to other sample means or μ, understanding their expected spread will allow us to identify means that likely don't reflect the same population parameter, just as knowledge of the spread of variates can help us determine which variates might be part of a different population than the others. There are a couple of ways to determine the measures of dispersion for sample means. One is to literally take a bunch of sample means and calculate the standard error for them using a modified form of Equation (4.3), the standard deviation. The only difference would be substituting $\bar{Y}_i$ for Y_i, and remembering the n now stands for the number of samples, not the size of each sample. This is a viable option, and underlies some "Monte Carlo"-style analyses. However, in most archaeological circumstances we can more easily calculate an *expected value* of the standard error based on the variation of the parent population and the sample size. This estimate is calculated using σ from the original distribution and n (sample size) from our sample. It perfectly approximates the values we would obtain if we were to take a bunch of samples from the same population and calculate the standard error for them. This estimate of standard error is calculated using the following equation, where σ is the population's standard deviation and n is the sample size of our sample.

The standard error

$$\sigma_{\bar{Y}} = \frac{\sigma}{\sqrt{n}} \tag{8.1}$$

Consider for a moment the distributions in Figures 8.3–8.5. Summary statistics for them are presented in Table 8.1. All three of the distributions provide reasonable estimates of μ (3.1 cm). Note also that the distributions' standard errors get smaller as the size of the samples increases. Further note that values obtained using Equation (8.1) very closely approximate the actual values calculated from the samples.

$$n = 3, \sigma_{\bar{Y}} = \frac{1.5}{\sqrt{3}} = .866$$

Table 8.1 Summary statistics for the three series of samples of the Timberlake Ruin flake lengths

Sample	Grand mean ($\bar{\bar{Y}}$) (cm)	Calculated standard error (cm)	Standard error derived using Equation (8.1) (cm)
15 samples of 3	3.1	.87	.87
20 samples of 10	3.0	.38	.34
30 samples of 20	3.1	.28	.27

$$n = 10, \sigma_{\bar{Y}} = \frac{1.5}{\sqrt{20}} = .335$$

$$n = 20, \sigma_{\bar{Y}} = \frac{1.5}{\sqrt{30}} = .274$$

If we continued to draw random samples from the population of flake lengths an innumerable number of times, the standard error calculated from the samples will exactly match the standard error determined using Equation (8.1). This means that if we know the standard deviation of a population and the size of our samples, we can determine the standard error without directly computing it from a group of samples. But why bother? Why not just calculate the standard error directly from the means of multiple samples? Well, the most obvious reason is that using Equation (8.1) saves us the trouble of trying to obtain multiple samples from the same population. Sometimes we may not have the money, time, opportunity, or interest to obtain 10 or 20 samples of the same population. There are also times when we may wish to use published summary information collected by other researchers for which it may be impossible to collect comparable data. A site can only be dug once, and it may be impossible to obtain multiple samples of its attributes. Finally, Equation (8.1) provides a more accurate estimate of the true standard error when compared to an estimate based on a series of samples. As with all samples, the samples used to calculate standard error will vary from the true population parameter to varying degrees. Equation (8.1) determines the population parameter $\sigma_{\bar{Y}}$, whereas the standard error estimates of the standard error in Table 8.1 only approximate it. Thus, the results of Equation (8.1) are both superior to and easier to calculate than directly determining the standard error from a series of samples. The utility of $\sigma_{\bar{Y}}$, whether it is calculated from a sample of means or using Equation (8.1), is most clear when we consider the use of standard errors in hypothesis testing through the establishment of confidence limits.

Comparing Sample Means to μ

The central limit theorem makes it easy to evaluate hypotheses such as $H_0 : \bar{Y} = \mu$ using the standard error and the standardized normal distribution. This can be done using all three of the methods for hypothesis testing outlined in Chapter 7 and above: direct determination of a probability, critical values, and confidence intervals. We will illustrate how this is done using a contrived example. Let us assume that maximum wall thickness does indeed reflect pottery use and that we wish to evaluate the proposition that 25 pots found in a store room are not meaningfully different from pottery used for storing food found in households throughout a very large site. The null hypothesis corresponding with this hypothesis of interest is $H_0 : \bar{Y} = \mu$, where $\bar{Y}$ is the average of the sample of 25 pots stockpiled in the storeroom and μ is the average maximum wall thickness for household food storage pots from the site. The alpha level is set at .05, and the alternate hypothesis is $H_a : \overline{Y} \neq \mu$. $\overline{Y} = 6.2$ mm, and the population parameters are $\mu = 7$ mm and $\sigma = 2.3$ mm.

We can construct 95% confidence limits around the parameter μ in the following way:

$L_1 = \mu - 1.96\sigma_{\overline{Y}}$

$L_2 = \mu + 1.96\sigma_{\overline{Y}}$.

The value 1.96 reflects the critical value of the standardized normal distribution associated with $\alpha = 0.05$. To complete the calculations, we need to calculate $\sigma_{\overline{Y}}$ as follows:

$$\sigma_{\overline{Y}} = \frac{\sigma}{\sqrt{n}} = \frac{2.3}{\sqrt{25}} = .46 \text{ mm.}$$

We can now calculate the confidence interval of the mean by calculating the upper and lower confidence limits.

$L_1 = 7 - 1.96(.46) = 6.10 \text{ mm}$

$L_2 = 7 + 1.96(.46) = 7.90 \text{ mm}$

The interval 6.10 mm to 7.90 mm contains the value of $\overline{Y} = 6.2$ mm. We therefore fail to reject the null hypothesis; the mean maximum wall thickness of the pottery from the store room is not significantly different than that from the site as a whole.

But what if there was a second store room with 25 pots that produced a mean of $\overline{Y}_2 = 5.9$ mm? Would this mean be encompassed by the confidence interval around μ? No, the interval 6.10 mm to 7.90 mm does not contain $\overline{Y}_2 = 5.9$ mm, so we would reject the null hypothesis. How about a 99% confidence interval around μ? We know before any calculations are completed that a 99% confidence interval will be broader than a 95% interval; after all, all possible values for $\overline{Y}_i$ contained in the 95% confidence interval will invariably be contained in the 99% confidence limit as well. Here is the calculation for the 99% confidence intervals:

$L_1 = 7 - 2.576(.46) = 5.82 \text{ mm}$

$L_2 = 7 + 2.576(.46) = 8.18 \text{ mm}$

This broad confidence interval does indeed contain $\overline{Y}_2$, causing us to be unable to reject the null hypothesis for $\alpha = .01$ despite the fact that we did reject it when $\alpha = .05$.

Given that sample means are normally distributed as specified by the central limit theorem, the Z-score can also be used to evaluate the null hypothesis $H_0 : \overline{Y} = \mu$, just as it could be used to evaluate the null hypothesis $H_0: Y_i = \mu$. We accomplish this by modifying Equation (6.1) as indicated in Equation (8.2).

The Z-score comparing a sample mean to μ

$$Z = \frac{\bar{Y} - \mu}{\sigma_{\bar{Y}}}, \text{which can also be stated as} \frac{\bar{Y} - \mu}{\sigma/\sqrt{n}} \qquad (8.2)$$

Equations (6.1) and (8.2) differ in that $\bar{Y}$ is substituted for Y_i and the standard error ($\sigma_{\bar{Y}}$) replaces σ, but both equations provide Z-scores that correspond with the standard deviation units of the standard normal distribution. Once we specify a level of rejection, we can test the null hypothesis $H_0: \bar{Y}_i = \mu$ using the probabilities in Appendix A.

Using the previous pottery example, the Z-score can be calculated as:

$$Z = \frac{\bar{Y} - \mu}{\sigma_{\bar{Y}}}$$

$$Z = \frac{6.2 - 7}{.46}$$

$$Z = -1.74$$

The negative sign reflects that $\bar{Y}$ is less than the mean, and can simply be compared to the critical value of ± 1.96 for $\alpha = .05$. Given that -1.74 is more than -1.96, we again cannot reject $H_0: \bar{Y}_i = \mu$ and we must conclude that the mean maximum wall thickness of the pottery in the store room does not significantly differ from μ for the entire site. The structure of these hypothesis tests perfectly mimic the structure we presented for evaluating the null hypothesis $H_0: Y_i = \mu$ except that $\sigma_{\bar{Y}}$ is used instead of σ. Remember, sample means are distributed according to $\sigma_{\bar{Y}}$ whereas variates are distributed according to σ. If you remember this, you will find evaluating both the null hypotheses $H_0: \bar{Y}_i = \mu$ and $H_0: Y_i = \mu$ very straightforward.

Statistical Inference and Confidence Limits

Having illustrated how confidence limits can be used as a means of hypothesis testing, we now want to discuss how they can be used to determine the accuracy of our statistics as estimates of population parameters. This use focuses on the process of *statistical inference*, which is using statistics describing samples to make inferences about population parameters. When population parameters are unknown, it is often useful to know how reliable our sample statistics are. For example, an archaeologist might wish to use a sample of atlatl dart points to characterize the maximum shoulder width for dart points in an effort to demarcate between dart, arrow, and thrusting spear points. The average maximum dart shoulder width provides an estimate of μ, but it should be obvious from Figures 8.3 through 8.5 that the sample mean probably will not be *exactly* equal to μ. We will therefore want to know what a reasonable range would be for the parameter that the statistic is estimating. Assessing this would intuitively seem to require us to already know what μ is, which would make the whole exercise of using $\overline{Y}$ to estimate μ moot. Fortunately, though, we can use the standard error to determine confidence limits that reflect the likely placement of μ.

The central limit theorem states that sample means will be normally distributed around μ. Recall that in a normal distribution:

50% of all deviates fall between $\mu \pm .674\sigma$
95% of all deviates fall between $\mu \pm 1.96\sigma$
99% of all deviates fall between $\mu \pm 2.58\sigma$

Since the means of repeated samples are normally distributed, the same relationship holds except that the standard error will replace σ:

50% of all sample means fall between $\mu \pm .674\sigma_{\overline{Y}}$
95% of all sample means fall between $\mu \pm 1.96\sigma_{\overline{Y}}$
99% of all sample means fall between $\mu \pm 2.58\sigma_{\overline{Y}}$

Using the standard error, then, we can place confidence limits around μ to determine the range in which most sample means will fall (e.g., 95% of the sample means will be between $\mu \pm 1.96\sigma_{\overline{Y}}$). The inverse of this relationship is also true, such that μ will be contained by $\overline{Y} \pm 1.96\sigma_{\overline{Y}}$ for 95% of the sample means. We can express this relationship symbolically as $P(\overline{Y} - 1.96\sigma_{\overline{Y}} \leq \mu \leq \overline{Y} + 1.96\sigma_{\overline{Y}}) = .95$, which indicates that the probability is .95 that $\overline{Y} - 1.96\sigma_{\overline{Y}}$ is less than or equal to μ and that $\overline{Y} + 1.96\sigma_{\overline{Y}}$ is greater than or equal to μ. This being the case, our lithic analyst could derive a "very good estimate" (i.e., 95%, 99%, or whatever probability the archaeologist selects) of the range that encompasses the true population parameter μ for the maximum shoulder widths of dart points using the sample mean she or he determines.

It is important to fully understand what we mean by saying we are 95% certain the confidence limits encompass μ. μ is a population parameter that does not change. It is fixed, and reflects the totality of the population defined. Sample means, however, change as the composition of the sample is altered. In the example presented above, 95% of the confidence intervals placed around the mean of a sample of atlatl darts will include this fixed point corresponding with μ, but 5% will not. As a consequence, it is inaccurate to state that μ is "95% likely to fall within the confidence limits" given that it is the confidence limits that change, not μ. This is a subtle, yet important, distinction.

Confidence intervals can reflect whatever alpha value is appropriate. If we desire confidence intervals that are 99% certain to encompass μ, 2.58 would be used instead of 1.96. Since 2.58 is a larger number than 1.96, multiplying $\sigma_{\overline{Y}}$ by 2.58

would yield a larger value, and our confidence limits would be broader. With 99% confidence limits, we would be more certain that μ is within the confidence interval, but we would be less certain of the exact value of μ because the confidence intervals would be over 30% broader. Given the implications of the central limit theorem, we can use the values in Appendix A for the standardized normal distribution to derive whatever probability is useful.

In most archaeological analyses, a 95% confidence interval is just fine, but researchers can change this level of confidence as they see fit given their theoretical and analytic structure. Typically, though, archaeologists would like to have the best estimate of μ that they can. There are several ways to make the confidence interval smaller. Consider that confidence intervals corresponding with an alpha of .05 are calculated as:

$$L_1 = \bar{Y} - 1.96\frac{\sigma}{\sqrt{n}}$$

$$L_2 = \bar{Y} + 1.96\frac{\sigma}{\sqrt{n}}$$

If we wish to decrease the confidence intervals, then we must make the term $1.96\frac{\sigma}{\sqrt{n}}$ smaller. One way to do this would be to increase the alpha value, which in turn reduces the reliability of our estimate. If we decrease the confidence intervals from 95% to say 50% certainty, we would substitute .674 for 1.96 and thereby decrease the size of our confidence intervals by over 60%. While this does substantially narrow the confidence intervals, it also decreases the reliability of our estimated range for μ. We would now expect that there is only a one in two chance that the confidence intervals actually encompass μ, instead of a 95% chance. Fifty percent odds aren't particularly good betting odds, and most archaeologists would find this unsatisfactory. We can live with being wrong 5% of the time, but being wrong 50% of the time has obvious drawbacks for most analyses, especially given that errors can compound throughout an analysis as one incorrect conclusion justifies more erroneous conclusions. Simply choosing whatever probability produces superficially appealing narrow confidence limits is analytically flawed, to be charitable.

Two other more useful options remain. First, even small increases in the sample size (n) can decrease the confidence limits' size by decreasing the standard error. As illustrated in Table 8.1, increasing our sample size from 3 to 10 led to a 61% decrease in the standard error (which in turn would cause the area encompassed by the confidence limits to decrease by 61%). Second, a reduction of σ will also tighten our confidence limits. This may seem like an odd statement, given that σ is a population parameter and is therefore fixed. However, at times it may be possible to redefine our hypothesis of interest so as to reduce a population to a more restricted number of individuals that may reflect less variation. For example, instead of treating atlatl dart points as a population, an archaeologist might redefine the population of interest to the dart points of a specific region or of a specific point type. This might cause σ to be reduced without sacrificing any aspect of the analysis's integrity. Another example is faunal analyses of sexually dimorphic species. The great overlap in size between males and females often creates a dispersed unimodal distribution, but the distributions of males and of females are narrower when considered in isolation. Redefining a population to differentiate between males and females of a given species might in turn create two tighter distributions with smaller standard deviations, instead of a single distribution with a large σ.

The t-Distribution

In the example above, our knowledge of the normal distribution allows us to predict with 95% certainty (or whatever probability we might select) the location of μ using confidence limits around a sample mean, but this requires us to know σ in order to calculate $\sigma_{\bar{Y}}$. Rarely in the real world, however, do archaeologists have knowledge of σ when they don't also know μ. Typically a researcher either is able to determine all of the population parameters, which makes estimating them unnecessary, or the archaeologist is forced to use a sample to approximate the parameters, which means he or she won't be able to calculate $\sigma_{\bar{Y}}$ using σ. Given that the standard deviation (s) is an estimate of σ, would it be possible to use it to estimate $\sigma_{\bar{Y}}$ so that we can go ahead and determine a range for μ even if we don't know σ? Yes, we can substitute s in the equation for standard error to derive an estimate of $\sigma_{\bar{Y}}$ as illustrated in Equation (8.3).

The standard error for a sample

$$s_{\bar{Y}} = \frac{s}{\sqrt{n}} \qquad (8.3)$$

$s_{\bar{Y}}$, the sample standard error, is the best estimator of the standard error $\sigma_{\bar{Y}}$ available when using a sample to estimate the population parameters. Unfortunately though, it is a biased estimate of $\sigma_{\bar{Y}}$. While repeated samples of $\dfrac{\bar{Y} - \mu}{\sigma_{\bar{Y}}}$ are distributed normally, repeated samples of $\dfrac{\bar{Y} - \mu}{s_{\bar{Y}}}$ are not. The reason for this is actually quite simple. Given that $\sigma_{\bar{Y}}$ is a population parameter based on σ, it is a constant. The central limit theorem tells us that the means of unbiased samples are normally distributed, so subtracting μ (another constant) from $\bar{Y}$ and dividing by the constant $\sigma_{\bar{Y}}$ doesn't change the shape of the distribution. After all, adding, subtracting, dividing, and multiplying values from all the members of any distribution doesn't change the shape of the distribution, it just changes the magnitude of the values.

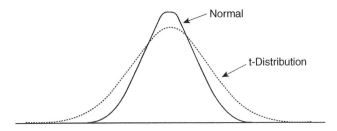

Figure 8.6 Comparisons of the t-distribution and the normal distribution

As an estimate of $\sigma_{\bar{Y}}$ based on a sample, the value of $s_{\bar{Y}}$ varies from sample to sample to varying degrees. Thus, $s_{\bar{Y}}$ isn't a constant any more than $\bar{Y}$ is; they will both change depending on each sample's variates. The relationship between $\dfrac{\bar{Y} - \mu}{s_{\bar{Y}}}$ will consequently change because of the variation in $\bar{Y}$ *and* $s_{\bar{Y}}$, causing the resulting distribution to not be normally distributed. The result is a distribution called Student's t-distribution that shares many characteristics with the standardized normal distribution. Like the standardized normal distribution, the t-distribution is symmetrical and asymptotic at both ends to infinity. However, it is somewhat more dispersed at the tails, and somewhat flatter (platykurtic) on top when compared to the standardized normal distribution (Figure 8.6). The consistency in shape, however, allows the distribution to be modeled statistically in a manner similar to the standardized normal distribution with one exception – the shape of the t-distribution is determined by the degrees of freedom, or v (pronounced "nu").

Degrees of freedom and the t-distribution

We will discuss the degrees of freedom here in the context of the t-distribution, but it is an important statistical concept that is pertinent to a great many quantitative methods. The degrees of freedom are required by the mathematical nature of statistics, because of the fact that many distributions change as the sample size changes (as illustrate by Figures 8.3 through 8.5). While the following definition is not necessarily intuitively meaningful, the degrees of freedom are defined as the number of quantities under consideration that are free to vary. In this case, this is $n - 1$. The implications of this definition, and the mathematical necessity of the degrees of freedom, can be illustrated using the Alyawara example initially presented in Chapter 4.

In Chapter 4, we determined that the average settlement size of the eight Alyawara camps is 23.4 people. We now know that the average of our eight variates is an estimate of μ and therefore it must, according to probability, equal (some value close to) that parameter. Thus, our eight variates must produce $\bar{Y} = 23.4$ people, which by extension necessitates that the sum of the variates must equal (23.4×8)

or 187. If we want to find eight variates that sum to 187, seven of the variates (Y_1 through Y_7) are completely free to vary. The final variate, Y_8, however, must be a variate that forces the sum of the eight variates to equal 187. In other words, it is mathematically necessary that Y_8 be the correct value such that the sum of the variates is equal to 187. Seven variates can vary, but the value of one variate is mathematically fixed. Therefore, the degrees of freedom in this case are $n - 1$, which is 7. Don't worry too much if this doesn't make sense. Another, more intuitive way to think of degrees of freedom is that they reflect the number of variates that can vary. A sample of one doesn't have any variation at all. A sample of two reflects variation caused by one variate (i.e., one degree of freedom), given that there must be one variate present before any variation can be present. A sample of three reflects two degrees of freedom, and so on. Only $n - 1$ of the variates are in fact free to vary.

Neither of these simple formulations are entirely adequate but both are defensible. They are certainly superior for most archaeologists when compared to the more statistically appropriate statement that degrees of freedom are "differences in dimensionalites of parameter spaces" (Good, 1973: 227). The point is that the shape of the t-distribution changes as a result of the impact of the sample size on both the variation in $\bar{Y}$ and $s_{\bar{Y}}$ (again as illustrated in Figures 8.3 through 8.5). The way we control this variation is through the degrees of freedom. When using most distributions including the t-distribution, the degrees of freedom is calculated using Equation (8.4), where n is the sample size.

The degrees of freedom

$$v = n - 1 \tag{8.4}$$

As previously discussed, sample means become more accurate and precise predictors of μ as sample size (n) increases. This is also true for s as an estimate of σ. As a result, $s_{\bar{Y}}$ becomes a better estimate of the population parameter $\sigma_{\bar{Y}}$ as n increases, which in turn causes $\dfrac{\bar{Y} - \mu}{s_{\bar{Y}}}$ (i.e., the t-distribution) to better approximate $\dfrac{\bar{Y} - \mu}{\sigma_{\bar{Y}}}$ (i.e., the standardized normal distribution) as the sample size n becomes close to the population size N. When n grows to be N, $s_{\bar{Y}}$ is in fact $\sigma_{\bar{Y}}$ because it reflects the entire population. Predictably, then, the t-distribution varies the most from the standardized normal distribution when $v = 1$, and is indistinguishable from the standardized normal distribution when $v = $ infinity. For our purposes, when $v > 120$, the t-distribution and standardized normal distribution are identical. Therefore, whenever we use $s_{\bar{Y}}$ to estimate $\sigma_{\bar{Y}}$ and when sample sizes are less than 121 variates, we should use the t-distribution to construct confidence intervals instead of the normal distribution. Figure 8.7 illustrates the different shapes of the t-distribution as determined by v. Critical values corresponding with various probability levels are presented in Appendix B for degrees of freedom up to $v = 120$.

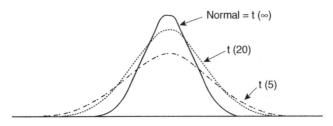

Figure 8.7   Different shapes of the t-distribution as determined by ν

Hypothesis Testing Using the t-Distribution

The t-distribution can be used to place confidence intervals around $\bar{Y}$ to estimate the likely placement of μ, but it is also useful for evaluating hypotheses such as $H_0: Y_i = \bar{Y}$, $H_0: \bar{Y}_i = \bar{Y}_2$, and $H_0: \bar{Y} = \mu$ using s to estimate σ. We will consider the comparatively complicated evaluation of $H_0: \bar{Y}_1 = \bar{Y}_2$ in detail in Chapter 9, but $H_0: Y_i = \bar{Y}$ and $H_0: \bar{Y} = \mu$ can be evaluated using either confidence intervals or critical values. Using critical values is accomplished by calculating a t-score using a modified version of the Z-score introduced in Equations (6.1) and (8.2). For the hypothesis $H_0: Y_i = \bar{Y}$ the t-score is calculated using Equation (8.5) (compare with Equation (6.1)) and for $H_0: \bar{Y} = \mu$ it is calculated using Equation (8.6) (compare with Equation (8.2)). The differences between Equations (8.5) and (8.6) hinge on whether the variation being considered is between a variate and $\bar{Y}$, which is quantified using s, or between μ and $\bar{Y}$, which is quantified using $s_{\bar{Y}}$. Again, simply remember that standard deviations quantify the variation in variates, whereas standard errors quantify the variation in sample means. If you do this, you will always be certain which of these two equations to use.

The t-score comparing a sample mean to a variate

$$t = \frac{Y_i - \bar{Y}}{s} \tag{8.5}$$

The t-score comparing a sample mean to μ using $s_{\bar{Y}}$ to approximate $\sigma_{\bar{Y}}$

$$t = \frac{\bar{Y} - \mu}{s_{\bar{Y}}} \tag{8.6}$$

The null hypothesis $H_0: Y_i = \bar{Y}$ is a common comparison in archaeological contexts, perhaps the most common comparison involving means. It is the question being asked whenever an archaeologist wonders if a particular variate is substantially different from the others in a sample. Does Burial 5 have more pots than the other burials? Is Room 3-B significantly larger than the other rooms in the structure? Is the pot found in the plaza larger than is typical of the other pots on the site? Questions such as these all focus on $H_0: Y_i = \bar{Y}$.

Consider for example the analysis of a fluted point from Sheriden Cave, Ohio, presented by Redmond and Tankersley (2005). Morphological comparisons intuitively suggested to the researchers that this point resembled the small fluted Paleoindian points recovered from the nearby Gainey Site. The researchers wanted to evaluate this hypothesis of interest using quantitative methods to determine if the point's length, basal width, and other measurements were statistically similar to the Gainey Site sample (Redmong and Tankersley 2005:518). This involved evaluating the null hypothesis $H_0 : \overline{Y}_{Sheriden} = \overline{Y}_{Gainey}$ for these various measurements. To illustrate this process, we consider the maximum length measurement, which for the Sheriden Cave point is 35.6 mm and for the Gainey Site sample ($n = 9$) is $\overline{Y} = 47.8$ mm with $s = 6.5$ mm. The alpha value is set at .05. Because we do not have σ, we will need to use the t-distribution to evaluate this hypothesis.

Using Equation (8.5), we can determine a t-score as:

$$t = \frac{Y_i - \overline{Y}}{s}$$

$$t = \frac{36.5 - 47.8}{6.5}$$

$$t = -1.738.$$

The negative sign indicates that the Sheriden point is smaller than the average Gainey Site small fluted point, but is this difference statistically significant? We can determine this using Appendix B to identify the critical value. Determining critical values using Appendix B is quite different from the table in Appendix A for the standard normal distribution. Appendix A provided areas under a single curve that allows us to calculate probabilities. This is possible because the standard normal distribution truly is a single distribution. However, the t-distribution is actually a bunch of similar, but slightly different distributions based on the degrees of freedom. As a result, the probabilities under the curve differ as v changes, meaning that it would be necessary to create a different probabilities table for $v = 1$, $v = 2$, $v = 3$, and so forth until $v > 120$, when we could simply revert to the standardized normal distribution.

Listing 120 different tables for the t-distribution seems excessive, so a more useful alternative is simply to provide the critical values for the t-score that correspond to given probabilities and degrees of freedom. This is what Appendix B does. Various alpha values are listed at the top of the table. Values for the degrees of freedom are listed on the left-hand side. The critical value for a given test is the value that corresponds with the degrees of freedom determined using Equation (8.4) and the appropriate α value. For this example, $\alpha = .05$ and $v = 9 - 1 = 8$. The corresponding critical value displayed in Appendix B is 2.306. As with the Z-score, the negative sign for our t-score reflects only the direction of difference, so we can discard it when comparing it to a critical value. Our calculated t-score of 1.738 is less than the critical value of 2.306, meaning that we cannot reject the null hypothesis. There is no statistically significant difference between the maximum length of the Sheriden Cave point and the average maximum length of the small fluted points from the Gainey Site, a conclusion that supports Redmond and Tankersley's (2005) hypothesis of interest. The authors then build on this relationship to make various conclusions concerning the Paleoindian occupation of Ohio.

The null hypothesis can also be easily evaluated using confidence intervals. The upper and lower confidence intervals for evaluating $H_0: Y_i = \bar{Y}$ are calculated as $\bar{Y} \pm t_{[\alpha,v]}s$. The term $t_{[\alpha,v]}$ simply means the t-value for a given alpha value and degrees of freedom. In this case, $t_{[.05,8]}$ equals 2.306, so $L_1 = 47.8 - (2.306 \times 6.5) = 32.8$ mm and $L_2 = 47.8 + (2.306 \times 6.5) = 62.8$ mm. Given that $Y_{Sheriden} = 35.6$ mm falls between these two limits, we again cannot reject the null hypothesis and consequently conclude that the Sheriden fluted point's maximum length is consistent with the small fluted points from the Gainey Site.

Archaeologists also perform comparisons between a sample mean and μ using $s_{\bar{Y}}$ to approximate $\sigma_{\bar{Y}}$, typically when they are predicting some μ value based on their knowledge of the world or are using previously derived means (especially from published reports based on unpublished data). In these cases, the archaeologist wishes to determine whether $\bar{Y}$ is substantially different than μ, but will need to evaluate this proposition using $s_{\bar{Y}}$ for the sample because $\sigma_{\bar{Y}}$ is not known. (Of course, when we know $\sigma_{\bar{Y}}$, we can simply calculate a Z-score as outlined above.)

We will illustrate such a case using Nelson et al.'s (2006) analysis of Mimbres sites from southwestern New Mexico. Nelson and her colleagues wished to study residential mobility strategies by examining differences in room use and repair through time. They present evidence concerning the density of postholes in rooms, as well as other variables that they argue reflect the presence and extent of room modification and repair (Nelson et al. 2006:416–17). Posthole density for the Classic Mimbres rooms ($n = 30$ rooms) averaged 1.16 postholes per m² with a standard deviation of .73 postholes per m². They are interested in determining if this is significantly different than the later Reorganization Phase rooms that averaged .78 postholes per m². Treating the Reorganization Phase sample as a population, this hypothesis of interest can be evaluated using the null hypothesis $H_0: \bar{Y}_{Mimbres} = \mu_{Re\,organization}$. In this case, the t-score can be calculated using Equation (8.6) as:

$$t = \frac{\bar{Y} - \mu}{s_{\bar{Y}}}$$

$$t = \frac{1.16 - .78}{\left(.73/\sqrt{30}\right)}$$

$$t = 2.851$$

The degrees of freedom for this comparison are $30 - 1 = 29$. The critical value (again from Appendix B) is $t_{[.05,29]} = 2.045$. The computed t-value is larger than the critical value, so we reject the null hypothesis, just as Nelson and her colleagues did.

There are in fact differences between the time periods in regards to room repair and modification. Nelson *et al.* (2006) use this information to construct arguments about changing settlement and movement patterns through time.

Of course, the null hypothesis could also be evaluated using confidence limits, determined as $\overline{Y} \pm t_{[\alpha, v]} s_{\overline{y}}$. Here, $L_1 = 1.16 - (2.045 \times .133) = .89$ postholes per m^2 and $L_2 = 1.16 + (2.045 \times .133) = 1.43$ postholes per m^2. This range does not encompass $\mu = .78$ postholes per m^2, again prompting us to reject the null hypothesis.

Note that we treated both of the Mimbres examples as two-tailed tests (see Figure 8.1), that is, as evaluating whether the mean of the Mimbres period rooms was significantly larger or smaller than μ for the Reorganization Phase. Appendix B is in fact organized for two-tailed tests, which are most common. However, the table can be easily adapted for one-tailed hypotheses tests, the topic to which we now turn.

Testing One-Tailed Null Hypotheses

Often the archaeologist's hypothesis of interest is focused on whether two or more properties are equal to one another. In these cases there is no initial concern about the direction of the difference and null hypotheses such as $H_0: \overline{Y} = \mu$ are appropriate. As just mentioned, such tests are two-tailed tests as illustrated in Figure 8.1. On occasion, though, an archaeologist may have a hypothesis of interest that specifies the direction of difference that is considered analytically significant. The null hypothesis will therefore need to consider whether a difference in one direction is present. For example, an archaeologist proposes that habitations built in elevated portions of the site are high status as reflected by larger room sizes. The null hypothesis $H_0: \overline{Y} = \mu$ is no longer adequate, because it does not reflect the hypothesis of interest, which instead focuses on whether $\overline{Y}$ for the elevated rooms is *larger* than μ, not *different* than μ. Rephrasing the null hypothesis as $H_0: \overline{Y} \le \mu$ better reflects the hypothesis of interest, in that the failure to reject the null hypothesis invalidates the hypothesis of interest, whereas rejecting the null hypothesis lends support to the archaeologist's hypothesis. This means though, that probabilities on only one side of the mean will be considered significant for rejecting the null hypothesis. Thus, instead of having areas of rejection in both tails of the distribution, the area of rejection is concentrated in only one tail (compare Figure 8.1 with Figure 8.8).

There is a problem using the critical values listed in Appendix B to evaluate the revised null hypothesis, however. This (and most such tables) assume that the hypothesis is a two-tailed test. As a result, the table actually reports a value corresponding to a probability that is *half* of the actual alpha value (again, consult Figure 8.1). How can the critical values be changed to reflect a one-tailed test, in which the entire area of rejection is focused in a single tail of the distribution? Quite easily, in that all we need to do is double the actual alpha value when consulting the table. Thus, the critical values for two-tailed tests with $\alpha = .10$ in Appendix B correspond exactly with the critical values for one-tailed tests with $\alpha = .05$.

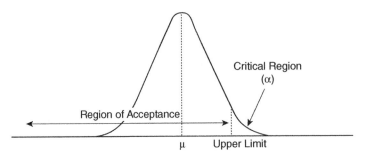

Figure 8.8   Region of rejection for a one-tailed test corresponding with $H_0 : \overline{Y} \leq \mu$

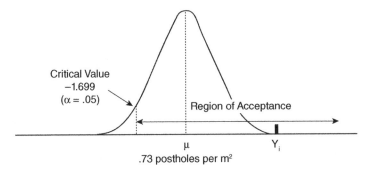

Figure 8.9   Critical value and area of rejection for $H_0 : \overline{Y} \geq \mu$

Using the Mimbres posthole analysis to demonstrate a one-tailed test, imagine that Nelson *et al.* (2006) were interested in determining if there was *an increase* in the number of postholes in the Reorganization Phase relative to the Mimbres Phase. The null hypothesis would become $H_0 : \overline{Y} \geq \mu$. Rejecting this null hypothesis would support the hypothesis of interest, whereas failing to reject the null hypothesis would undermine the hypothesis of interest. We would go about calculating the t-score just as we did above, again concluding that the value was 2.851. However, the null hypothesis now specifies a direction of difference, reflecting a one-tailed test. Here, it will only be rejected if μ is greater than $\overline{Y}$, meaning that the critical t-Value *must* be negative (Figure 8.9). Further, all of the region of rejection must now be focused in that lower tail of the distribution. Thus, the critical value for $\alpha = .05$ with $v = 29$ corresponds to the listed critical value of $\alpha = .10$ with $v = 29$ in Appendix B. The corresponding critical value is thus −1.699. The t-value of 2.851 is larger than −1.699, preventing us from rejecting the null hypothesis, despite the fact that we did reject in for $H_0 : \overline{Y} = \mu$ (see Figure 8.9).

As the above example illustrates, the devil does indeed live in the statistical details, and the difference between rejecting and not rejecting a null hypothesis often is contingent on how the hypothesis is phrased. Every analyst should take the time to clearly state the hypothesis of interest and then be certain that their null hypothesis adequately reflects it. Only by doing this can you be certain of the rigor of your quantitative analysis and thereby use this powerful tool to help build the archaeologically relevant argument you seek to pursue. In the next chapter, we will further expand on some of these themes, especially as they relate to the concept of power.

Practice Exercises

1 Define and differentiate between the following:

  (a) standard deviation and standard error

  (b) the grand mean, μ, and $\bar{Y}$.

  (c) standardized normal distribution and the t-distribution

  (d) confidence limits and critical values

  (e) one-tailed and two-tailed tests.

2 When would you wish to use the t-distribution instead of the standardized normal distribution? When would you wish to use the standard error instead of the standard deviation?

3 Determine whether you should use the standard error or the standard deviation to evaluate the following null hypotheses:

  (a) $H_0: Y_i = \mu$

  (b) $H_0: \bar{Y} = \mu$

  (c) $H_0: Y_i = \bar{Y}$

4 Determine whether you should use the standardized normal distribution or the t-distribution to evaluate the following null hypotheses:

  (a) $H_0: Y_i = \mu$ where you know σ

  (b) $H_0: \bar{Y} = \mu$ where you are using s to estimate σ

  (c) $H_0: Y_i = \mu$ where you are using s to estimate σ

  (d) $H_0: \bar{Y} = \mu$ where you know σ

  (e) $H_0: Y_i = \bar{Y}$ where you are using s to estimate σ

5   Prasciunas (2007) presents the results of an experimental comparison of
    the efficiency in terms of the amount of usable cutting edges for flakes
    produced using bifacial and generalized cores. Below is the data for the
    bifacial cores.

    (a)   Evaluate the null hypothesis ($\alpha = .05$) that the average weight of an
          exhausted bifacial core is statistically identical to the average weight
          of exhausted generalized cores ($\mu_{gen} = 57.65$).

    (b)   Evaluate the null hypothesis ($\alpha = .05$) that the average weight of an
          exhausted bifacial core is equal to or greater than the average weight
          of exhausted generalized cores ($\mu_{gen.} = 57.65$).

    (c)   How did shifting the null hypothesis from a two-tailed to a one-
          tailed test impact the outcome?

Core	End weight
B1	131.41
B3	150.35
B4	125.41
B5	110.32
B6	170.27
B7	131.88
B8	160.69
B9	135.66
B10	132.48
B11	150.44

6   Evaluate the null hypothesis that a historic settlement where the average
    house size for 11 homes is $\bar{Y} = 312 \, m^2$ and $s = 43 \, m^2$ is consistent with a
    mean house size of $325 \, m^2$ using a critical value derived from the
    t-distribution ($\alpha = .05$). Re-evaluate the null hypothesis using confidence
    intervals. Did you reach the same conclusion in both cases?

7   A sample of 13 ceramic pipes has an average weight of 43 g with a standard
    deviation of 4 g. Using confidence intervals ($\alpha = .05$), determine which of
    the following ceramic pipes are consistent with the sample: 29 g, 50 g, 47 g,
    64 g, 22 g, 33 g, and 56 g.

9

Hypothesis Testing III: Power

Now that you understand the fundamentals of hypothesis testing, it is possible to fully examine the relationship between Type I and Type II errors. As previously discussed in Chapter 7, β or Type II errors occur when we fail to reject the null hypothesis when it is false and another hypothesis is true. A clear relationship exists between Type I and Type II errors. As we reduce the probability of committing a Type I error by expanding our confidence limits, we increase the probability of accepting a null hypothesis when it is false – making a Type II error. Likewise, if we decrease the probability of making a Type II error by shrinking our confidence limits, we increase the probability of rejecting a null hypothesis when it is true – a Type I error. This relationship will be easier to understand with an example.

Let us begin with two hypothetical distributions of maximum rim thickness for pots, both of which have $\sigma = 0.5\,mm$ but one which has $\mu_1 = 7\,mm$ and the other $\mu_2 = 8\,mm$. Given $\mu_1 = 7\,mm$, let's say we wish to evaluate the null hypothesis $H_0 : Y_i = \mu_1$, where the alternate hypothesis is $H_a : Y_i = \mu_2$. The null hypothesis can be tested by placing confidence limits around μ_1 as follows:

Lower confidence limit $= \mu_1 - 1.96\sigma = 7 - 1.96(.5) = 6.02\,mm$
Upper confidence limit $= \mu_1 + 1.96\sigma = 7 + 1.96(.5) = 7.98\,mm$

Ninety-five percent of the distribution associated with $\mu_1 = 7\,mm$ falls within this range, but what about the alternate hypothesis $H_a : Y_i = \mu_2$? As illustrated in Figure 9.1, there is a considerable amount of the distribution corresponding with $\mu_2 = 8\,mm$ in the acceptance region for $H_0 : Y_i = \mu_1$. This means that it is very likely that we will not reject the null hypothesis when $H_a : Y_i = \mu_2$ is true. The area of the distribution μ_2 inside the confidence limits of $H_0 : Y_i = \mu_1$ is β, the probability of failing to reject

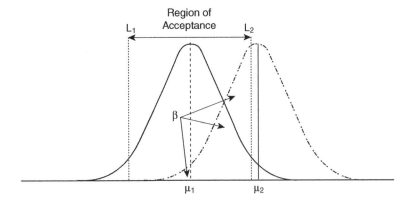

Figure 9.1 β associated with $H_a : Y_i = \mu_2$

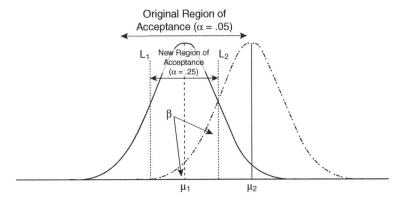

Figure 9.2 Illustration of β decreasing as α increases

the null hypothesis when it is false and $H_a : Y_i = \mu_2$ is true (Figure 9.1). When rejecting the null hypothesis, we would conclude that Y_i is associated with some alternate distribution such as μ_2, so the probability of committing a β error doesn't matter. However, if we fail to reject $H_0 : Y_i = \mu_1$, then we must be concerned about the likelihood that Y_i is actually a part of some alternate distribution (in this case $\mu_2 = 8\,\text{mm}$) but happens to fall within the region of acceptance by chance. As illustrated in Figure 9.1, there is indeed a pretty good chance that we will not reject the null hypothesis when it is actually false and $H_a : Y_i = \mu_2$ is true. Of course, saying that there is "a pretty good chance" of committing a Type II error is ambiguous, and we will demonstrate how we can determine this probability directly.

It should be clear, though, that the probability of committing a Type II error is a significant issue. Just because we fail to reject the null hypothesis doesn't mean it is true. One way to decrease β is of course to increase α so that the confidence limits are narrower. This will decrease the area of overlap between the second distribution and the acceptance region (Figure 9.2). Increasing α also makes it more likely to commit a Type I error, though, which presents its own problem in evaluating null

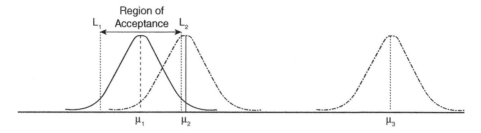

Figure 9.3 β associated with evaluating $H_0: Y_i = \mu_1$ where $H_1: Y_i = \mu_2$ and $H_2: Y_i = \mu_3$

hypotheses. Thus, we are caught between the proverbial "rock" and the "hard place"; setting α so that we have a good probability of not rejecting a null hypothesis when it is true increases the likelihood of mistakenly failing to reject the null hypothesis when in fact it is false. Setting α so as to decrease the likelihood of mistakenly failing to reject a null hypothesis increases the likelihood that we will reject the null hypothesis when it is true. Which is better, minimizing α or β? There is no good rule of thumb. It depends on the research design and the implications that making each of these errors has on the analysis. Ideally, we would like the probability of making either a Type I or Type II error to be quite small. However, this is not always possible. We must therefore weigh the results of making both types of errors within the context of our specific analysis.

An additional point to keep in mind is that β is unique for each alternate hypothesis. While each null hypothesis has only one alpha level, β must be considered for each and every alternate hypothesis. For example, there may be a large amount of overlap between the acceptance region for $H_0: Y_i = \mu_1$ and the alternate hypothesis $H_1: Y_i = \mu_2$, but little overlap between the null hypothesis and $H_2: Y_i = \mu_3$, where $\mu_3 = 16\,mm$ and $\sigma = .5\,mm$. β then may be very large between $H_0: Y_i = \mu_1$ and $H_1: Y_i = \mu_2$, but almost non-existent between $H_0: Y_i = \mu_1$ and $H_2: Y_i = \mu_3$ (Figure 9.3). In fact, the only reason we consider it only "almost" non-existent is because the two distributions are hypothetically asymptotic to infinity – an assumption of the normal distribution.

Calculating β

Calculating β is simple, once the null hypothesis's region of acceptance and one or more specific alternate hypotheses have been defined. To calculate β, use the standardized normal distribution or the t-distributions to determine the area of the alternate distribution that falls within the confidence limits. For example, all three of our hypothetical distributions above ($\mu_1 = 7\,mm$, $\mu_2 = 8\,mm$, and $\mu_3 = 16\,mm$) have standard deviations of 0.5 mm. As demonstrated, we can calculate the upper and lower confidence limits for $H_0: Y_i = \mu_1$ using $\alpha = 0.05$ as $7 \pm 1.96\sigma$, which produces a lower limit of 6.02 mm and an upper limit of 7.96 mm. Figure 9.3 illustrates these confidence limits, and their relationships to the means of the other two distributions being considered. We now have both α (specified at .05) and the area of acceptance for $H_0: Y_i = \mu_1$.

To determine our two β values for $H_1: Y_i = \mu_2$ and $H_2: Y_i = \mu_3$ we must identify the areas of the distributions associated with μ_2 and μ_3 that fall within the acceptance region for the null hypothesis. This can be calculated by computing the Z-score for each distribution corresponding with the upper and lower limits of the acceptance region of $H_0: Y_i = \mu_1$. β will be the area under the two alternate distributions between the two values.

For $H_1: Y_i = \mu_2$, the Z-scores are:

$$Z = \frac{L_2 - \mu_2}{\sigma} \text{ or } Z = \frac{7.98 - 8}{.5} = -.04 \text{ and } Z = \frac{L_1 - \mu_2}{\sigma} \text{ or } Z = \frac{6.02 - 8}{.5} = -3.96$$

For $H_2: Y_i = \mu_3$, the Z-scores are:

$$Z = \frac{7.98 - 16}{.5} = -16.04 \text{ and } Z = \frac{6.02 - 16}{.5} = -19.96$$

Remember that positive and negative Z-values reflect only whether the variates are greater or lesser than μ. Using Appendix A (areas under the normal distribution), we find that the tabled values of $-.04$ and -3.96 correspond to areas under the curve of .016 and .500, respectively. These values reflect the distance from $\mu_2 = 8$ mm, so β, the distance between them, is calculated as $\beta = .50 - .016 = .486$ (Figure 9.4). This value reflects that 48.6% of the variates drawn from the population $\mu_2 = 8$ mm will be in the acceptance region, which makes them indistinguishable from the variates that are members of $\mu_1 = 7$ mm when evaluating the null hypothesis $H_0: Y_i = \mu_1$ using $\alpha = .05$. Put another way, we will fail to reject the null hypothesis roughly 49% of the time we encounter a member of distribution $\mu_2 = 8$ mm. If we

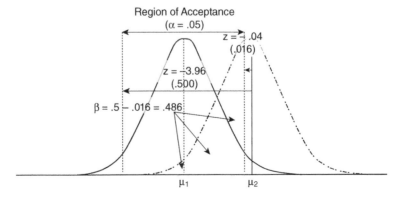

Figure 9.4 Calculation of β for the alternate hypothesis $H_1: Y_i = \mu_2$

fail to reject $H_0 : Y_i = \mu_1$, then it is possible that Y_i is from the population $\mu_1 = 7$ mm or $\mu_2 = 8$ mm.

In contrast, for $H_2 : Y_i = \mu_3$, the values from Appendix A both correspond to .500. This means that $\beta < .001$, and the small portion of the asymptotic left tail that does fall within the acceptance region for $H_0 : Y_i = \mu_1$ is too small to meaningfully measure for our purposes. The likelihood of accepting the null hypothesis $H_0 : Y_i = \mu_1$ when in fact $H_2 : Y_i = \mu_3$ is true is extremely small.

What do the β values mean in real-world terms? Well, archaeologists are often interested in both differences and similarities within their data. When a null hypothesis is rejected, then it can be concluded, at a previously specified level of certainty, that Y_i *is not* a member of the distribution specified in the null hypothesis. But if we "accept" the null hypothesis, can we conclude that Y_i *is* a member of the distribution specified in the null hypothesis? In this case, we can be quite certain that $H_2 : Y_i = \mu_3$ is unlikely to be true, given that very few members of $\mu = 16$ mm fall within the region of acceptance. However, nearly half of the distribution associated with $\mu_2 = 8$ mm does. This means that a poorly trained or lazy archaeologist who simply does accept that $H_0 : Y_i = \mu_1$ is true every time he or she fails to reject the null hypothesis will be wrong 48.6% of the time when $H_1 : Y_i = \mu_2$ is actually true. We don't know about you, but that is a higher level of error than we find comfortable when making statements in press. Failing to acknowledge and measure the probability of committing a Type II error will lead (and has led) fellow archaeologists to make the mistake of concluding that all variates within the acceptance region are members of the same population when they likely are not, which in turn can lead to poor reasoning and faulty conclusions as analytically distinct distributions are collapsed into each other. Again, "failing to reject" the null hypothesis does not indicate that the null hypothesis is true, but simply indicates it remains plausible. We should also determine which alternate hypotheses are also plausible. While the example above used the Z-score, power can just as easily be determined using many of the other distributions such as the t-distribution we introduce in this text.

Returning to our measurements of β, we found that β for $H_2 : Y_i = \mu_3$ is effectively zero, but that β for $H_1 : Y_i = \mu_2$ is roughly 49%. It intuitively seems that testing $H_0 : Y_i = \mu_1$ is a powerful means of differentiating between the members of the distributions μ_1 and μ_3, but isn't a powerful test for differentiating between μ_1 and μ_2. Further, knowing the power of the test could be analytically useful when our hypothesis of interest is phrased such that some differences between means are important, but others aren't. For example, if the three means reflect some attribute of cultural historical types, there could easily be a situation where the archaeological context and other temporally diagnostic materials indicate that $H_2 : Y_i = \mu_3$ is a plausible alternate hypothesis but that $H_1 : Y_i = \mu_2$ is not. This in turn means that our analysis of β indicates that the significant overlap between distributions associated with $H_0 : Y_i = \mu_1$ and $H_1 : Y_i = \mu_2$ is in practicality insignificant. Statisticians and other scientists have recognized this aspect of β, and have consequently developed a formalized means of stating a test's *power*, which will be illustrated below.

Statistical Power

The power of a statistical test is derived using the formula: Power $= 1 - \beta$, and is a probabilistic statement of the ability of a particular statistical test to correctly differentiate between two distributions, samples, or populations. Formally, the power of the test is defined as *the probability of rejecting the null hypothesis when it is false and the alternate hypothesis is true*. The greater the power, the more certainty there is that a statistical test can accurately differentiate between the members of the different distributions.

The preceding example considering $H_0: Y_i = \mu_1$ and $H_1: Y_i = \mu_2$ illustrates the relationship between β and power (Figure 9.5). Using the formula Power $= 1 - \beta$, we find that the power between the hypotheses $H_0: Y_i = \mu_1$ and $H_1: Y_i = \mu_2$ is .514. Given that $\beta < .001$, the power between the hypotheses $H_0: Y_i = \mu_1$ and $H_2: Y_i = \mu_3$ is $>.999$. Thus, we can accurately differentiate only 51.6% of the variates belonging to the distribution $\mu_2 = 8\,\mathrm{mm}$ when evaluating the null hypothesis $H_0: Y_i = \mu_1$. The rest of the time we will be unable to determine if the variates belong to $\mu_1 = 7\,\mathrm{mm}$ or $\mu_2 = 8\,\mathrm{mm}$. In contrast, we can easily differentiate between effectively all of the variates associated with $\mu_1 = 7\,\mathrm{mm}$ and $\mu_3 = 16\,\mathrm{mm}$.

Increasing the power of a test

As you can see from the example, if the degree of the overlap of the two distributions is large, the power of a statistical test differentiating between the alternate hypotheses is small (e.g., $H_0: Y_i = \mu_1$ and $H_1: Y_i = \mu_2$). If the degree of the overlap of the two distributions is small, the power of a statistical test differentiating between the alternate hypotheses is large (e.g., $H_0: Y_i = \mu_1$ and $H_2: Y_i = \mu_3$).

Ideally, we would like our tests to be as powerful as possible. A test's power can be increased using the same means for tightening our confidence limits as previ-

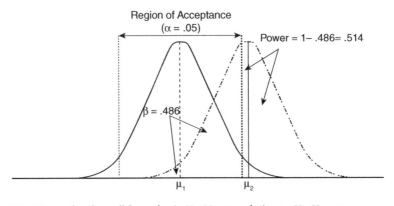

Figure 9.5　Power for the null hypothesis $H_0: Y_i = \mu_1$ relative to $H_1: Y_i = \mu_2$

ously discussed, because the test's power is directly controlled by the size of the confidence limits. The easiest way is to increase α, but this solution has the disadvantage of increasing the frequency of Type I errors. Another useful strategy, especially when dealing with distributions of means, is to increase the sample size, which often results in a tighter distribution. As a result, the alternate distributions will include less overlap, which in turn increases the power of the test. We can also occasionally increase a test's power by redefining the null hypothesis (e.g., differentiating between sexes for sexually dimorphic species). However, it is always important to remember that the null hypothesis must be clearly related to the hypothesis of interest. Redefining a null hypothesis so that it results in a powerful test, but no longer reflects the hypothesis of interest, doesn't provide any analytic benefit at all.

Calculating Power: An Archaeological Example

An archaeological example will help demonstrate the utility of the concept of power and the impact that β can have on archaeological research. Anthropologists have noted a relationship between the organization of craft production and various cultural attributes related to political and economic complexity (e.g., Costin and Hagstrum, 1995). Craft production in politically complex groups tends to be dominated by specialists making items for use by people outside of their immediate household. Craft production in politically simple groups tends to be dominated by generalized, household production. Likewise, ethnoarchaeologists have found that the products of specialists tend to be highly morphologically standardized when compared with the products of generalized (non-specialist) producers. This relationship, dubbed the standardization hypothesis, allows archaeologists to study the organization of craft production using measures of morphological and compositional variation in artifact assemblages.

The following table (Table 9.1) provides a list of the coefficient of variation (*CV*) for the maximum ceramic vessel diameter in samples from ethnographically studies of modern groups with specialized and generalist ceramic production (Crown, 1995: 150–1). These data have been used to argue that ceramics produced by specialists will be morphologically characterized by attributes producing *CV*s equal to or smaller than 10% whereas generalized producers manufacture ceramics characterized by morphological attributes associated with *CV*s larger than 10% (e.g., Benco, 1988; Crown, 1994: 116, 1995: 148–9; Longacre *et al.*, 1988). This 10% breaking point does seem intuitively meaningful; most pottery samples made by specialists have coefficients of variation less than 10% while those made by non-specialists tend to have coefficients of variation greater than 10%. Sill, how certain are archaeologists about our conclusions derived using this rule of thumb? What is the likelihood that the researcher is incorrect about the organization of production? We don't know by simply looking at the table. It *looks like* the 10% cutoff is very powerful for differentiating between the two groups of potters, but there are obvious

Table 9.1 Coefficients of variation for ceramics produced by specialists and non-specialists

Corrected CV of maximum diameter	Group	Form
Specialists		
.04	Paradijon	Small–medium cooking vessels
.03	Paradijon	Medium cooking vessels
.04	Paradijon	Medium–large cooking vessels
.11	Paradijon	Small flower pots
.10	Paradijon	Medium flower pots
.08	Paradijon	Large flower pots
.07	Paradijon	Extra large flower pots
.15	Amphlett Island	Small household cooking vessels
.08	Amphlett Island	Ceremonial cooking vessels
.06	Amphlett Island	Ceremonial cooking vessels
.18	Amphlett Island	Large household cooking vessels
.15	Sacoj Grande	Medium cooking vessels
.06	Sacoj Grande	Medium cooking vessels
.07	Sacojito	Medium water containers
.05	Sacojito	Large water containers
.02	Durazno	Small water containers
.03	Durazno	Medium–large water containers
.05	Duranzo	Medium–large water containers
.14	Ticul	Plant pots
.06	Ticul	Decorative vessels
.18	Ticul	Small food bowl
Generalist (household production)		
.12	Kalinga	Medium vegetable pots
.10	Kalinga	Medium rice bowls
.13	Goodenough Island	Small cooking vessels
.12	Goodenough Island	Small cooking vessels
.12	Goodenough Island	Small cooking vessels
.16	Shipibo-Conibo	Small cooking vessels
.22	Shipibo-Conibo	Medium cooking vessels
.12	Shipibo-Conibo	Large cooking vessels
.16	Shipibo-Conibo	Water containers
.18	Shipibo-Conibo	Water containers
.18	Shipibo-Conibo	Water containers

Source: Crown, Patricia (1995). The production of the Salado polychromes in the American Southwest. In B.J. Mills and P.L. Crown (eds.), *Ceramic Production in the American Southwest* (pp. 150–1). Tucson: The University of Arizona Press.

exceptions in Table 9.1. Thus, we know that there is at least some likelihood that assemblages made by generalists and specialists will be misclassified when using 10% as a demarcation. Shouldn't archaeologists using these data to help determine the organization of production know what this likelihood is? An error created while studying the nature of craft specialization can lead to errors in reconstructing a culture's social differentiation, political complexity, economic organization, resource distribution, class distinctions, and so forth. The likelihood of committing a Type II error when using the standardization hypothesis consequently is not a trivial issue. Archaeologists using this methodology need to take the time to calculate β in order to ascertain the probability that they incorrectly classify a sample made by specialists as the product of non-specialists and vice versa.

We can easily determine the probability of committing a β error using the Z-score. The β error is the portion of the distribution of specialists with a *CV* greater than 10% and the proportion of generalized producers with a *CV* equal to or less than 10% (i.e., the area of each distribution in the acceptance region of the other distribution). For ceramics made by specialists, the average *CV* is 8.33% (μ_{spec} = 8.33%) and the standard deviation is 4.97% (σ_{spec} = 4.97%). For generalized producers, μ_{gen} = 14.63% and σ_{gen} = 3.64%. The relationships between the two distributions and the 10% cutoff value is depicted in Figure 9.6.

To use the Z-score to calculate the power of the tests, let us begin with the null hypothesis $H_0: Y_i = \mu_{spec}$ where $H_a: Y_i = \mu_{gen}$. Determining β requires that we determine the probability of concluding that a ceramic was made by specialists, when in actuality it had been made by generalized producers (Figure 9.7). To do so, we use the Z-score to calculate the proportion of the generalist producer distribution that is below 10%. The Z-score is calculated as:

$$Z = \frac{Y_i - \mu_{gen.}}{\sigma_{gen.}} = \frac{10 - 14.63}{3.64} = -1.27$$

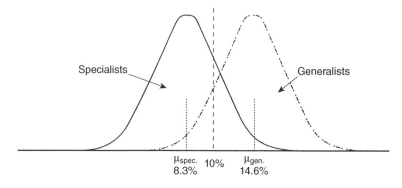

Figure 9.6 Distribution of coefficients of variation associated with pottery assemblages made by specialists and generalized producers

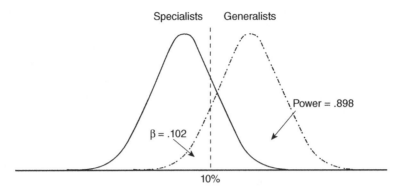

Figure 9.7　Illustration of β and the power associated with determining that an assemblage was made by specialists $(H_0 : Y_i = \mu_{spec.})$

Appendix A shows us that a Z-score of −1.27 corresponds to an area of 0.3980. β is therefore $0.5 - 0.398 = 0.102$. The power of the test is $1 - \beta = 1 - 0.102 = 0.898$ (Figure 9.7). Clearly, the cutoff point of a *CV* of 10% is fairly powerful when distinguishing specialist from non-specialist production. Only 10% of the time will pottery produced by generalists be incorrectly identified as the product of specialists meaning that 90% of the time the test will properly reject the null hypothesis when in fact it is false. Is this power sufficient? That is up to the investigator conducting the analysis, but 90% odds is pretty good, and we suspect that most archaeologists would feel comfortable with this result. At least now the researcher can directly address the issue.

But what about the power for the null hypothesis $H_0 : Y_i = \mu_{gen}$ where $H_a : Y_i = \mu_{spec}$. We can also use the Z-score to determine the probability of concluding that a ceramic was made by generalist producers, when in actuality it had been made by specialists. Now we use the Z-score to determine the portion of the distribution of ceramics manufactured by specialists with a coefficient of variation greater than 10% (i.e., the area of the specialist distribution in the acceptance region of the generalized producers) (Figure 9.8). The Z-score is calculated as:

$$Z = \frac{10\% - \mu_{spec.}}{\sigma_{spec.}} = \frac{10 - 8.33}{4.97} = 0.34$$

This Z-score corresponds to an area of 0.1331 from the mean.

To determine the area in the distribution of specialists that is greater than 10% then we must subtract 0.1331 from 0.5. The resultant β error, 0.3669, is the probability of incorrectly concluding that generalized producers made a sample of pots when they were actually manufactured by specialists. In other words, 36.7% of the time we would expect to incorrectly conclude ceramic vessels made by specialists were made by generalists. The power of the test is $1 - \beta = 1 - 0.3669 = 0.6331$

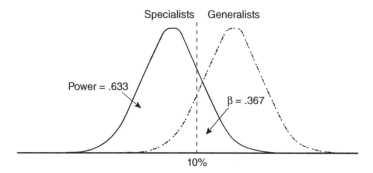

Figure 9.8 Illustration of β and the power associated with determining that an assemblage was made by generalists ($H_0 : Y_i = \mu_{\text{gen.}}$)

(Figure 9.8). Only about 63% of the time will we correctly conclude that specialists were responsible for the assemblages they indeed did produce. In this instance, we suspect that many archaeologists would find this test not sufficiently powerful for drawing substantial conclusions about past craft organization.

Taking the powers we calculated above, the implications for the use of coefficients of variation to differentiate between ceramics made by specialists and generalists are quite straightforward. First, we are 90% certain given these data that a coefficient of variation below the cutoff point of 10% represents ceramics made by specialists. Most archaeologists would likely find this to be an appreciable, but acceptable, likelihood for error, and would be willing to accept the argument that assemblages with CVs less than 10% were made by specialists. This argument would become even more persuasive if there were additional lines of evidence indicating craft specialization (e.g., craft barrios, expansive workshops). We are far less certain that ceramics with coefficients of variation greater than 10% were made by generalized producers, though. This area does contain 90% of the generalized producers but it also includes 37% of the assemblages manufactured by specialists. Thus, we expect to misclassify roughly two-fifths of the assemblages made by specialists. This large β error is probably greater than most researchers would find acceptable, and would probably cause many archaeologists to have little confidence when concluding that generalists made an assemblage. At the least, a rigorous argument would need additional lines of evidence supporting generalized craft production (e.g., kilns within individual households).

In addition to relying on corroborating, independent evidence, researchers interested in using the standardization hypothesis to identify the presence of generalized production could potentially construct a more powerful test in a number of ways. First, we could increase the cutoff value from 10% to some higher number such as 12%. This would increase the power of the test for determining generalized production, but it would decrease the power of the test when evaluating specialists' production considerably. In other words, increasing the cutoff point would decrease the probability that we would incorrectly conclude ceramics made by specialists were made by generalists, but would increase the probability of incorrectly conclud-ing ceramics made by generalists were made by specialists.

We could also define an "indeterminate" area of CV values that may represent either specialist or generalist production. For example, we could continue to use a 10% cutoff to classify assemblages as the product of specialists, but classify assem-blages with CVs of between 10% and 14% as "uncertain" or "indeterminate" in that they cannot be definitively associated with either specialists or non-specialists. Ceramic assemblages with CVs greater than 14% could then be classified as the products of generalists. Doing so would maintain the power of the test when con-cluding that ceramics were made by specialists at 10%, but would increase the power of the test when concluding that ceramics were made by generalists to 87%, which is a considerable improvement over the original 63%. However, this would cause about 24% of the distribution associated with specialists and 33% of the generalist distribution to be considered indeterminate (i.e., 24% of the specialist distribution and 33% of the generalist distribution fall between 10% and 14%). (We didn't bother including the Z-score calculations for the numbers presented above, but you can calculate them easily, if you care to do so.)

Either of these alternatives are reasonable, and we will leave it to individual researchers to decide how best to maximize the power of their statistical tests. However, the consideration of the rigor associated with the 10% cutoff is only pos-sible when the power of the statistical test is determined. Archaeologists (and all social scientists) should include such considerations as a general feature of their quantitative analyses. In this example, the consideration of power is quite straight-forward, given that there are two, and only two, competing hypotheses (a ceramic assemblage was made by specialists, or it was made by generalists). What do we do when no alternate hypotheses are explicitly defined? In such cases, power curves, our next topic, are useful.

Power Curves

Quantitative methods are an essential tool for archaeological comparisons, so much so that archaeologists give tremendous stock to arguments based on quantitative comparisons. If you really want to show archaeologists that site density didn't change between two periods, or that projectile points A and B really are functionally equivalent, or that bedrock mortars throughout a region were used to process the same resource, the surest way to build an argument that others will accept is to use quantitative methods to demonstrate similarities. This approach is especially per-suasive when there are explicit alternate hypotheses that can be considered (e.g., projectile points A and B are similar to each other and different from projectile point C). However, in archaeology, explicitly defined alternate hypotheses are not always known, especially given that something like projectile point morphology or settlement density can vary substantially *within analytic groups*. This variation can be as much as or more than the variation *among analytic groups*. Thus, quantitative comparisons showing similarities can be useful, but what if the range of variation was so great that it is impossible to reliably differentiate between members of different distributions? The lack of power prevents any meaningful conclusions about similarity because of the excessive probability of committing a Type II error. For example, what if a projectile point type is so broadly defined that virtually every corner notched point, whether it is long or short, wide or narrow, could fit within it? Can an archaeologist put much stock in a quantitative analysis demonstrating that point A's length is statistically consistent with the typical length of the type? Probably not, given that this result would be expected for nearly every "typically sized" projectile point. The failure to reject the null hypothesis just means that point A *could* be classified as a member of the type, not that it is more similar to that type than other possible types.

Unfortunately, archaeologists rarely consider such factors, and it seems to us often draw rather sweeping conclusions about similarities when in reality it would be very difficult to detect actual differences that might be present. This is fundamentally a problem of power, in that the tests used to evaluate the null hypotheses are not adequately powerful to justify the archaeologists' conclusions. In the example provided above, we could directly determine the power, which in turn suggested problems with the uncritical application of the standardization hypothesis. Is it possible to do something similar when considering a null hypothesis without clear alternatives? Yes, it is. Even in these cases power can provide insight into the sensitivity and usefulness of our statistical tests. Using *hypothetical* alternate distributions, archaeologists can calculate a *power curve* that describes the relationship between the null hypothesis being tested and a range of plausible alternate hypotheses. This is extremely useful, especially when dealing with certain types of data that have realistic limits to the values that are possible.

We will illustrate power curves using the rim thickness example presented above where $H_0: Y_i = \mu_1$; $\mu_1 = 7\,\text{mm}$, $\sigma_1 = .5\,\text{mm}$, and $\alpha = .05$. We have already determined that β may be a significant issue for alternate hypotheses such as $H_a: Y_i = \mu_2$ where $\mu_2 = 8\,\text{mm}$. As a result, we can ask whether evaluating $H_0: Y_i = \mu_1$ is analytically useful at all, given the realistic limits on the average rim thickness of pottery assemblages. Regardless of any specific alternate hypotheses, it is unlikely that a pottery assemblage in most archaeological cultures would produce an average maximum rim thickness less than 2 mm or significantly larger than 30 mm. Although we suppose it is possible to have assemblages exceed these limits, we doubt that they are common. We could ask, then, given the range of generally plausible alternate hypotheses (which are somewhere between $\mu = 2\,\text{mm}$ and $\mu = 30\,\text{mm}$), could we reliably identify when $H_0: Y_i = \mu_1$ is true, as opposed to some alternate hypothesis?

Table 9.2 presents the power for alternate hypotheses between $\mu = 2\,\text{mm}$ and $\mu = 30\,\text{mm}$ for every whole millimeter. The power for each of the contrived alternate values is computed using the Z-score as outlined above. Because the alternate means are simply invented, we do not have a standard deviation for each to use in the equation. We can expect, however, that the variation in other assemblages will

Table 9.2   Power curve for $H_0: Y_i = \mu_1$, where $\sigma = .5\,mm$

μ_{alt} (mm)	Derivation	Z-value	β	Power $(1 - \beta)$
2	$(6.02 - \mu_a)/\sigma$	8.04	0.00	1.00
3	$(6.02 - \mu_a)/\sigma$	6.04	0.00	1.00
4	$(6.02 - \mu_a)/\sigma$	4.04	0.00	1.00
5	$(6.02 - \mu_a)/\sigma$	2.04	0.02	0.98
6	$(6.02 - \mu_a)/\sigma$	0.04	0.49	0.51
7	$(6.02 - \mu_a)/\sigma$	−1.96	0.95	0.05
	AND	AND		
	$(7.98 - \mu_a)/\sigma$	1.96		
8	$(7.98 - \mu_a)/\sigma$	−0.04	0.49	0.51
9	$(7.98 - \mu_a)/\sigma$	−2.04	0.02	0.98
10	$(7.98 - \mu_a)/\sigma$	−4.04	0.00	1.00
11	$(7.98 - \mu_a)/\sigma$	−6.04	0.00	1.00
12	$(7.98 - \mu_a)/\sigma$	−8.04	0.00	1.00
13	$(7.98 - \mu_a)/\sigma$	−10.04	0.00	1.00
14	$(7.98 - \mu_a)/\sigma$	−12.04	0.00	1.00
15	$(7.98 - \mu_a)/\sigma$	−14.04	0.00	1.00
16	$(7.98 - \mu_a)/\sigma$	−16.04	0.00	1.00
17	$(7.98 - \mu_a)/\sigma$	−18.04	0.00	1.00
18	$(7.98 - \mu_a)/\sigma$	−20.04	0.00	1.00
19	$(7.98 - \mu_a)/\sigma$	−22.04	0.00	1.00
20	$(7.98 - \mu_a)/\sigma$	−24.04	0.00	1.00
21	$(7.98 - \mu_a)/\sigma$	−26.04	0.00	1.00
22	$(7.98 - \mu_a)/\sigma$	−28.04	0.00	1.00
23	$(7.98 - \mu_a)/\sigma$	−30.04	0.00	1.00
24	$(7.98 - \mu_a)/\sigma$	−32.04	0.00	1.00
25	$(7.98 - \mu_a)/\sigma$	−34.04	0.00	1.00
26	$(7.98 - \mu_a)/\sigma$	−36.04	0.00	1.00
27	$(7.98 - \mu_a)/\sigma$	−38.04	0.00	1.00
28	$(7.98 - \mu_a)/\sigma$	−40.04	0.00	1.00
29	$(7.98 - \mu_a)/\sigma$	−42.04	0.00	1.00
30	$(7.98 - \mu_a)/\sigma$	−44.04	0.00	1.00

be (more or less) comparable to that in the assemblage specified in the null hypothesis, so we estimate $\sigma = .5\,mm$ for the range of alternate hypotheses. For hypothetical means less than the lower confidence limit for $H_0: Y_i = \mu_1$, we use the lower limit $(L_1 = 6.02\,mm)$ when calculating power, because β will be the area of the hypothetical distribution that is greater than the lower acceptance limit. For values larger than the upper confidence interval, we use $L_2 = 7.98\,mm$ because β will be the area of the distribution that is smaller than this value. Both of these are straightforward applications of the methods we presented in the previous section, as illustrated in Figures 9.9 and 9.10. But what about alternate means that fall within the confidence

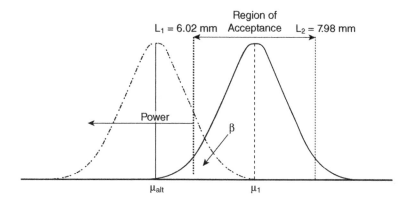

Figure 9.9 β and power for alternate means (μ_{alt}) less than the lower confidence limit

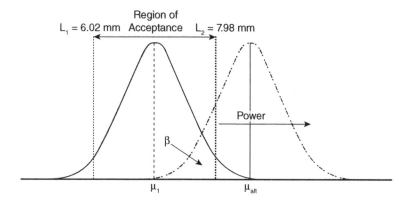

Figure 9.10 β and power for alternate means (μ_{alt}) larger than the upper confidence limit

intervals (Figure 9.11)? This creates a more complicated issue that we wish to explore in more detail.

The reason why the calculation of power becomes more complicated when the alternate mean is within the region of acceptance is that the area corresponding with β includes areas on both sides of the alternate distribution's mean. As a result, we must calculate two Z-scores to quantify this area. Further, power continues to be calculated as $1 - \beta$, but the region corresponding with power also includes substantial areas on both sides of the confidence limits (Figure 9.11). Again, we encourage you to draw the relationships between your distributions every time you calculate power or otherwise calculate areas under the normal or t-distributions.

The calculations of all of the powers for the alternate distributions are presented in Table 9.2. Please take the time to draw the relationship between several of these alternate hypotheses and the power of $H_0: Y_i = \mu_1$ to be sure you understand both how power is calculated and how the values in Table 9.2 are derived. The results of the power calculations indicate that the evaluation of $H_0: Y_i = \mu_1$ is quite powerful,

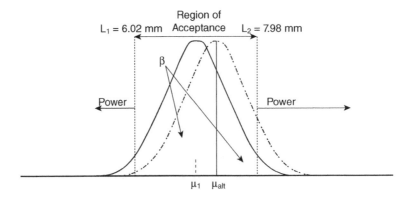

Figure 9.11 β and power for an alternate mean (μ_{alt}) contained within the confidence limits

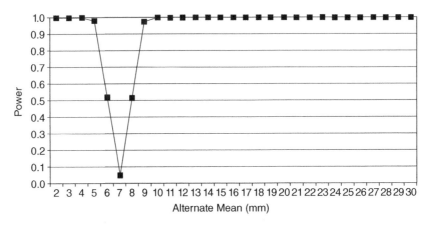

Figure 9.12 The power curve for the hypothetical alternative distributions to $H_0 : Y_i = \mu_1$

except for alternate distributions with means between 5 mm and 9 mm. Within this range, but especially between 6 mm and 8 mm, the test's power is quite low. These results are graphically represented in Figure 9.12. Are these results good? It depends on the analytic context. If we can reasonably expect that the most likely alternate distributions will have mean rim thickness less than 5 mm or greater than 9 mm, then we can make a compelling argument that the failure to reject $H_0 : Y_i = \mu_1$ is meaningful in that it indicates Y_i likely is a member of the distribution characterized by μ_1. However, if there are likely alternate distributions that fall within the range characterized by low power, we should not conclude that $H_0 : Y_i = \mu_1$ is true simply because we failed to reject the null hypothesis. Doing so will likely result in a Type II error.

The power curve provides valuable insights into the strength of our statistical tests. If a test is weak over the entire range of probable alternate hypotheses, then the test is prone to Type II errors and is consequently analytically suspect. Failing to reject a null hypothesis is not significant evidence that the null hypothesis is true, because the variate could be a member of many different distributions other than the one specified in the null hypothesis. This won't be a problem if the null hypothesis is rejected, but does seriously hamper any attempt to identify similarities in the archaeological record. If the power is weak, we can again increase the test's power in three ways: increase the sample size; redefine the null hypothesis; or increase the alpha level (with the associated consequences of increasing the likelihood of committing a Type I error).

Putting It All Together: A Final Overview of Hypothesis Testing

Understanding the relationship between probability, the null hypothesis, and Type I and Type II errors provides all of the tools necessary to efficiently evaluate hypotheses. The process of using these tools and reporting their results can be broken into six steps as follows:

Steps to hypothesis testing

1   State the statistical hypothesis in English, and as a formal statistical hypothesis. Be clear in the presentation of what is being measured, how it is being measured, and the validity of the measurement. Be sure to clearly specify the relationship between the hypothesis of interest and the statistical hypothesis.
2   Consider the consequences of making Type I and Type II errors and the relationship of these errors to power. With these in mind, set alpha.
3   Select the appropriate statistical procedure.
4   Define the region of rejection, or critical value.
5   Perform the computations and make the appropriate statistical decision (reject or fail to reject the null hypothesis).
6   State the decision regarding the null hypothesis in statistical terms and in English. If you fail to reject the null hypothesis, assess the power of the test. If the test is not powerful, come to no conclusions regarding the original hypothesis being tested, other than they are plausible.

Evaluating common hypotheses

You are now able to complete powerful statistical analyses. Below are common hypotheses that archaeologists test, and the calculation formula used to do so. This summary builds on the discussions in the preceding three chapters, and is intended as a quick reference guide to aid you in your work.

$H_0: Y_i = \mu$. **Determining if a single observation comes from a population**

To test if a single observation Y_i comes from a normally distributed population with a mean of μ, use the familiar formula to calculate a Z-score:

$$Z = \frac{Y_i - \mu}{\sigma}$$

$H_0 : Y_i = \bar{Y}$. **Determining if a single observation comes from a sample**
If σ is known, use a Z-score to evaluate this null hypothesis:

$$Z = \frac{Y_i - \bar{Y}}{\sigma}$$

If σ isn't known, substitute s for σ to calculate a t-score as follows:

$$t = \frac{Y_i - \bar{Y}}{s}$$

$H_0 : \bar{Y} = \mu$. **Determining if a sample mean comes from a population**
When σ is known, calculate $\sigma_{\bar{Y}}$ and use a Z-score to evaluate the null hypothesis as follows:

$$Z = \frac{\bar{Y} - \mu}{\sigma_{\bar{Y}}}$$

If the population parameter $\sigma_{\bar{Y}}$ is unknown, substitute $s_{\bar{Y}}$ and use the t-distribution:

$$t = \frac{\bar{Y} - \mu}{s_{\bar{Y}}}$$

$H_0 : \bar{Y}_1 = \bar{Y}_2$. **Comparing two sample means**
The comparison of two sample means is a common statistical comparison in archaeology, and is commonly called the t-test, although (as we have seen) we can use the t-distribution to evaluate other relationships as well. It is a bit more complicated than simply using the formula for the t-score presented in Equation (8.6), so we will discuss it in more detail here.

Because σ is rarely known when dealing with samples, s is typically used to estimate it. However, there typically are two sample standard deviations, which correspond with the two sample means. We can use s to estimate σ for calculating the standard error $\sigma_{\bar{Y}}$, but which of these standard deviations is a better estimate of the population parameter σ? Simply put, there is no *a priori* way to know. As a result, the best approach is to "pool" them into a single measure of dispersion called the "pooled standard error" symbolized as $s_{\bar{Y}_1 - \bar{Y}_2}$. The pooled standard error is based on a pooled standard deviation, symbolized as s_p. Substituting these terms into Equation (8.6) results in the following formula:

$$t = \frac{\bar{Y}_1 - \bar{Y}_2}{s_{\bar{Y}_1 - \bar{Y}_2}}$$

where

$$s_{\bar{Y}_1 - \bar{Y}_2} = \sqrt{\frac{s_p^2}{n_1} + \frac{s_p^2}{n_2}}$$

and

$$s_p = \sqrt{\frac{\sum y_1^2 + \sum y_2^2}{n_1 + n_2 - 2}}$$

Please remember from Chapter 4 that $\sum y^2$ is the sum of squares for a distribution calculated as $\sum (Y_i - \bar{Y})^2$. The value for n is simply the sample size of the respective sample.

$H_0: \bar{Y}_1 = \bar{Y}_2 = \bar{Y}_3 = \bar{Y}_n \dots$. Comparing multiple sample means using paired t-tests and the Bonferroni correction

We will discuss this in the next chapter, but ANOVA (analysis of variance) is the best means of evaluating this null hypothesis. However, many archaeologists (and researchers in other fields) use a technique called paired t-tests, which, while not inherently flawed, has significant drawbacks that are typically overlooked. Paired t-tests work by reducing null hypotheses such as $H_0: \bar{Y}_1 = \bar{Y}_2 = \bar{Y}_3$ into a series of binary comparisons such as $H_0: \bar{Y}_1 = \bar{Y}_2$, $H_0: \bar{Y}_1 = \bar{Y}_3$, and $H_0: \bar{Y}_2 = \bar{Y}_3$. Each of these are evaluated independently using the formula of comparing sample means presented above, with the original null hypothesis $H_0: \bar{Y}_1 = \bar{Y}_2 = \bar{Y}_3$ being rejected if any of the subsidiary hypotheses are rejected. Although it can be made to work, this is a fundamentally flawed way to evaluate the parent null hypothesis, because it exponentially increases the likelihood of committing a Type I error. The t-test is designed to determine if there is a significant difference between two means. While the likelihood of committing a Type I error (incorrectly rejecting the null hypothesis) in an individual t-test perfectly corresponds with alpha, the probability of committing a Type I error increases as the number of tests increases. This means that you are more likely to commit at least one Type I error when evaluating three null hypotheses when compared to evaluating a single null hypothesis with the same alpha level. This is the same process that we illustrated using the Psychic's trick. You could commit a Type I error when evaluating the first null hypothesis, and the second one, and the third one. Thus, the true probability of committing a Type I error when using paired t-tests does not correspond with .05 or whatever alpha level you specified – you are much more likely to commit a Type I error when evaluating the null hypothesis $H_0: \bar{Y}_1 = \bar{Y}_2 = \bar{Y}_3 = \bar{Y}_n$ using paired t-tests then the stated alpha level indicates.

Table 9.3 illustrates the rapid distortion of the true error rate for $H_0: \bar{Y}_1 = \bar{Y}_2 = \bar{Y}_3 = \bar{Y}_n$ by calculating the increasing probability of making at least one Type I error for at least one of the derivative paired t-tests. (You can replicate these results using a binomial analysis if you care to. Set $p = 1 - \alpha$ and then solve for all successes.) If we compare two means at $\alpha = .05$, the true probability of making a Type I error is five times out of 100, which is what we would expect. However, if we test three means, two at a time, the number of paired t-tests is three and the probability of committing a Type I error increases to .14. By the time 20 means are compared, we are evaluating 190 different paired t-tests, causing us to be virtually assured of committing a Type I error even when using a conservative alpha value of .02 (the cumulative probability of committing a Type I error is .98). Regardless of whether $H_0: \bar{Y}_1 = \bar{Y}_2 = \bar{Y}_3 = \bar{Y}_n$ is true or not, we are likely to reject it for no other reason than a Type I error. At the very least, the likelihood of doing so is not the same as the alpha that was selected.

One solution to the problem of compounding Type I error rate (also called the familywise error rate) is called the Bonferroni correction. It is a simple, yet effective solution created by dividing the alpha by the number of paired t-tests (α/n tests). When used to evaluating the difference between three means ($H_0: \bar{Y}_1 = \bar{Y}_2 = \bar{Y}_3$) at a *cumulative alpha* of .05, the alpha level for the three paired t-tests ($H_0: \bar{Y}_1 = \bar{Y}_2$, $H_0: \bar{Y}_1 = \bar{Y}_3$, and $H_0: \bar{Y}_2 = \bar{Y}_3$) is .05/3 = .0167. This allows the cumulative alpha to be constant so that the familywise error rate is in fact our stated α. Although the Bonferroni correction allows archaeologists to use paired t-tests, ANOVA, which is the subject of the next chapter, is a much easier and methodologically eloquent approach. We would never dream of forbidding people from using pair t-tests, but we recommend that you take care to consider the familywise error rate and use them properly.

Table 9.3 The true probability of committing a Type I error using paired t-tests

Number of means	Number of paired t-tests	Stated α level					
		0.2	*0.1*	*0.05*	*0.02*	*0.01*	*0.001*
2	1	0.20	0.10	0.05	0.02	0.01	0.00
3	3	0.49	0.27	0.14	0.06	0.03	0.00
4	6	0.74	0.47	0.26	0.11	0.06	0.01
5	10	0.89	0.65	0.40	0.18	0.10	0.01
10	45	1.00	0.99	0.90	0.60	0.36	0.04
20	190	1.00	1.00	1.00	0.98	0.85	0.17
∞	∞	1.00	1.00	1.00	1.00	1.00	1.00

Practice Exercises

1   What is the relationship between Type I errors, Type II errors, and power? How can we increase the power of a test?

2   Please answer the following questions:

(a)   How much of a distribution is contained in the area of acceptance when using confidence limits reflecting $\alpha = .05$?

(b)   What is the likelihood of committing a Type I error when $\alpha = .05$?

(c)   How much of a distribution is outside of the area of acceptance when using confidence limits reflecting $\alpha = .10$?

(d)   What is the change in the likelihood of committing a Type I error if a researcher increases α from .05 to .10?

3   A sample of ceramic vessels has a mean opening diameter of 12 cm, $s = 4$ cm, $n = 10$. Another sample of ceramic vessels has a mean diameter of 19 cm, $s = 3$ cm, $n = 7$. Use a t-test to determine if the sample means are significantly different.

4   Treat both distributions mentioned in Question 3 above as populations. Using $\alpha = .05$, determine if a vessel with an opening 17 cm in diameter is consistent with the population with an average opening diameter of 12 cm. Treat the second distribution ($\mu = 19$ cm) as an alternate distribution and determine the power of the test. How does knowledge of the power impact the interpretation of the hypothesis test?

5   In a study of weasels from archaeological faunal materials, Lyman (2004) compares the mandible lengths of long-tailed weasels with a sample of the modern ermine living in the central Columbia Basin of northeast United States. Evaluate the following hypotheses:

(a)   The long-tailed weasels have an average mandible length of 5.16 mm with a standard deviation of .48 mm ($n = 30$). The ermine have an average mandible length of 4.47 mm with a standard deviation of .33 ($n = 20$). Determine if the means of the two samples are the same.

(b) Following are mandible lengths for weasels of unknown species from various archaeological assemblages: 4.88 mm, 4.80 mm, 4.66 mm, 4.64 mm, 4.62 mm, and 4.56 mm. Determine which are consistent with the sample of long-tailed weasels ($\alpha = .10$). Determine if these same individuals are consistent with the ermine ($\alpha = .10$). Which, if any, of the individuals can be clearly assigned to either group?

(c) Does the probability of committing a Type II error pose a significant problem in determining the species of any of the archaeological specimens?

10

Analysis of Variance and the F-Distribution

Given archaeology's comparative nature, it is not surprising that archaeologists often wish to determine if there are meaningful differences in the central tendencies, especially means, of more than two samples. Does the mean room size in a Hallstatt village ($\bar{Y}_1$) differ significantly from an earlier Urnfield culture village ($\bar{Y}_2$) and a later La Tène settlement ($\bar{Y}_3$)? Given a suitable analytic and theoretical approach, this could be an interesting question that could further our understanding of social differentiation, corporate group size, craft production, household organization, and a host of other variables. We would like to test the null hypothesis about the average room size $H_0: \bar{Y}_1 = \bar{Y}_2 = \bar{Y}_3$, and its alternate hypothesis $H_a: \bar{Y}_1 \neq \bar{Y}_2 \neq \bar{Y}_3$. Such comparisons are at the core of archaeological analysis.

As described at the end of Chapter 9, we could use paired t-tests to evaluate null hypotheses such as $H_0: \bar{Y}_1 = \bar{Y}_2 = \bar{Y}_3$, but, for the reasons outlined there, doing so is problematic. The ANalysis Of VAriance (ANOVA) is a much easier and robust alternative. ANOVA is one of the most powerful tools in the statistician's toolkit, yet it hasn't been applied as widely as it deserves to be in archaeological analysis. This is unfortunate, given that it is mathematically quite simple and its eloquent structure allows you to think about and examine the organization of the world in new and exciting ways. Further, ANOVA is built on the F-distribution, which is useful in a variety of additional contexts.

One of the strengths of ANOVA is its conceptual structure. It isn't just a mathematical formula, but is instead a conceptually explicit framework for deriving meaning from the comparison of means. ANOVA can be used for EDA to inductively identify differences and similarities that otherwise might go unnoticed in analyses, but ANOVA's real strength is its potency for evaluating well-defined hypotheses of interests. When there are some empirical or theoretical reasons for

Quantitative Analysis in Archaeology, Todd L. VanPool and Robert D. Leonard
© 2011 Todd L. VanPool and Robert D. Leonard

expecting that either the null hypothesis $(H_0: \mu_1 = \mu_2 = \mu_3 = \mu_n)$ or the alternate hypothesis $(H_a: \mu_1 \neq \mu_2 \neq \mu_3 \neq \mu_n)$ is true, ANOVA can be used to test the underlying models and premises. There are two basic contexts for doing this, which are codified as *fixed effects ANOVA (Model I)* and *random effects ANOVA (Model II)*. These ANOVA models differ conceptually, not computationally; they are calculated using the same formulas, but the implications of the results are different. In short, the two models reflect how the source of variation among the means is explained.

In Model I, the sources of variation are introduced (generally experimentally) by the investigator, and the ANOVA analysis determines whether or not the investigator's treatment results in a significant difference in means. For example, imagine that an archaeologist wants to conduct an experiment to determine if the surface treatments of pots impact heat transfer to the pots' contents. Using the same clay and temper, the researcher makes morphologically identical vessels that differ only in their surface treatment. Some are corrugated, some are slipped, and some are plain. The archaeologist then subjects them to the same heat, and measures the time it takes to bring water to a boil. Given that all other variables are held constant, the only factor that should create differences in heat transfer is differences in the pots' surface textures. Thus, differences in the time to reach boiling reflect the variation introduced by the researcher (i.e., the researcher's fixed effects). This being the case, the ANOVA quantifies the impacts of the analytically introduced variation, and thereby *explains* the differences in the means (e.g., corrugated surface treatment increases the rate of thermal transfer).

In Model II ANOVA, the investigator wants to explain differences that can be observed, but the source of variation among means is beyond the investigator's direct control. Except in experimental studies, this is typical of most archaeological research given that human behavior (especially past human behavior) is outside of the researcher's direct control. (We frequently hear archaeologists wish it were otherwise.) The sources of variation reflected in differences among means could be the result of taphonomic processes, differences in artifact/feature composition and construction, differences in artifact/feature use or discard, different symbolic or conceptual associations, or any of the host of factors archaeologists invoke to explain the variation we observe. Unlike Model I ANOVA, in which identifying the differences allowed them to be clearly linked to previously specified, explanatorily relevant variation, identifying differences using Model II ANOVA is a starting point for the identification of the factors that underlie differences. To consider the differences in Model I and Model II ANOVA, consider the following archaeological examples.

Model II ANOVA: Identifying the Impacts of Random Effects

We start by outlining Model II ANOVA, given that its application will be more intuitive for most archaeologists. Table 10.1 presents the maximum length in millimeters of ten unbroken flakes of four raw material types recovered from Cerro

Table 10.1 Maximum flake length (mm) for 40 flakes from Cerro del Diablo, Chihuahua, Mexico

	Obsidian	Chert	Rhyolite	Silicified wood
	30	41	135	113
	53	110	141	111
	45	73	138	97
	34	52	175	70
	105	176	143	117
	102	61	132	48
	51	69	130	134
	47	40	109	115
	71	64	125	103
	58	48	120	106
$\sum\limits^{n} Y$	596	734	1348	1014
$\bar{Y}$	59.6	73.4	134.8	101.4
$\sum\limits^{n} Y^2$	8603.62	8520.31	8573.22	8656.87
$\sum\limits^{n} y^2$	6012.4	15396.4	2743.6	5598.4
$\bar{\bar{Y}} = 92.3$				

del Diablo, a Late Archaic site in northern Chihuahua, Mexico. It also presents ΣY_i (the sum of the individual variates calculated within each group), $\bar{Y}$ (the mean for each group), ΣY^2 (the sum of all of the squared variates within each group), Σy^2 (the sum of the squared deviations of each variate from its group mean), and $\bar{\bar{y}}$ (the grand mean of all members of all four groups). These values will be used below to calculate the ANOVA.

Table 10.1 represents a sample of the flaked stone artifacts from Cerro del Diablo that can be used to estimate μ and σ^2. A quick glance at the data causes us to suspect that the average length of flakes differs according to raw material (the average length of rhyolite flakes is over twice that of obsidian flakes), which could reflect differences in flaked stone reduction, use, and discard. We can test the hypothesis of interest that there are differences in flake length according to raw material type using the null hypothesis $H_0: \bar{Y}_{obs} = \bar{Y}_{chert} = \bar{Y}_{rhy} = \bar{Y}_{sw}$, but the way we will do so may seem counterintuitive. As the name analysis of *variance* suggests, we will actually be comparing various estimates of σ^2, the population variance. These are then tied to μ through a cleaver (yet simple) mathematical relationship. To illustrate how this can be done, consider the various estimates of σ^2 that can be calculated using the data in Table 10.1.

The most obvious estimates of σ^2 are s^2 for each group. With four groups, A (obsidian), B (chert), C (rhyolite), and D (silicified wood), we can calculate four estimates s_A^2, s_B^2, s_C^2, and s_D^2 where we use our usual formula (Equation (4.2)) for

the sample variance: $s^2 = \dfrac{\sum y^2}{n-1}$.

Our estimates are as follows:

$$s_A^2 = \frac{6012.4}{9} = 668.0 \text{ mm}$$

$$s_B^2 = \frac{15396.4}{9} = 1710.7 \text{ mm}$$

$$s_C^2 = \frac{2743.3}{9} = 304.8 \text{ mm}$$

$$s_D^2 = \frac{5598.4}{9} = 622.0 \text{ mm}$$

Assuming that the null hypothesis is correct, the starting assumption of all statistical analyses, each of these is an estimate of σ^2, but which is the best estimate? We don't know, so instead we can calculate a pooled estimate using (Equation (10.1)), where the numerator is the sum of the sum of squares for each raw material group, and the denominator is the sum of the degrees of freedom for each group.

The pooled variance

$$s^2 = \frac{\sum y_A^2 + \sum y_B^2 + \sum y_C^2 + \sum y_D^2}{(n_A - 1) + (n_B - 1) + (n_C - 1) + (n_D - 1)} \tag{10.1}$$

For our example:

$$s^2 = \frac{6012.4 + 15396.4 + 2743.3 + 5598.4}{9 + 9 + 9 + 9} = 826.5 \text{ mm}$$

This value is called the average variance within groups, or more commonly, the *variance within groups*, because the sum of squares is calculated *within* the groups (i.e., using the difference from each variate to its own group's mean). If the four groups are drawn from the same population (i.e., if the null hypothesis is true), probability suggests that the average variance within groups will be a better estimate of the population variance than would any single group variance.

Another estimate of the population variance σ^2 can be calculated using $s_{\bar{Y}}^2$ to estimate $\sigma_{\bar{Y}}^2$ (the variance among means, which is the squared standard error discussed in Chapter 8). The variance among means is calculated using the variation of the four group means $\bar{Y}_i$ from the grand mean $\bar{\bar{Y}}$ as indicated in Equation (10.2).

The variance among means

$$s_{\bar{Y}}^2 = \frac{\sum_{n=1}^{n=a}(\bar{Y}_i - \bar{\bar{Y}})^2}{a-1} \tag{10.2}$$

a is the number of groups, in this case the four raw material types, and $\sum(\bar{Y}_i - \bar{\bar{Y}})^2$ is called the *sum of squares of means*. We can calculate the sum of squares of means by literally subtracting the grand mean from each sample mean, squaring these values, and then adding them together as indicated by Equation (10.2), but it can also be calculated using Equation (10.3).

Calculation formula for the sum of squares of means

$$\sum \bar{Y}^2 - \frac{\left(\sum \bar{Y}_i\right)^2}{a} \tag{10.3}$$

Equation (10.3) is computationally equivalent to $\sum(\bar{Y}_i - \bar{\bar{Y}})^2$ but often is easier to calculate. Using the data from Table 10.1, the sum of squares of means is:

$$\sum_{n=1}^{n=a}\bar{Y}^2 - \frac{\left(\sum_{n=1}^{n=a}\bar{Y}\right)^2}{a} = 37392.72 - \frac{136308.60}{4} = 3315.56\,\text{mm}^2$$

After the sum of squares of means is calculated, we can calculate $s_{\bar{Y}}^2$ (the variance among means) as follows:

$$s_{\bar{Y}}^2 = \frac{\sum_{n=1}^{n=a}(\bar{Y} - \bar{\bar{Y}})^2}{a-1} = \frac{3315.56}{3} = 1105.19\,\text{mm}^2$$

Our goal in calculating $s_{\bar{Y}}^2$ is to provide an estimate of σ^2, but one is the squared standard error, reflecting the variation of means around the grand mean, whereas the other is a variance, reflecting the variation in variates around their means. They aren't equivalent; $\sigma^2 = \frac{\sum(Y_i - \mu)^2}{n}$ whereas $s_{\bar{Y}}^2 = \frac{\sum(\bar{Y} - \bar{\bar{Y}})^2}{a-1}$. Given that $\frac{\sum(Y_i - \mu)^2}{n} \neq \frac{\sum(\bar{Y}_i - \bar{\bar{Y}})^2}{a-1}$ how can we use $s_{\bar{Y}}^2$ to derive s^2 so that we can estimate σ^2? Recall Equation (8.1), which presents the relationship between the standard error and the standard deviation as $\sigma_{\bar{Y}} = \frac{\sigma}{\sqrt{n}}$. Squaring both sides of the equation causes it to be $\sigma_{\bar{Y}}^2 = \frac{\sigma^2}{n}$. Multiplying both sides of the equation by n we

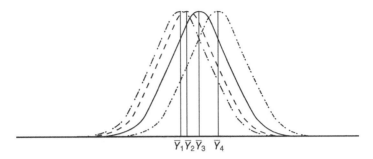

$$\overline{Y}_1\overline{Y}_2\overline{Y}_3\ \ \overline{Y}_4$$

Figure 10.1   Distribution of means in which the *among-group variance* is comparable to the *within-group variance*

obtain: $n\sigma_{\overline{Y}}^2 = \sigma^2$. This same relationship holds for $s_{\overline{Y}}^2$ and s^2, given that they are simply estimates of $\sigma_{\overline{Y}}^2$ and σ^2. Thus, $s^2 = n(s_{\overline{Y}}^2)$. Given this relationship, once we have calculated the variance among means $s_{\overline{Y}}^2$, multiplying this value by n provides us with s^2, an additional estimate of our population parameter σ^2. For our example:

$$s^2 = n(s_{\overline{Y}}^2) = 10(1105.19) = 11051.9 \,\text{mm}^2$$

This estimate of the population variance σ^2 is called the *variance among groups*, because it reflects the variation of the group means from the grand mean.

We now have created two independent estimates of the population variance σ^2: *variance within groups* = 826.4 mm, and *variance among groups* = 11,051.9. Here is the core of ANOVA: assuming the null hypothesis is true, these estimates of σ^2 ought to be (roughly) the same, because they are likely equally good estimates of the *same* population parameter. In other words, if $H_0: \overline{Y}_{\text{obs}} = \overline{Y}_{\text{chert}} = \overline{Y}_{\text{rhy}} = \overline{Y}_{\text{sw}}$ is indeed true, then both approaches to estimating σ^2 should produce (roughly) identical results. In contrast, if $H_0: \overline{Y}_{\text{obs}} = \overline{Y}_{\text{chert}} = \overline{Y}_{\text{rhy}} = \overline{Y}_{\text{sw}}$ is not true, then the estimates of σ^2 will likely differ appreciably. Compare the distributions reflected in Figures 10.1 and 10.2.

Figure 10.1 reflects the expected spread of distributions if the null hypothesis $H_0: \overline{Y}_1 = \overline{Y}_2 = \overline{Y}_3 = \overline{Y}_4$ were true. The four distributions have very similar spreads, indicating that their variances (s^2) are similar. The within-group variance thus should be an excellent estimate of σ^2. Likewise, the sample means are closely clustered as would be expected if they reflected the same population. Their distribution is consistent with the expected variation of sample means around μ as measured by the standard error ($\sigma_{\overline{Y}}$). As a result, the estimate s^2 calculated from the variance among groups should be an excellent estimate of σ^2 and should be roughly equal to the within-group variance.

Figure 10.2 is quite different, in that one of the distributions is distant from the others. In this case, all four distributions have similar spreads, causing the within-group variance to be very close to s^2 for each of the individual distributions.

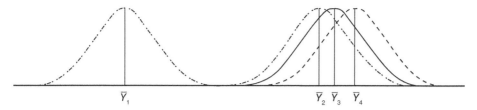

Figure 10.2 Distribution of means in which the variance among means exceeds the variance within means

Table 10.2 Maximum flake lengths (mm) of four raw materials from Cerro del Diablo, Chihuahua, Mexico

Obsidian	Chert	Rhyolite	Silicified wood
30	41	135	113
53	110	141	111
45	73	138	97
34	52	175	70
105	176	143	117
102	61	132	48
51	69	130	134
47	40	109	115
71	64	125	103
58	48	120	106

However, the difference between $\bar{Y}_1$ and the other three means will result in a comparatively large variance among means that will reflected by a large $s_{\bar{Y}}^2$. The subsequently derived estimate of s^2 will be considerably larger than the variance within groups. ANOVA is based on comparing the variance estimates. If the among-group variance and within-group variances are similar, then the null hypothesis $H_0 : \bar{Y}_1 = \bar{Y}_2 = \bar{Y}_3 = \bar{Y}_4$ is plausible (i.e., it cannot be rejected). However, if the among-group variance is considerably larger than the within-group variance, the sample means are more widely disbursed than expected according to the standard error, and the null hypothesis is unlikely to be true. Thus, even though we are comparing measures of variance, we are testing the relationship among the means.

Calculating the actual ANOVA comparison is quite easy, but by now you know that statistical analysis relies upon precise and straightforward notation. We consequently need to introduce some additional symbolism. To begin with, we need symbols to represent the location of individual observations as members of each group. Table 10.2 presents only the data from Table 10.1. Table 10.3 shows that each observation can be specified by the group to which it belongs (j), and which observation it is within that group (i). In Table 10.2, $Y_{1,1} = 30$ mm and $Y_{2,4} = 111$ mm.

Integrating the previously outlined conceptual structure with Table 10.3, the variance within groups can be symbolized as illustrated in Equation (10.4). The

Table 10.3   Matrix illustrating Y_{ij}

		a groups (*j*)			
		1	2	3	4
	1	1,1	1,2	1,3	1,4
	2	2,1	2,2	2,3	2,4
n items (*i*)	3	3,1	3,2	3,3	3,4
	4	4,1	4,2	4,3	4,4
	5	5,1	5,2	5,3	5,4
	6	6,1	6,2	6,3	6,4

inclusion of the double summation $\sum\limits_{j=1}^{j=a}\sum\limits_{i=1}^{i=n}$ makes this formula look peculiar, but the formula remains straightforward; it says to subtract the group mean from each individual variate, square the resulting values, sum them, and then divide the total by the degrees of freedom ($n-1$ for each group). This formula is merely an expanded form of the standard variance presented in Chapter 4 calculated as

$$s^2 = \frac{\sum\limits_{i=1}^{i=n}(Y_i - \bar{Y})^2}{n-1}$$ to include multiple groups instead of just one.

Variance within groups

$$s^2_{\text{within}} = \frac{1}{a(n-1)}\sum_{j=1}^{j=a}\sum_{i=1}^{i=n}\left(Y_{i,j} - \bar{Y}_j\right)^2 \tag{10.4}$$

The variance *among* groups can be calculated using Equation (10.5). These instructions state to subtract the grand mean from each group mean, square these values, sum them, multiply the summed value by the number of variates in each group, and then divide by the degrees of freedom. *Note that n here refers to the number of variates in each group (assuming equal sample sizes for all groups), in our example 10, not to the total number of variates for all groups.*

Variance among groups

$$s^2_{\text{among}} = \frac{n}{a-1}\sum_{j=1}^{j=a}\left(\bar{Y}_j - \bar{\bar{Y}}\right)^2 \tag{10.5}$$

Returning to our example of the Cerro del Diablo flaked stone assemblage, we need to test whether s^2_{among} and s^2_{within} reflect the same population parameter σ^2. The F-distribution provides the means of evaluating this. It is built on the relationship created by dividing one sample variance by another, i.e., $F_s = \dfrac{s_1^2}{s_2^2}$.

Obviously dividing σ^2 by itself will produce a result of 1 because σ^2 never changes. It is a population parameter defined in time and space that cannot change; $\dfrac{\sigma^2}{\sigma^2}$ always equals 1. However, s^2 should be close to the σ^2 it estimates, but most likely varies a bit from the true value. Taking two variances that estimate the same σ^2 and dividing one by the other will produce a value close to but probably not exactly 1, but probability also implies that sometimes s^2 will vary quite a bit from σ^2. Thus, although we expect $F_s = \dfrac{s_1^2}{s_2^2}$ to be (roughly) equal to 1, when the numerator is by happenstance substantially larger and/or the denominator is substantially smaller than σ^2, the F-value will be quite a bit larger than 1. Likewise, if the numerator is smaller and/or the denominator is larger than expected, F_s will be considerably smaller than 1 although it must always be larger than 0. The resulting distribution of sample variances divided by each other produces the F-distribution which has an average of 1, a lower limit almost but never equal to 0, and an asymptotic upper tail causing it to be skewed to the right (Figure 10.3).

As with the standardized normal distribution or the t-distribution, the F-distribution allows us to measure probabilities, in this case probabilities associated with the relationship between two sample variances. We can consequently use the F-distribution to determine if the ratio of two variances is larger than expected assuming they estimate the same parameter. If the two sample variances s_{among}^2 and s_{within}^2 estimate the same parameter, the values of each should be approximately equal and the ratio $\dfrac{s_{among}^2}{s_{within}^2}$ should be roughly equal to 1. If s_{among}^2 is considerably larger than s_{within}^2, the ratio will be larger than 1, suggesting that the various sample means do not reflect the same μ. The F-distribution allows us to determine if the ratio deviates sufficiently from 1 to cause the null hypothesis to be rejected at a given α. By convention, s_{among}^2 is always placed in the numerator and can be symbolized as s_a^2. s_{within}^2 is the denominator and can be symbolized as s_w^2.

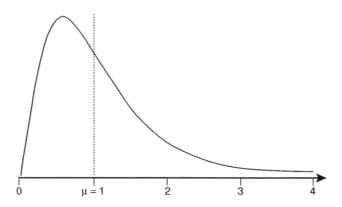

Figure 10.3 The F-distribution

The shape of the F-distribution changes depending on the degrees of freedom, because the accuracy of s^2 as an estimate of σ^2 improves as the sample size increases. Thus, $F_s = \dfrac{s_a^2}{s_w^2}$ should be constrained much closer to 1 if both sample variances are based on large samples than if one or both are based on smaller samples. Since there are two sample variances, there are also two degrees of freedom: one associated with the numerator and one for the denominator. Appendix C presents critical values of the F-distribution for various degrees of freedom for v_1 (numerator) and v_2 (denominator).

For the Cerro del Diablo example $F_s = \dfrac{s_a^2}{s_w^2} = \dfrac{11051.9}{826.4} = 13.37$, where $v_1 = a - 1 = 4 - 1 = 3$ degrees of freedom, and $v_2 = a(n-1) = 4(10-1) = 36$ degrees of freedom. Using $\alpha = .05$ and Appendix C, the critical value for F is 2.87. Given that $13.37 > 2.87$, we reject the null hypothesis and instead conclude that $H_0 : \bar{Y}_{obs} = \bar{Y}_{chert} = \bar{Y}_{rhy} = \bar{Y}_{sw}$ is unlikely to be true. The variation among the means is much greater than would be expected if in fact the average flake length did not differ as a result of lithic raw material.

What caused the differences among the means? This is a Model II ANOVA. We, as researchers, did not in any way control the variation in flake length according to lithic raw material, so we know the source of the variation lays elsewhere. Several possible reasons for the observed variation come quickly to mind. Given that flaked stone modification is a reductive technology, it is physically impossible to produce large flakes from a small cobble. Perhaps the raw materials differ in their original core size, allowing large flakes of some raw materials but only small flakes of other raw materials. The raw materials also differ in the sharpness and durability of their edges with obsidian and fine-grained chert producing sharper but less durable edges than the silicified wood and rhyolite. These differences could have prompted past peoples to use them for different purposes, which might have caused them to produce differently sized flakes. Or perhaps some materials are from more distant sources, and therefore are represented by smaller, more heavily reduced cores that produced smaller flakes. A host of other possibilities could be suggested; the results of the ANOVA give us the impetuous to pursue relevant explanatory variables (in conjunction with our theoretical and analytic structures). Put another way, it gives us a great starting point to refine our understanding of the archaeological record.

Model I ANOVA: The Analysis of Treatment Effects

To illustrate Model I ANOVA, consider a study of inter-observer measurement error performed by the senior author and his colleague, R. Lee Lyman at the University of Missouri (Lyman and VanPool, 2009). Archaeologists have long noted the potential of inter-observer variation when recording data from artifacts and features (e.g., Fish, 1978). This is an especially pressing issue when an archaeologist wishes to compare data collected by different people from different assemblages using different tools. Various studies have shown that problems of consistency can be severe (Fish, 1978; Gnaden and Holdaway, 2000), especially when using classifications to type artifacts (Whittaker *et al.*, 1998). Lyman and VanPool (2009) seek to evaluate the significance of this problem when taking and comparing "simple" dimensional measurements of artifacts and bones. As part of the study, the authors independently measured attributes of the same 30 projectile points twice, thereby creating four data sets reflecting the same artifacts. At least one month passed between the times when the analysts remeasured the points, and VanPool used two different calipers to introduce possible variation related to measurement tool accuracy and precision.

Table 10.4 presents the data for projectile point length from the study. Given that the points measured for each sample are the same, the only source of variation between samples should be differences introduced because of analyst error (e.g., failure to measure an attribute consistently (intraobserve error) or systematic differences between the analysts in how they measured the attributes (interobserver error)) or instrument error. If there is no substantial analyst or instrument error, then we should not be able to reject the null hypothesis $H_0 : \bar{Y}_{L1} = \bar{Y}_{L2} = \bar{Y}_{VP1} = \bar{Y}_{VP2}$. In contrast, the rejection of the null hypothesis will indicate the presence of significant analyst/instrument error. This is a perfect Model I ANOVA, in that any differences that might be present among the measurements are attributable to the fixed effects introduced by the researchers (e.g., different measurement tools, different researchers taking the measurements).

To evaluate if the variation among groups exceeds the variation within groups, we proceed with our ANOVA by calculating the variance *within groups* as follows:

$$s_w^2 = \frac{1}{a(n-1)} \sum_{j=1}^{j=a} \sum_{i=1}^{i=n} \left(Y_{ij} - \bar{Y}_j\right)^2$$

$$s_w^2 = \frac{1}{4(30-1)} \sum_{j=1}^{j=a} (2238.78 + 2259.24 + 2062.61 + 2252.20)$$

$$s_w^2 = \frac{8812.83}{116} = 75.97 \text{ mm}^2$$

We do not bother to show the actual values for each $(Y_{ij} - \bar{Y}_j)^2$, but you can check our work, if you wish.

We next calculate the values required to calculate the sum of squares of means (Equation (10.3)). These are:

$$\sum \bar{Y}_j = 104.10 \text{ mm}$$

$$\bar{\bar{Y}} = 26.025 \text{ mm}$$

$$\sum \bar{Y}_j^2 = 2709.221 \text{ mm}^2$$

Table 10.4   Measurements of the length of the same 30 points measured independently four times

Specimen	Lyman 1	Lyman 2	VanPool 1	VanPool 2
1	23.78	23.76	23.84	23.77
2	20.42	20.40	20.88	20.43
3	20.88	20.86	20.44	20.86
4	24.96	24.94	24.95	24.95
5	17.06	16.94	17.02	16.99
6	15.40	15.46	15.69	15.54
7	19.04	19.00	20.01	19.06
8	23.10	23.08	22.92	23.05
9	25.18	25.24	25.18	25.20
10	21.54	21.52	21.40	21.56
11	17.66	17.66	17.64	17.66
12	16.94	16.92	16.92	16.83
13	13.58	13.56	13.53	13.56
14	18.00	17.94	17.26	17.28
15	14.68	14.66	14.66	14.59
16	21.70	21.60	22.09	21.44
17	23.26	23.24	23.24	23.07
18	42.38	47.36	47.36	47.34
19	42.34	42.30	42.28	42.26
20	27.86	27.82	27.84	27.83
21	29.94	29.96	29.88	29.80
22	30.92	30.90	30.90	30.90
23	25.92	25.90	26.34	25.94
24	33.54	33.52	33.52	33.74
25	33.60	33.60	33.61	33.02
26	30.26	30.24	30.25	30.24
27	27.54	27.52	27.54	27.50
28	37.04	37.02	37.04	36.89
29	38.00	37.94	37.76	37.97
30	41.04	41.04	41.08	41.20

Putting these values into Equation (10.3), we can calculate the mean sum of squares as:

$$\sum\left(\bar{Y}_j-\bar{\bar{Y}}\right)^2=\sum\bar{Y}_j^2-\frac{\left(\sum\bar{Y}_j\right)^2}{a}=2709.221-\frac{(104.10)^2}{4}=.0188$$

The *variance among means* can then be calculated as:

$$s_{\bar{Y}}^2=\frac{\sum\left(\bar{Y}-\bar{\bar{Y}}\right)^2}{a-1}$$

$$s_{\bar{Y}}^2 = \frac{.0188}{3} = .006$$

The *variance among groups* is derived as:

$$s_a^2 = n\left(s_{\bar{Y}}^2\right) = 30(.006) = .18$$

Now we can examine the relationship between the variance among groups and the variance within groups by using the F-distribution:

$$F_s = \frac{s_a^2}{s_w^2} = \frac{.18}{75.97} = .002$$

This ratio shows that the variance among groups is considerably less than the variance reflected in the groups. Looking up the critical value corresponding with our degrees of freedom ($v_a = 4 - 1 = 3$ and $v_w = 4(30 - 1) = 116$) in Appendix C for $\alpha = .05$, we see that the critical value is 2.68. Obviously, .002 is smaller than 2.68, so we cannot reject the null hypothesis. There is no evidence that the differences in the analyst or the instruments used systematically introduced variation into the measurement of the points' lengths. This indicates that the measurements are precise, even when different researchers record the data using different tools. Archaeologists wishing to compare different samples using point length should be pleased by this result.

Let us revise this example using a contrived situation. As mentioned above, VanPool 2 was taken using a different set of calipers than the other measurements. Suppose that the calipers were biased such that they systematically produce measurements that were 6 mm too large, a small difference that might not be noticed by a rushed analyst. The revised data are illustrated in Table 10.5. The variance within groups will remain exactly the same; the difference between the original and the contrived data is the change in magnitude of a single group, but the variance *within each group* isn't changed at all. Each variate maintains the same relationship with its group mean that it had before.

However, the *variance among groups* increases substantially. It is calculated as follows:

$$\sum \bar{Y}_j = 110.10 \text{ mm}$$

$$\bar{\bar{Y}} = 27.53 \text{ mm}$$

$$\sum \bar{Y}_j^2 = 3057.41 \text{ mm}^2$$

Using Equation (10.3), we can calculate the sum of square of means as:

Table 10.5   Contrived measurements of the length of the 30 points listed in Table 10.4

Specimen	VanPool 1	Modified VanPool 2 (VanPool 2 +6 mm)	Lyman 1	Lyman 2
1	23.84	29.77	23.78	23.76
2	20.88	26.43	20.42	20.40
3	20.44	26.86	20.88	20.86
4	24.95	30.95	24.96	24.94
5	17.02	22.99	17.06	16.94
6	15.69	21.54	15.40	15.46
7	20.01	25.06	19.04	19.00
8	22.92	29.05	23.10	23.08
9	25.18	31.20	25.18	25.24
10	21.40	27.56	21.54	21.52
11	17.64	23.66	17.66	17.66
12	16.92	22.83	16.94	16.92
13	13.53	19.56	13.58	13.56
14	17.26	23.28	18.00	17.94
15	14.66	20.59	14.68	14.66
16	22.09	27.44	21.70	21.60
17	23.24	29.07	23.26	23.24
18	47.36	53.34	42.38	47.36
19	42.28	48.26	42.34	42.30
20	27.84	33.83	27.86	27.82
21	29.88	35.80	29.94	29.96
22	30.90	36.90	30.92	30.90
23	26.34	31.94	25.92	25.90
24	33.52	39.74	33.54	33.52
25	33.61	39.02	33.60	33.60
26	30.25	36.24	30.26	30.24
27	27.54	33.50	27.54	27.52
28	37.04	42.89	37.04	37.02
29	37.76	43.97	38.00	37.94
30	41.08	47.20	41.04	41.04

$$\sum\left(\bar{Y}_j-\bar{\bar{Y}}\right)^2=\sum\bar{Y}_j^2-\frac{\left(\sum\bar{Y}_j\right)^2}{a}=3057.41-\frac{(110.10)^2}{4}=26.91\,\mathrm{mm}^2$$

The *variance among means* can then be calculated as:

$$s_{\bar{Y}}^2=\frac{\sum\left(\bar{Y}-\bar{\bar{Y}}\right)^2}{a-1}=\frac{26.91}{3}=8.97\,\mathrm{mm}^2$$

The *variance among groups* is:

$$s_a^2 = n\left(s_{\bar{Y}}^2\right) = 30(8.97) = 269.07$$

The new F-value is:

$$F = \frac{s_a^2}{s_w^2} = \frac{269.07}{75.97} = 3.54$$

This value exceeds the critical value for $F_{[.05,3,116]} = 2.68$, prompting us to reject the null hypothesis in this contrived example. In this case, our biased instrument did introduce a significant treatment effect that caused the null hypothesis $H_0: \bar{Y}_{L1} = \bar{Y}_{L2} = \bar{Y}_{VP1} = \bar{Y}_{VP2}$ to no longer be true. While this did not happen in Lyman and VanPool's (2009) study, it does demonstrate how biased measurements, even when the bias might intuitively appear small (about half a centimeter), can lead to serious problems of accuracy that can undermine quantitative analyses. This again illustrates how quantitative methods are only meaningful in the context of a larger analytic structure that must account for factors such as measurement accuracy and precision, the clear definition of variables measured, and consistent measurement methods.

The impact that the treatment effect had on the point length measurements can be mathematically (and conceptually) understood by realizing that the ratio between the two variance estimates (variance within groups and variance among groups) actually estimates the following parameters in Model I ANOVA: $\frac{s_a^2}{s_w^2}$ estimates $\frac{\sigma^2 + \alpha}{\sigma^2}$ where α in the numerator represents the treatment effects (not Type I error). If there are no treatment effects, $\frac{s_a^2}{s_w^2}$ estimates the ratio $\frac{\sigma^2}{\sigma^2}$ because $\alpha = 0$. However, if there are significant treatment effects as was the case in our contrived example, $\alpha \neq 0$ causing $\frac{s_a^2}{s_w^2}$ to quickly become larger than 1 as their magnitude grows.

A Final Summary of Model I and Model II ANOVA

The preceding examples lay the utility of and differences between Model I and Model II ANOVA in sharp relief while also reflecting their potential for archaeological analysis. The results of the Model I ANOVA allow us to determine if the treatment effects had an impact on samples that otherwise reflect the same population. The presence of a difference is therefore directly attributable to the experimentally introduced (fixed) effect, and hence the difference (or lack thereof) is directly tied to the experimental design. The experimental design thus *explains* the results of the Model I ANOVA when the null hypothesis is rejected. Others may repeat our experiment, verify our conclusions, and establish to their satisfaction that we have adequately *explained* the differences in means through our treatments.

In contrast, the results of the Model II ANOVA indicate whether a difference that is beyond the researcher's direct control is present. This difference may or may not be expected. If there is some theoretical or empirical reason to expect a particular outcome (either rejecting or failing to reject the null hypothesis), Model II ANOVA tests whether the underlying conceptual or empirical structure is plausible. This can be a very powerful tool for evaluating our understanding of the factors shaping the archaeological record. At the very least, rejecting the null hypothesis helps archaeologists detect differences worthy of additional investigation. However, the explanation of the differences relies on factors outside of the researcher's direct control, and their explanation will therefore require the detailed theoretical and empirical frameworks that connect the ANOVA analysis to other causal factors. Any differences that are present are *random effects* (relative to the researcher) as opposed to the *fixed effects* associated with Model I ANOVAs. Model II ANOVAs are consequently generally a starting point for continued analysis as researchers attempt to understand the nature of any random effects that structure the variation between means, whereas Model I ANOVAs are the ending point that links observed differences to the experimental design.

ANOVA Calculation Procedure

Now that you understand how ANOVA works, we provide you with a more rapid but less intuitive way to calculate them. Also, unlike the methods presented above, which are applicable only for groups with the same sample size, this computational method remains the same regardless of whether the sizes of each sample are the same. We use the data in Table 10.4 to illustrate this computation.

Quantity 1. The grand total, which is calculated by adding all Y_{ij} values.

$$\sum_{j=1}^{j=a}\sum_{i=1}^{i=n} Y_{ij} = 783.07 + 780.47 + 777.56 + 781.90 = 3123\,\text{mm}$$

Quantity 2. The sum of the squared individual observations, which is calculated by squaring all of the Y_{ij} values then adding them together.

$$\sum_{j=1}^{j=a}\sum_{i=1}^{i=n} Y^2 = 22678.73 + 22563.69 + 22215.93 + 22631.12 = 90089.47\,\text{mm}^2$$

Quantity 3. The sum of the squared group totals, each divided by its sample size.

$$\sum_{j=1}^{j=a} \frac{\left(\sum Y_i\right)^2}{n_i} = \frac{(783.07)^2}{30} + \frac{(780.47)^2}{30} + \frac{(777.56)^2}{30} + \frac{(781.90)^2}{30} = 81276.64 \text{ mm}^2$$

Quantity 4. The grand total squared and divided by the total sample size. This is also called the *correction term*, or CT.

$$CT = \frac{\left(\sum_{j=1}^{j=a} \sum_{n=1}^{n=i} Y\right)^2}{\sum_{j=1}^{j=a} n_i} = \frac{(3123)^2}{120} = 81276.07 \text{ mm}^2$$

Quantity 5. The total sum of squares.

$$SS_{total} = \sum_{j=1}^{j=a} \sum_{n=1}^{n=i} Y_i^2 - CT$$

$$SS_{total} = \text{Quantity 2} - \text{Quantity 4} = 90089.47 - 81276.08 = 8813.39$$

Quantity 6. The group sum of squares.

$$SS_{groups} = \sum_{j=1}^{j=a} \frac{\left(\sum Y_i\right)}{n_i} - CT$$

$$SS_{groups} = \text{Quantity 3} - \text{Quantity 4} = 81276.64 - 81276.07 = .57$$

Quantity 7. The sum of squares within.

$$SS_{within} = SS_{total} - SS_{groups}$$

$$SS_{within} = \text{Quantity 5} - \text{Quantity 6} = 8813.39 - .57 = 8812.82$$

Although less intuitive than the equations for the variance among and within groups we originally presented, these quantities produce exactly the same results. The calculations' results are then customarily presented in an ANOVA table that shows the sources of variation, the associated degrees of freedom, the sum of squares, the mean squares, and F_s (Table 10.6). The complete ANOVA analysis for the projectile point data is presented in Table 10.7.

The ANOVA table contains all of the information necessary to evaluate whether there is a difference in group means using the comparison of the variation *among* groups and *within* groups. Let us examine the ANOVA table in some detail.

Table 10.6 Generalized ANOVA table

Source of variation	SS	df	MS	F_S
$\bar{Y} - \bar{\bar{Y}}$ Among group	Quantity 6	$a - 1$	$\dfrac{\text{Quantity 6}}{a-1}$	$\dfrac{MS_{\text{groups}}}{MS_{\text{within}}}$
$Y - \bar{Y}$ Within group	Quantity 7	$\displaystyle\sum_{j=1}^{j=a} n_i - a$	$\dfrac{\text{Quantity 7}}{\displaystyle\sum_{j=1}^{j=a} n_i - a}$	
$Y - \bar{\bar{Y}}$ Total	Quantity 5	$\displaystyle\sum_{j=1}^{j=a} n_i - 1$		

Table 10.7 ANOVA analysis comparing the mean length of the 30 projectile points presented in Table 10.4

Source of variation	SS	Df	MS	F_S
$\bar{Y} - \bar{\bar{Y}}$ Among group	.56	3	.18	.002
$Y - \bar{Y}$ Within group	8812.82	116	75.97	
$Y - \bar{\bar{Y}}$ Total	8813.39	119		

In Tables 10.6 and 10.7 you will notice a new term, the total sum of squares. This is a new source of variation, one that demonstrates the sum of the squared deviations of each variate from the grand mean. Notice that the among-group sum of squares and the within-group sum of squares are additive to the total sum of squares. As a result, we have the following sources of variation:

$\bar{Y} - \bar{\bar{Y}}$ *among groups*: the variation attributed to the differences of the group means from the grand mean.

$Y - \bar{Y}$ *within groups*: the variation attributed to the differences of each variate from its own mean.

$Y - \bar{\bar{Y}}$ *total*: the variation attributed to the differences of each variate from the grand mean.

Each of these sources of variation has an associated degrees of freedom. The degrees of freedom are necessary to calculate a variance, or a mean square (i.e., the specific estimate of σ^2). The among-group degrees of freedom and within-group degrees of freedom are additive to the total degrees of freedom.

Each source of variation also has a *sum of squares* (SS), the sum of the deviations squared. The sum of squares is also necessary to calculate the variance, or mean square. As is true for the degrees of freedom, the among-group sum of squares and the within-group sum of squares are additive to the total sum of squares.

The *mean square* (MS) for each source of variation is the sum of squares for each source of variation divided by the appropriate degrees of freedom. The *among-group mean square* describes the dispersion of the group means around the grand mean. If there are no random or fixed effects, it estimates σ^2. If there are random or fixed effects, the *among-group mean square* estimates σ^2 plus the influence of the random or fixed effects. The *within-group mean square* describes the average dispersion of the observations in each group around the group's mean. The *within-group mean square* estimates σ^2 if the groups are representative samples from the same population.

The relationship among the three sums of squares can be illustrated as $(\bar{Y}-\bar{\bar{Y}})+(Y-\bar{Y})=(Y-\bar{\bar{Y}})$. Consider the variate $Y_{3,2}$ in Table 10.4, which is 20.86 where $\bar{Y}_2 = 26.06$ and $\bar{\bar{Y}} = 26.02$. Placing these values into the equation $(\bar{Y}-\bar{\bar{Y}})+(Y-\bar{Y})=(Y-\bar{\bar{Y}})$ produces:

$$(26.06-26.02)+(20.86-26.06)=(20.86-26.02)$$

$$.04+(-5.2)=-5.16$$

$$-5.16=-5.16$$

Why do we care about this relationship? The reason is that it demonstrates the additive effects of the deviation that are the foundation of ANOVA. A careful consideration of this equation illustrates that the value of each variate Y_i is caused by $\bar{\bar{Y}}$ plus the difference between $\bar{Y}$ and $\bar{\bar{Y}}$ plus the difference between $\bar{Y}$ and the variate Y_i. ANOVA, especially when presented in the format of Table 10.7, identifies each of these sources of variation and is in fact a formal comparison of them. If the variation between each group's $\bar{Y}$ and the individual variates is greater than the variation reflected between $\bar{\bar{Y}}$ and the sample means, then the variance within groups is larger than the variance among groups, and we will not be able to reject the null hypothesis. However, if the variance between the sample means and $\bar{\bar{Y}}$ is greater, then one or more fixed effects (Model I) or random effects (Model II) are contributing disproportionately to the total variation between the individual variates and $\bar{\bar{Y}}$.

We can consequently build a straightforward model of the sources of variation using ANOVA. If there was no variation, then all $Y_{ij} = \bar{\bar{Y}}$ (which would actually be μ). In the real world we do expect variation within a population for a whole host of reasons. In archaeological contexts, variation in artifact data can be caused by differences in artifact life history, a plethora of taphonomic processes, differential breakage and repair, imprecise cultural transmission, isochrestic variation in manufacture, and measurement error, just to name a few possibilities. Some of these differences will be systematic, and will therefore be uniform within a sample. These are the fixed effects in Model I ANOVA or the random effects in Model II ANOVA. We can symbolize this relationship as $Y_{ij} = \mu + a_j$ for Model I ANOVA and $Y_{ij} = \mu + A_j$ for Model II ANOVA. We use the lower case a and capital A to differentiate between fixed and random effects in order to clearly differentiate between the two ANOVA models. For Model I ANOVA, *a* is controlled by the researcher whereas *A* in Model II reflects factors outside the researcher's control.

Adding the fixed/random effects reflects the addition of the variation among groups that impacts the value of Y_{ij}, but if this was all of the variation present within a data set, every Y_{ij} would equal its group mean. As just mentioned, archaeologists expect variation within our samples for various reasons such as various taphonomic processes, differences in preservation, variation in the original artifact morphology, differences in use and repair, etc. Some of these sources of variation will not uniformly impact all members of a sample the same way, meaning that they create variation *within* the sample. (If they had a uniform impact on all members of the sample, they would be random or fixed effects). The influence of such variables is symbolized as e_{ij}, which reflects all sources of variation (typically considered "error" or "noise") that are not the result of the random/fixed effects. This "error" constitutes the within-group variation, which is then compared with the fixed/random effects to determine if they contribute disproportionately to the total variation. This conceptual link between within-group variation and the idea of error is why some statisticians refer to the *within-group MS* as the *error MS*. The model for Model I ANOVA is thus $Y_{ij} = \mu + a_j + e_{ij}$ and for Model II ANOVA is $Y_{ij} = \mu + A_j + e_{ij}$.

When rejecting the null hypothesis $H_0: \bar{Y}_1 = \bar{Y}_2 = \bar{Y}_a$, the analyst is formally concluding that random or fixed effects are different for the various samples. To put it another way, ANOVA is actually a means of testing for differences in treatment effects. We could therefore just as easily have phrased the null hypothesis as $H_0: a_1 = a_2 = a_n$ for Model I ANOVA and $H_0: A_1 = A_2 = A_n$ for Model II ANOVA (i.e., are the effects impacting each sample the same).

Given the nature of the "random" (with respect to the fixed/random effects) error, plots of e_{ij} from the same sample should approximate the normal distribution. If the distribution is not normal, the variation is likely non-random (with respect to the fixed/random effects) and we likely have an additional treatment effect that needs to be identified and controlled by redesigning the experiment or through using one of the more advanced forms of ANOVA presented in Chapter 14 that consider multiple treatment effects. More significantly, ANOVA analysis requires two assumptions related to this random error. First, the variances within each group must be roughly comparable. If there are vast differences, then it is not reasonable to assume that the variances of each group estimate the same σ^2. Second, all of the samples considered in the ANOVA analysis must be roughly normally distributed. Heavily skewed distributions produce large variances relative to normal distributions because the difference between the mean and its outliers is large. As a result, the variances within groups will be disproportionately large when one or more of the distributions is heavily skewed. When either of these assumptions is not met, we encourage you to use the Median Test introduced in Chapter 13 or the Kruskal-Wallis nonparametric ANOVA discussed in Chapter 14.

Identifying the Sources of Significant Variation in Model I and Model II ANOVA

In many cases, the exact source of significant variation may be evident when rejecting the null hypothesis $H_0: \bar{Y}_1 = \bar{Y}_2 = \bar{Y}_a$. For example, a quick glance at the sample means associated with Table 10.5 will indicate that the average for VanPool 2 is quite different from the other three averages and is consequently the main source of the variance among groups. However, it is often not so clearly cut and dry, and the researcher will want to formally establish which means are different. This can be done in a variety of ways, but the easiest method is to use paired t-tests. Yes, we know that we just got through trying to convince you that paired t-tests were methodologically flawed at the end of Chapter 9, but hear us out for a moment. The problem with paired t-tests is that they make it increasingly likely that we will incorrectly reject the null hypothesis $H_0: \bar{Y}_1 = \bar{Y}_2 = \bar{Y}_a$ as the number of paired tests increases. However, once a difference has been identified using ANOVA, this problem is less severe, given that *a difference already has been established at the stated α level.* Using the paired t-tests consequently won't cause us to reject the entire null hypothesis based on a Type I error. Further, if the Bonferroni method (discussed at the end of Chapter 9) is used to control for the impact on α, then we will be less prone to make Type I errors for the individual comparisons.

Consider the data for the maximum sherd thickness of ceramic types presented in Table 10.8. These three ceramic types are quite prominent in archaeological sites in western New Mexico and eastern Arizona from the 12th and 13th centuries. We know that they are different in one way – two are polychromes and the third is a bichrome. Intrigued that they co-occur so commonly, we may ask the question, "Do these ceramics serve similar or different functions?" If similar, we would expect there to be no significant differences in mean ceramic width. If dissimilar, we expect there may be differences. Our null hypothesis is $H_0: \bar{Y}_{\text{wingate}} = \bar{Y}_{\text{tularosa}} = \bar{Y}_{\text{St.Johns}}$ and the alternative hypothesis is $H_a: \bar{Y}_{\text{wingate}} \neq \bar{Y}_{\text{tularosa}} \neq \bar{Y}_{\text{St.Johns}}$. Table 10.9 presents the results of our Model II ANOVA. (In case you are wondering why we do not present all of the individual variates, the answer is that we don't need to. We are instead using the reported standard deviation to determine the sums of squares (Σy^2) for each sample, which in turn let us calculate the variance within the groups. You can

Table 10.8 Summary information for the maximum sherd thickness of a sample of three southwestern pottery types

Pottery type	n	$\bar{Y}$	Standard deviation
Wingate polychrome	25	5.6 mm	.661 mm
Tularosa bichrome	25	5.0 mm	.924 mm
St. Johns polychrome	25	5.8 mm	.707 mm

Table 10.9   ANOVA analysis comparing the sherd thickness of three pottery types

Source of variation	SS	dF	MS	F_s
$\bar{Y} - \bar{\bar{Y}}$ Among group	8.67	2	4.33	7.26
$Y - \bar{Y}$ Within group	43.00	72	.597	
$Y - \bar{\bar{Y}}$ Total	51.67	74		

Table 10.10   Confidence intervals for the maximum sherd thickness of three pottery types

Pottery type	n	$\bar{Y}$	Standard error	Lower limit	Upper limit
Wingate polychrome	25	5.6	0.13	5.26	5.94
Tularosa bichrome	25	5.0	0.18	4.52	5.48
St. Johns polychrome	25	5.8	0.14	5.45	6.16

do the same with published summary statistics even when you may not have the raw data, so long as you know the sample size.)

Using Appendix C to determine that the critical value for $F_{[.05,2,72]} = 3.12$ prompts us to reject the null hypothesis. There is a significant difference among the means. This is of course a Model II ANOVA given that the variation in the prehistoric pottery is out of our control. As a result, an interested ceramic analyst would likely wish to determine which mean(s) are different so that the random effects underlying the differences can be explored. This can be done easily by calculating confidence intervals around each mean to determine which encompass the means of the other samples. These confidence limits are presented in Table 10.10. We use the Bonferroni correction to maintain our α level at .05. In this case, the corrected significance level is $\alpha/3 = .05/3 = .0167$; $t_{[.0167,24]} = 2.57$.

Inspecting the means and confidence intervals, we conclude that the polychromes are statistically indistinguishable, given that their confidence intervals encompass each other's mean. However, the Tularosa bichrome mean lies outside of the other confidence intervals, and its confidence intervals do not encompass the other means. We therefore conclude that the large among-group variation is a result of the small mean value for the Tularosa ceramics. We now must seek to explain the random effects contributing to the thinner Tularosa sherds (or thicker polychrome sherds). Given that the variation is outside of the researcher's control, answering this question will likely require the evaluation of a number of other hypotheses using various lines of additional evidence (e.g., archaeological context, residue analysis, etc.).

Comparing Variances

There is a final use of the F-distribution we wish to introduce before turning our attention to regression analysis in the next chapter. It is the comparison of sample variances. Archaeologists are often interested in comparing the amount of variation within assemblages or samples as well as their central tendencies. Roux (2003) for example uses ethnoarchaeological data of pottery production to evaluate the intensity of craft production in Mesopotamia. She reports the standard deviations for pottery assemblages produced by several different potters from Andhra Pradesh, India, but notes that some attributes appear to be more standardized than others for the same potter depending on the use of the pots. For example the mean aperture size for *pedda baba* (storage jars) and *ralla catti* (vegetable cooking vessels) are comparable ($\bar{Y}_{\text{pedda baba}} = 17.02$ cm and $\bar{Y}_{\text{ralla catti}} = 16.93$ cm) but the variances intuitively seem different ($s^2_{\text{pedda baba}} = 2.37$ cm and $s^2_{\text{ralla catti}} = 1.54$ cm) (Roux 2003:777). She suggests that such variation reflects emic conceptions of acceptable variation, which in turn impacts archaeologists' ability to use variation to monitor craft production.

The hypothesis of interest regarding the potential difference between variances can be evaluated using the F-distribution, given that it is based on the comparison of variances. An F-score can be calculated by dividing the larger variance by the smaller one to derive an F value. This is called an F-test, which is presented in Equation (10.6). The null hypothesis is $H_0 : s^2_{\text{pedda baba}} = s^2_{\text{ralla catti}}$, and the alternate hypothesis is $H_a : s^2_{\text{pedda baba}} \neq s^2_{\text{ralla catti}}$. Here, $F = 2.37/1.54 = 1.54$. The sample of *pedda baba* vessels is based on 84 vessels whereas the *ralla catti* sample used 166 vessels. The degrees of freedom are thus 83 and 165, respectively. There is a bit of a difference in determining the critical value in this case. ANOVA as outlined above is a one-tailed test, in that we are testing to see if the variance among groups is larger than the variance within groups. The null hypothesis $H_0 : s^2_{\text{pedda baba}} = s^2_{\text{ralla catti}}$ is a two-tailed test, however, given that $s^2_{\text{pedda baba}}$ could be larger or smaller than $s^2_{\text{ralla catti}}$. The probabilities in Appendix C are for one-tailed tests, so the correct critical value for the two-tailed test will be the value listed for $\alpha/2$; the critical value for a two-tailed test at $\alpha = .05$ is the value listed for .025. The critical value of $F_{[.05,83,165]} = 1.44$. Given that 1.54 is larger than the critical value, we reject the null hypothesis and conclude that the variation in aperture size is different for the two functional groups of artifacts.

The F-test

$$F_s = \frac{s^2_1}{s^2_2} \qquad (10.6)$$

The F-test can be used to compare any two variances so long as two general assumptions are met. First, the means must be roughly equal. As noted during the discussion of the corrected *CV*, the size of the standard deviation (and hence the variance) correlates with the size of the distribution's mean. If the means are significantly different, the variance corresponding with the large mean will naturally be larger. An overly conservative but useful rule of thumb is that if there is no statistical difference between the means, then the variances can be compared. However, statistically different means can be compared so long as the differences between them are not too great.

Second, the underlying distributions must be normally distributed. As previously mentioned when outlining the assumptions of ANOVA, heavily skewed distributions produce large variances relative to normal distributions. They therefore cannot be compared to other distributions using the F-distribution. Distributions that are clearly not normally distributed should not be compared using an F-test.

Practice Exercises

1   Explain why an ANOVA analysis is preferable to a series of paired t-tests for evaluating the null hypothesis $H_0: \bar{Y}_1 = \bar{Y}_2 = \bar{Y}_a$. What are the assumptions of an ANOVA analysis?

2   Differentiate between a Model I and Model II ANOVA. Provide an example of a potential Model I ANOVA analysis that might interest an archaeologist. Provide an example of a potential Model II ANOVA analysis that might interest an archaeologist. How would the difference between the models impact the interpretation of the results?

3   Data for the posthole spacing (cm) for circular structures at the Oak Hill village site are presented below (Perttula and Rogers 2007:77). Use an ANOVA analysis to evaluate the null hypothesis that the average posthole spacing among the groups is the same ($\alpha = .05$). Is this a Model I or Model II ANOVA?

Structure group		
A	G	E & H
56	70	50
41	88	67
55	60	59
73	60	49
67	61	61
65	80	53
50	52	51

4   Following is the maximum thickness (mm) of Middle Archaic projectile points from the Ryan site in Alabama (Baker and Hazel 2007:47).

   (a)   Use an ANOVA to evaluate the null hypothesis that all of the average thicknesses of the points are the same ($\alpha = .01$). If they differ, use paired t-tests to determine which means are different.

   (b)   Is this a Model I or Model II ANOVA?

Benton knife/spear burial offerings	Benton projectile points/darts	Side notched projectile points
9.4	6.3	6.5
9.5	6.2	5.1
9.7	8.1	5.4
8.7	8.5	7.2

5   Below are data for the diameters (m) of pit structure clustered into three spatially distinct groups.

   (a)   Use paired t-tests ($\alpha = .05$) without using the Bonferroni correction to control for familywise error (see the end of Chapter 9) to evaluate the null hypothesis that the three groups have the same average diameter.

   (b)   Re-evaluate the null hypothesis using paired t-tests modified using the Bonferroni correction for a cumulative error of $\alpha = .05$.

   (c)   Re-evaluate the null hypothesis using an ANOVA ($\alpha = .05$). Did the results of your analysis change? Which method(s) are preferable and why?

Cluster 1	Cluster 2	Cluster 3
6.8	8.3	7.2
7.5	8.2	6.8
9.5	9.0	8.0
7.8	7.5	6.7
9	8.2	8.7
9.5	7.5	7.1
6.7	8.0	6.5

6   Crawford (1993: 9) presents the blade length of 27 British and Irish Iron Age swords. The standard deviation for the British swords is 10.07 cm ($n = 11$) while the standard deviation of the Irish swords is 4.51 cm ($n = 10$). Use the F-distribution to determine if the variation in the two distributions is significantly different ($\alpha = .05$).

11

Linear Regression and Multivariate Analysis

As useful as measures of central tendency and dispersion are, they cannot characterize all of the relationships that interest archaeologists. We may ask questions about the structure of a single variable of an assemblage (e.g., rim angles of different pottery types), but archaeological analysis often focuses on the relationships among two or more variables. Does rim angle change as maximum vessel height changes? Are longer projectile points also wider, or does the hafting element constrain a point's maximum width? Do settlements in an area get bigger through time? Do they get bigger moving down slope towards river flood plains? All of these questions might be interesting to an archaeologist, but they require the analyst to consider the relationships evident among two or more variables. As helpful as the mean, standard deviation, and other measurements we have discussed thus far are, they are not adequate for such tasks. The analyst needs additional statistical tools that can be called, as a group, multivariate analyses.

Perhaps the simplest and most straightforward multivariate method is linear regression, a technique that is one of the most widely used in archaeological (and other) analyses. Most people are familiar with it, at least in an abstract sense. It is useful in so many different contexts that we find it impossible to read the newspaper, a financial report, or listen to the daily news without encountering it (or at least the newsworthy results of its application). As common as it is, however, it does have certain limitations and assumptions, which are unfortunately often ignored, leading to hidden, but severe, analytic difficulties. Still, it is a powerful and flexible tool that is indispensible for archaeological analysis.

Simply put, linear regression allows us to examine the relationship between two continuously measured variables where we believe one variable influences the values of another. For example, we might expect the absolute number of hearths to increase in large habitation sites relative to small sites simply because more hearths are needed for cooking and heating in larger settlement. If this is so, then settlement size and hearth frequency share a *functional relationship*, in the sense that one of the variables (number of hearths) is *dependent* on the other (site size). As a tool, regression offers several benefits to archaeologists. It can be used to determine the presence and shape of functional relationships, which are often central to archaeological explanations of both similarities and differences in the archaeological record. We can also use the knowledge of the functional relationship to predict specific outcomes for cases where data are missing or to identify "odd", incongruent results that are atypical when compared to other cases.

In mathematical terms, the functional relationship allows us to use the independent variable (symbolized as X) to predict the values of the dependent variable (symbolized by the letter Y). Such relationships are expressed as $Y = f(X)$, which simply reads as Y is a function of X. Be sure to understand, though, that this is not an equal relationship of codependence between the two variables; the value of the independent variable (X) is not determined in any way by the value of the dependent variable (Y), but Y is at least to some degree controlled by X. Using our previous example, settlement size controls the number of hearths, not the other way around.

The simplest form of a functional expression is $Y = X$. The number of tree rings as a reflection of a tree's age is an excellent example of this relationship, as illustrated in Figure 11.1. Trees typically add a single ring every year they grow. We can therefore count the number of tree rings to determine a tree's age, an insight that remains crucial to dendrochronology. A more common and more complex relationship is $Y = bX$, where the coefficient b is a *slope* factor. If in our hypothetical example there

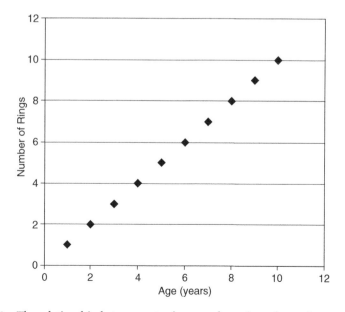

Figure 11.1 The relationship between a tree's age and number of tree rings

tends to be one hearth for every five habitation rooms, then the relationship between hearth and room frequencies would be $Y = .2X$.

As previously stated, regression is used when and only when there is a reason to believe (to hypothesize) that there is a relationship such that the variable X actually causes the value associated with Y to change. Let us take a moment to explore the structure of this causality. Is it fair to say that there is a causal connection between age and tree rings as illustrated in Figure 11.1 such that age causes the number of tree rings? The answer seems to us to be, "yes." Tree rings reflect the growing cycle of rapid growth during warm seasons and slow growth during winter dormancy. This cycle is measured as a year, so the causal connection between years and tree rings is clear. But what about the relationship between the exchange rates of the US dollar and the Mexican peso illustrated in Figure 11.2?

In the spring of 2010, the exchange rate was around 9.5 pesos for every US dollar ($Y = 9.5X$). The relationship follows the form of $Y = f(X)$ and looks like the relationship illustrated in Figure 11.1, but is it truly a functional relationship? Put another way, is the value of the Mexican peso *determined* by the value of the US dollar? The Mexican and US economies are certainly integrated with each other, but no, the dollar does not directly control the peso. Instead, variation in the value of each, which reflects underlying economic factors, causes them to correspond with each other at a fluctuating rate. Such nonfunctional correlations will be the subject of the next chapter, but regression requires a stronger causal relationship than such interdependency provides. For regression to work, the independent variable must in fact *control* the variation in the dependent variable in some way.

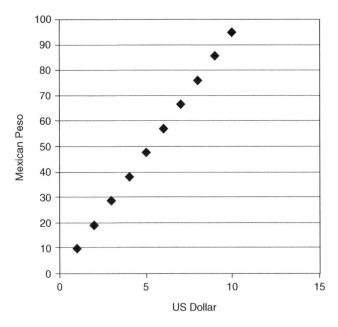

Figure 11.2   The relationship between the US dollar and Mexican peso in the spring of 2010

Let us consider a non-archaeological but intuitively meaningful example to further illustrate both the mathematical and conceptual structure of regression. Figure 11.3 illustrates the relationship between age and the average diastolic blood pressure in humans. Given our knowledge of human physiology and the effects of aging, we might very well expect there to be some relationship between age and blood pressure such that an individual's age affects his or her blood pressure. This hypothesis appears to be supported in Figure 11.3, in which the average blood pressure increases according to the individuals' ages.

While increases in X and Y in Figure 11.1 were uniformly linear, notice that this is not the case in Figure 11.3. Notice also that in Figure 11.1, when $X = 0$, then also $Y = 0$. That is not the case here. As illustrated in Figure 11.4, we can draw a line through the data points toward the y-axis to estimate that it would *intercept* the y-axis near 60. This makes sense; newborns have blood pressure. A regression line (the line illustrating the functional relationship) has both an *intercept* (the point at which it crosses the y-axis) and a *slope* (the rate at which Y changes in accordance with X). For any given relationship we can have a potentially infinite number of intercepts and slopes. These relationships take the form $Y = a + bX$. This formula is called the general linear regression equation; a is the *intercept*, and b is the *regression coefficient* or *slope*. Using our knowledge of age (X), the intercept (a), and the regression coefficient (b), we can predict an individual's blood pressure (a value of Y).

Data points are typically scattered about the regression line as illustrated in Figure 11.4 as opposed to falling on it perfectly as illustrated in Figure 11.1. As with the variation around means discussed in the last chapter, this variation from the regression line can be a product of measurement error, differences in artifact

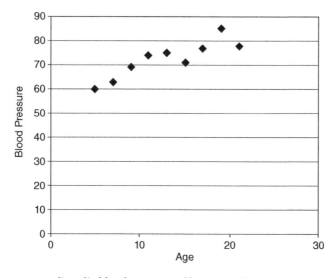

Figure 11.3   Average diastolic blood pressure of humans of various ages

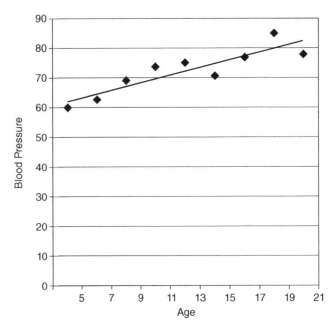

Figure 11.4   Regression line describing the relationship between age and diastolic blood pressure

manufacture and use, differences in preservation and recovery, and the influence of other factors impacting the dependent variable. Consider the data in Figure 11.5, which reflect the maximum length and weight of a random sample of 10 projectile points from Ventana Cave, Arizona (VanPool, 2003). It seems intuitive that there should be a causal relationship between point length and weight, in that a point's weight is controlled by its length. Figure 11.5 supports this self-evident proposition with the weight of points increasing with maximum length. However, the relationship is not perfect. Sometimes longer points weigh less than shorter points. Why?

The reason is that weight is also affected by maximum width, maximum thickness, basal width, basal thickness, raw material type, notch size, and a host of other morphological attributes. Maximum length does impact weight, but so do other variables meaning that only a portion of the variation in weight can be *explained* by variation in point length. The influence of the other variables also will be reflected in the weight values. Does this mean that regression is inappropriate for analyzing these data? Fortunately, no. Regression requires that Y be a product of X, but it does not require that Y be a product of *only* X. Other variables can affect the dependent variable too. As a result, the functional relationship between X and Y does not necessitate that the value of Y *must* be exactly equal to $a + bX$, but rather that the *mean* of Y for a given value of X is at $a + bX$. There will be variation in the data as a result of any other factors that influence the dependent variable. As a result, a regression line only reflects the average value of Y for a given X (which is, after

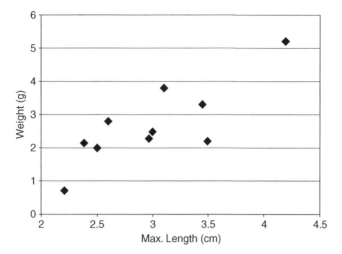

Figure 11.5 Scatter plot of the maximum length compared with weight for 10 projectile points from Ventana Cave, Arizona

all, the most likely value for Y), not the actual value that Y will take. (See, even here, the measures of central tendency and dispersion remain important.)

Realizing that regression analysis focuses on the average value of Y helps eluci-date the reasons for the four assumptions that are required for regression analysis. These assumptions are:

1 X is measured without error, or, in fancy statistical terms, it is fixed. It must be directly measured using adequately precise and accurate tools, and cannot be something like an average value or an estimate (e.g., it can't be something like a radiocarbon date). Otherwise, the relationship between the dependent and independent variables cannot be adequately quantified to allow the distribution of the mean values of the dependent variable to be represented by the regression line.

2 The expected (mean) values for Y fall on a straight (as opposed to curvilinear) line. Using more complicated jargon, this means that they can be described by the linear function $\mu_Y = \alpha + \beta X$.

3 For any given value of X (here symbolized as X_i), each Y value is independent of other Y values. This means that the value of one particular Y (e.g., a projectile point's weight) doesn't influence the values of other Ys (other point weights). Archaeological data typically meet this assumption, but there are cases where the value of one Y_i will influence other values. For example, the length of walls in square-shaped rooms are not free to vary independent of each other in that the length of one helps to determine the length of the others.

4 For each X_i, the distribution of Y is: (i) normally distributed and (ii) homo-scedastic (meaning that they have similar variances such that the amount of variation in Y around the regression line cannot increase or decrease as the value

of *X* increases). These related premises are essential for both reliably constructing a regression line and for using it to make accurate predictions. Figure 11.6
represents data that meet the four assumptions presented here. A sample of
these data will facilitate the creation of a useful estimate of the underlying
regression relationship. Compare this with Figures 11.7 and 11.8, which reflect
data that are not homoscedastic or normally distributed. A sample from these
data would quite likely produce regression lines that are quite different from

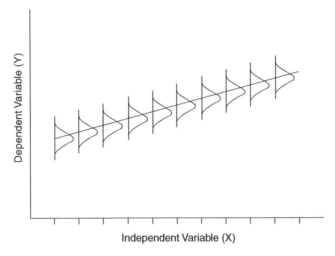

Figure 11.6   An example of a functional relationship that meets the assumptions necessary
for regression analysis

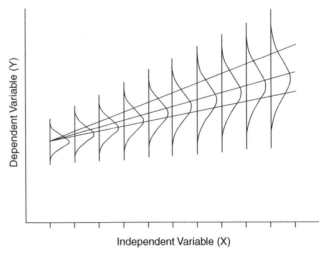

Figure 11.7   The effect on the likely range of regression lines as the inconsistent variation
resulting from non-homoscedastic data allows the analysis to stray excessively from the
mean for each X_i

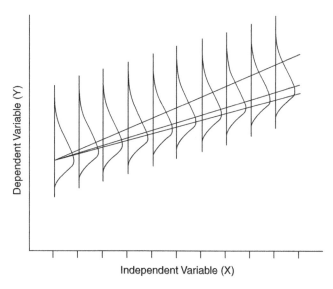

y-axis: Dependent Variable (Y)

x-axis: Independent Variable (X)

Figure 11.8 An example of the effect on the likely range of regression lines resulting from skewed distributions

the "true" regression line characterizing the means. Further, predictions for Y at each X_i based on samples from these populations would be very poor indeed.

If you understand these assumptions, you will find regression an easy tool to use and interpret. If not, you will be forever uncertain whether you should be using regression analysis, correlation analysis (discussed in the next chapter), or some other technique all together.

When all four assumptions are met, the formula for the average Y_i at a given X_i is $\hat{Y} = a + bX_i$, but this will likely differ from the observed Y_i values at each X by some small degree. ($\hat{Y}$(vocalized as Y-hat) is used to symbolize the average Y_i at a given X_i instead of $\bar{Y}$ in order to prevent confusing the average of all of the values for Y_i ($\bar{Y}$) with the average Y at a give X_i.) $\hat{Y}$ is an excellent estimate for Y_i, in that the mean value is the most common value in a normally distribution, but it cannot account for the influences of additional independent variables or other sources of error. From the perspective of X_i, this influence will be random (hence the assumption of a normal distribution). As a result, each Y_i is a product of $Y_i = a + bX_i + e_i$, where e_i is the error term reflecting variation caused by factors other than X (this should remind you of the conceptual model for ANOVA). In the dendrochronology example we used at the start of this chapter, e_i is virtually zero, given that the number of tree rings is nearly perfectly controlled by the years a tree has been growing. (We say "nearly" because on occasion harsh environmental conditions can disrupt the growing cycle and cause a tree to put on two or more rings in a single year or produce no ring at all.) In other cases, e_i may be comparatively large, as illustrated in Figure 11.5. Now let's turn our attention towards actually completing a regression analysis.

Table 11.1 Information for the regression analysis of 10 projectile points from Ventana Cave, Arizona

Proj. point	1 Total length X	2 weight Y	3 x $X_i - \bar{X}$	4 y $Y_i - \bar{Y}$	5 x^2 $(X_i - \bar{X})^2$	6 xy (Sum of products)	7 y^2 $(Y_i - \bar{Y})^2$	8 $\hat{Y}$	9 $d_{Y \cdot X}$ (Deviation of Y_i at X_i)	10 $d^2_{Y \cdot X}$ (Unexplained sum of squares)	11 $\hat{Y} - \bar{Y}$	12 $\hat{y}$ (Explained sum of squares)
1	4.2	5.2	1.21	2.51	1.46	3.03	6.30	4.66	0.54	0.29	1.97	3.88
2	3	2.5	0.01	−0.19	0.00	0.00	0.04	2.70	−0.20	0.04	0.01	0.00
3	2.5	2	−0.49	−0.69	0.24	0.34	0.48	1.89	0.11	0.01	−0.80	0.64
4	3.5	2.2	0.51	−0.49	0.26	−0.25	0.24	3.52	−1.32	1.74	0.83	0.69
5	2.6	2.8	−0.39	0.11	0.15	−0.04	0.01	2.05	0.75	0.56	−0.64	0.41
6	3.1	3.8	0.11	1.11	0.01	0.12	1.23	2.87	0.93	0.87	0.18	0.03
7	2.97	2.3	−0.02	−0.39	0.00	0.01	0.15	2.65	−0.35	0.13	−0.04	0.00
8	2.21	0.7	−0.78	−1.99	0.61	1.56	3.96	1.41	−0.71	0.51	−1.28	1.63
9	2.39	2.1	−0.60	−0.59	0.36	0.36	0.35	1.71	0.39	0.15	−0.98	0.96
10	3.45	3.3	0.46	0.61	0.21	0.28	0.37	3.44	−0.14	0.02	0.75	0.56
Σ	29.92	26.9	0.00	0.00	3.31	5.40	13.13	26.90	0.00	4.33	0.00	8.80

Constructing a Regression Equation

Given that the regression equation is $Y = a + bX$, we must determine a and b to solve for Y at a given X_i. To begin figuring out these values, we need the average for both the dependent and independent variables. For the Ventana Cave projectile point data, $\bar{X} = 2.99$ cm and $\bar{Y} = 2.69$ g. We also need the information presented in Table 11.1. Much of this will look familiar to you. For example x, y, x^2, and y^2 are the exact same deviations and squared deviations we would calculate when computing the standard deviation for X and Y.

The multiplication of the deviations of x and y in Column 4 to form xy may seem a bit odd at first glance, but it is actually quite easily understood. Squaring y gives y^2 (i.e., $[Y_i - \bar{Y}] \times [Y_i - \bar{Y}]$) that you will recall is used to determine the sums of squares, which reflects how much variation there is between $\bar{Y}$ and Y_i. (If you don't recall this, please refer back to the discussion of the standard deviation in Chapter 4.) This is useful for quantifying the variation in variable Y, but with regression we are interested in the covariance between variables X *and* Y, not just the variation in variable Y or X. Taking y and multiplying it by x (i.e., $[Y_i - \bar{Y}] \times [X_i - \bar{X}]$) actually provides a measure of the variation in variables X and Y relative to each other. If $Y_i - \bar{Y}$ gets larger as $X_i - \bar{X}$ get larger, then we know that they positively covary (Figure 11.9). If $Y_i - \bar{Y}$ gets smaller as $X_i - \bar{X}$ gets larger, then we know that the variables negatively covary (Figure 11.10). If $Y_i - \bar{Y}$ does not reliably get larger or smaller as $X_i - \bar{X}$ increases, then variables X and Y do not covary at all (Figure 11.11). Thus, using xy as reflected in Column 6, we can determine the degree of covariance between the two variables.

The amount of covariance between the dependent and independent variables in turn allows us to determine the amount of variation in variable Y that is *explainable*

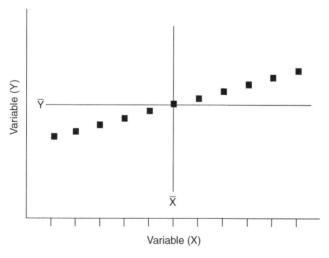

Figure 11.9 A relationship in which $(X_i - \bar{X})$ and $(Y_i - \bar{Y})$ positively covary, in that as $X_i - \bar{X}$ becomes large, so does $Y_i - \bar{Y}$

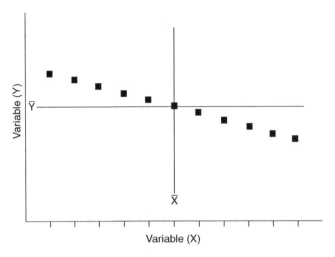

Figure 11.10 A relationship in which $(X_i - \bar{X})$ and $(Y_i - \bar{Y})$ negatively covary, in that as $X_i - \bar{X}$ becomes larger, $Y_i - \bar{Y}$ becomes smaller

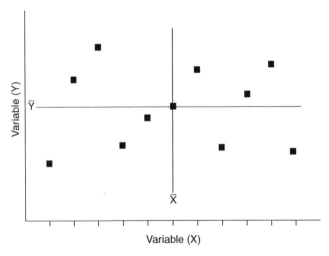

Figure 11.11 A relationship in which $(X_i - \bar{X})$ and $(Y_i - \bar{Y})$ do not covary, in that as $X_i - \bar{X}$ becomes larger, $Y_i - \bar{Y}$ does not consistently become larger or smaller

in terms of X, that is, how much of the variation in Y is caused by the change in X. The portion of the variation in Y that cannot be explained by X is *unexplainable* in terms of X and is the result of the influence of other variables or measurement error previously discussed. While building our regression equation, we will consequently also build an explained sum of squares, which measures the portion of the variation in Y caused by X, and an unexplained sum of squares, which measures the variation in Y caused by all other sources.

Calculating the regression equation and the explained/unexplained sums of squares is illustrated in Table 11.1 and is completed as follows:

Columns 1 and 2 reflect the maximum length and weight of the 10 Ventana Cave projectile points.
Column 3 presents x, the deviation of each X_i from its mean. Notice that this sums to zero.
Column 4 presents y, the deviation of each Y_i from its mean. This too sums to zero.
Column 5 presents the sum of squares for x, which is used in the denominator of the calculation of the regression coefficient presented in Equation (11.1).

The regression coefficient

$$b = \sum xy / \sum x^2 \qquad (11.1)$$

Column 6 presents the sum of products, that is, the product of x and y discussed in the preceding paragraphs. The sum of these products is used in the numerator of our calculation of the regression coefficient (Equation (11.1)).
Column 7 presents the sum of squares for y.
Column 8 presents our predicted value of Y_i for each X_i. Remember that this value in fact reflects the expected average of all Y_i at a given X_i. Again, to help prevent confusion between this predicted value and actual observed values (or $\bar{Y}$ for that matter), the predicted value is represented by $\hat{Y}$. The line connecting all of the expected values is the regression line. Remember that $\hat{Y}$ *always* refers to the average value of Y at a given X_i, and never to an actual observed Y_i. As Figure 11.4 illustrates, it is not necessary for even a single Y_i to fall perfectly on the regression line.

$\hat{Y}$ is calculated in the following manner. First, calculate the regression coefficient (or slope) using Equation (11.1).

$$b = \sum xy / \sum x^2$$

$$b = 5.40 / 3.31$$

$$b = 1.63$$

Now plug the slope into the regression equation and solve for a. We know that $Y = a + bX$. Calculating a therefore simply requires us to choose appropriate values for Y and X. There are a couple of ways to do this, but the most common method, which is associated with the "least squares" regression method we introduce here, is to use the values for $\bar{X}$ and $\bar{Y}$. We will briefly discuss the differences between least squares regression and its alternatives at the end of this chapter, but it is necessary to note here that this method mathematically requires the regression line to pass directly through $\bar{X}$ and $\bar{Y}$, so these values are ideal for determining a. Thus, a is calculated as follows.

$$\bar{Y} = a + b\bar{X}$$

$$a = \bar{Y} - b\bar{X}$$

$$a = 2.69 - 1.63(2.99)$$

$$a = -2.19$$

The fact that a is a negative number isn't a problem.

Now that we know a, we can present the regression line as $\hat{Y} = -2.19 + 1.63(X_i)$. We can then solve for the expected values in Column 8 of Table 11.1. As we would expect, these values are close to, but not identical to, the observed values Y_i for each given X_i (Figure 11.12). We can even calculate a measure of the variation between $\hat{Y}$ and Y_i using the difference between them, as reflected in Column 9. This *residual* is the deviation from the point on the regression line and the actual value of Y for each X_i illustrated in Figure 11.4. Column 10 is the square of the residuals in Column 9, which creates a sum of squares for the variation between $\hat{Y}$ and Y_i. This is the variation that *is not* explained by the variation in variable X, and is therefore called the *unexplained sum of squares*.

All of the variation in Y_i that is not reflected in the unexplained sum of squares is in fact the variation that is controlled by the independent variable (variable X). This probably sounds confusing unless you have a good mathematical background, so think about it this way. All of the variation in Y_i can be quantified using $Y_i - \bar{Y}$ as we do when calculating the variance. If a regression relationship exists, some of this variation will reflect the influence that the independent variable exerts over the dependent variable. Some of it will be caused by the influence of other variables or factors such as measurement error. This "extra" variation is the variation between Y_i and $\hat{Y}$, which is what is measured using the unexplained sum of squares. The

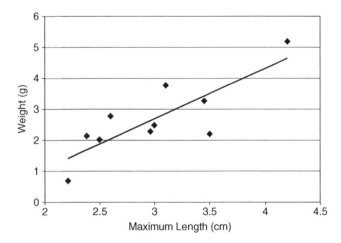

Figure 11.12   Regression line for the projectile point data presented in Table 11.1

remaining portion of the variation reflected in $Y_i - \bar{Y}$ is the result of the influence of the independent variable. It can be easily quantified using the variation between $\bar{Y}$ and $\hat{Y}_i$ (Column 11). Squaring these values in Column 12 provides a new sum of squares called the explained sum of squares. This term directly reflects the variation in the dependent variable (in this case, point weight) caused by the influence of independent variable (maximum point length). Notice that the explained sum of squares and the unexplained sum of squares do in fact equal the total sum of squares reflected in Column 7.

Figure 11.13 summarizes all of the sources of variation discussed above. It is critical to understand the relationships presented there, so humor us as we explore the various sources of variation one last time. The deviation of an individual observation (X_i, Y_i) from $\bar{Y}$ is calculated as $(Y_i - \bar{Y})$, and is symbolized by y. These are the deviations represented by the total sum of squares. Some of this deviation can be explained by the influence of variable X. This explainable variation is the difference between $\bar{Y}$ and the expected value for Y_i, which is $\hat{Y}$. Thus, the explained sum of squares is based on $(Y_i - \bar{Y})$, which is symbolized as $\hat{y}$. The remainder of the variation is not the result of the influence of variable X, and is therefore unexplained. This variation is the difference between the expected value and the corresponding y_i, which is quantified as $(Y_i - \hat{Y})$ and is symbolized as d_{Y*X} (the deviation of Y at X). d_{Y*X} is used to calculate the unexplained sum of squares, which reflects the influence of other variables and measurement error. By comparing the explained and the unexplained sum of squares, we can determine if variable X really has a significant influence on variable Y (reflected by a large explained sum of squares) or not. Thus, calculating the regression line is just the first step. The next is quantifying the strength of the association between a dependent and independent variable. We illustrate how to do this in the following section.

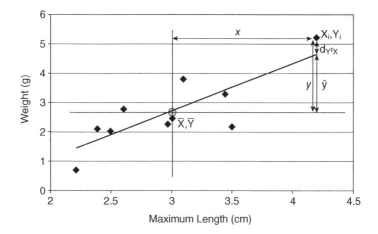

Figure 11.13 Sources of variation at X_i, Y_i

Evaluating the Statistical Significance of Regression

Here we will use data provided by Dohm (1990) to illustrate the application of regression analysis and its use to quantify both the explained and unexplained variation in archaeological data. Archaeologists often would like to know the demographic structure of the people who once occupied the settlements that we study. This often requires us to develop population estimates using some archaeologically measurable attribute (e.g., size of habitation area, number of rooms, etc.). Dohm's (1990) study focuses on understanding the effects of population nucleation and dispersion on settlement size in American Southwest. She quite reasonably proposes that the number of rooms and total roofed area in a settlement should be a function of the number of people who lived there, and measures this relationship using a regression analysis. However, she then goes on to explore the unexplained portion of the variation in the number of rooms to identify the impact of other factors such as regional population density on the number of rooms in a community. Dohm (1990) consequently perfectly illustrates the utility of using regression to identify the explanatory importance of a particular variable, and how this knowledge can help focus our research to provide a deeper understanding of the archaeological record. For this example, we wish to focus only on the first portion of Dohm's analysis, the relationship between population and the number of rooms in a settlement.

More people will need more rooms to live in and conduct all of their activities, all other variables being equal, and therefore the number of rooms should reflect past population size. Expressed more formally, this functional relationship can be written in the form of $Y = f(X)$ as (number of rooms) $= f$ (the number of people in a settlement).

Dohm's premise seems intuitively reasonable, and should be applicable to the prehistoric record, but archaeologists unfortunately don't know the independent variable (the number of people who once occupied a settlement). Dohm solves this problem by using historical and ethnographic evidence of settlement room frequencies and population size to: (i) establish that there is a relationship between population size and the number of rooms, and (ii) develop a means of estimating population sizes using architectural information. She recognizes that there is tremendous diversity in the way that different cultures use the built environment, so she used information from groups who are historically related to the people she is studying archaeologically (the Pueblo Indians of the American Southwest). She chose these groups because they live in similar buildings to those she is studying and they conduct many of the same activities as reflected in the archaeological and ethnographic records, suggesting that their use of built space may be similar.

Dohm's data are presented in Table 11.2 and Figure 11.14. There does intuitively appear to be a relationship between the two variables such that as X increases, so does Y. This suggests that population size does influence the number of rooms in a settlement. However, for archaeologists to be able to use this relationship to estimate the relationship between population size and room totals effectively, they will

Table 11.2 Historic pueblo room counts

Pueblo	Map date	Total population	Total rooms
Acoma	1948	879	387
Cochiti	1952	444	225
Isleta	1948	1470	804
Jemez	1948	883	459
Laguna	1948	711	189
Nambe	1948	155	34
Picuris	1948	130	59
San Felipe	1948	784	276
San Ildefonso	1948	170	207
San Ildefonso	1973	413	189
San Juan	1948	768	176
Sandia	1948	139	80
Santa Ana	1948	288	136
Santa Ana	1975	498	152
Santa Clara	1948	573	144
Santo Domingo	1948	1106	500
Shipaulovi	1882	113	131
Shongopavi	1882	216	253
Sichomovi	1882	104	105
Taos	1948	907	543
Taos	1973	1463	627
Tesuque	1948	160	116
Tewa Village	1882	175	158
Walpi	1882	270	363
Zia	1948	267	126

Source: Dohm, K. (1990). Effect of population nucleation on house size for pueblos in the American Southwest. *Journal of Anthropological Archaeology*, **9**: 201–39.

need to be sure that population size exerts a substantial control over the number of rooms. If total population is only one of many variables that weakly influence the dependent variable of number of rooms, then any explanation for differences in settlement size based on population estimates derived from room frequencies will be quite poor.

The least squares regression line illustrated in Figure 11.15 was calculated by solving for a and b using the method illustrated previously (Table 11.3).

$n = 25$

$\bar{X} = 523.44$

$\bar{Y} = 257.56$

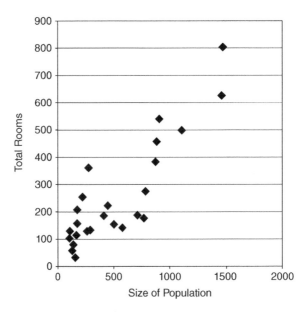

Figure 11.14   The relationship between site population and the total number of rooms

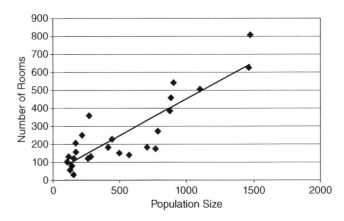

Figure 11.15   Regression relationship between population size and the total number of rooms

The regression coefficient is:

$$b = \frac{\sum xy}{\sum x^2} = \frac{1698469}{4146532} = .41$$

The Y intercept is:

$$a = \bar{Y} - b(\bar{X}) = 257.56 - .41(523.44) = 42.95$$

Table 11.3 Regression analysis of historic pueblo population size compared to room counts

Pueblo	Total population X	Total rooms Y	x ($X_i - \bar{X}$)	$\hat{x}$ ($X_i - \bar{X})^2$	y ($Y_i - \bar{Y}$)	xy (Sum of products)	y^2 ($Y_i - \bar{Y})^2$	$\hat{Y}$	$d_{Y\cdot X}$ (Deviation of Y_i at X_i)	$d^2_{Y\cdot X}$ (Unexplained sum of squares)	$\hat{y}$ ($\hat{Y} - \bar{Y}$)	$\hat{y}^2$ (Explained sum of squares)
Acoma	879	387	355.56	126422.91	129.44	46023.69	16754.71	403.20	−16.20	262.49	145.64	21211.48
Cochiti	444	225	−79.44	6310.71	−32.56	2586.57	1060.15	225.02	−0.02	0.00	−32.54	1058.82
Isleta	1470	804	946.56	895975.83	546.44	517238.25	298596.67	645.28	158.72	25191.33	387.72	150328.52
Jemez	883	459	359.56	129283.39	201.44	72429.77	40578.07	404.84	54.16	2933.30	147.28	21691.41
Laguna	711	189	187.56	35178.75	−68.56	−12859.11	4700.47	334.39	−145.39	21137.32	76.83	5902.36
Nambe	155	34	−368.44	135748.03	−223.56	82368.45	49979.07	106.64	−72.64	5276.95	−150.92	22776.06
Picuris	130	59	−393.44	154795.03	−198.56	78121.45	39426.07	96.40	−37.40	1398.93	−161.16	25971.80
San Felipe	784	276	260.56	67891.51	18.44	4804.73	340.03	364.29	−88.29	7794.85	106.73	11390.97
San Ildefonso	170	207	−353.44	124919.83	−50.56	17869.93	2556.31	112.79	94.21	8876.13	−144.77	20959.29
San Ildefonso	413	189	−110.44	12196.99	−68.56	7571.77	4700.47	212.32	−23.32	543.94	−45.24	2046.43
San Juan	768	176	244.56	59809.59	−81.56	−19946.31	6652.03	357.73	−181.73	33027.49	100.17	10034.97
Sandia	139	80	−384.44	147794.11	−177.56	68261.17	31527.55	100.09	−20.09	403.56	−157.47	24797.18
Santa Ana	288	136	−235.44	55431.99	−121.56	28620.09	14776.83	161.12	−25.12	631.06	−96.44	9300.48
Santa Ana	498	152	−25.44	647.19	−105.56	2685.45	11142.91	247.14	−95.14	9051.52	−10.42	108.59
Santa Clara	573	144	49.56	2456.19	−113.56	−5628.03	12895.87	277.86	−133.86	17918.60	20.30	412.10
Santo Domingo	1106	500	582.56	339376.15	242.44	141235.85	58777.15	496.18	3.82	14.57	238.62	56941.17
Shipaulovi	113	131	−410.44	168460.99	−126.56	51945.29	16017.43	89.44	41.56	1727.33	−168.12	28264.70
Shongopavi	216	253	−307.44	94519.35	−4.56	1401.93	20.79	131.63	121.37	14730.94	−125.93	15858.64
Sichomovi	104	105	−419.44	175929.91	−152.56	63989.77	23274.55	85.75	19.25	370.47	−171.81	29517.85
Taos	907	543	383.56	147118.27	285.44	109483.37	81475.99	414.67	128.33	16468.40	157.11	24483.78
Taos	1463	627	939.56	882772.99	369.44	347111.05	136485.91	642.41	−15.41	237.62	384.85	148113.32
Tesuque	160	116	−363.44	132088.63	−141.56	51448.57	20039.23	108.69	7.31	53.43	−148.87	22162.08
Tewa Village	175	158	−348.44	121410.43	−99.56	34690.69	9912.19	114.83	43.17	1863.23	−142.73	20370.47
Walpi	270	363	−253.44	64231.83	105.44	−26722.71	11117.59	153.75	209.25	43786.41	−103.81	10776.94
Zia	267	126	−256.44	65761.47	−131.56	33737.25	17308.03	152.52	−26.52	703.26	−105.04	11033.58
Σ	13086	6439	0.00	4146532.16	0.00	1698468.84	910116.16	6439.00	0.00	214403.14	0.00	695713.02

The explained sums of squares is:

$$\sum \hat{y}^2 = 695713.02$$

The unexplained sum of squares is:

$$\sum d_{Y*X}^2 = 214402.14$$

If variable X (population size) controls much of the variation in variable Y (number of rooms), then the explained sums of squares should reflect much more of the total variation than the unexplained sums of squares. Think back to Chapter 10's discussion of ANOVA in which we used the F-distribution to evaluate the ratio between two variances to determine if the variance among groups was substantially larger than the variance within groups. With regression we are faced with solving a similar problem, in that we wish to compare whether the explained variance is substantially larger than the unexplained variance. Fortunately, the F-distribution can be used in this context too.

Table 11.4 presents the test of significance of the regression relationship using the F-distribution in a manner that is directly comparable to an ANOVA table. Here we take the various sums of squares, transform them into mean sum of squares by dividing the sums of squares by their degrees of freedom, and then divide the $MS_{explained}$ by the $MS_{unexplained}$ to create an F-value. If the F-value is significant, then the independent variable does control the dependent variable in a meaningful way.

As with ANOVA, the null hypothesis being tested doesn't have an intuitive connection with the comparisons of variances. With this regression table, we are actually testing the null hypothesis $H_0 : b = 0$ (the slope of the regression line is zero). The connection between the null hypothesis and the comparisons between the explained and unexplained sums of squares rests on the realization that the only way the explained mean sum of squares can be substantially larger than the unexplained mean sum of squares is if the regression coefficient is significantly different than zero (i.e., if the variables covary). As illustrated in Figure 11.11, if the slope is zero, then the dependent variable varies irrespective of the independent variable producing a small value for the covariance based on xy. As a result, the explained mean sum of squares will not be larger than the unexplained mean sum of squares.

Table 11.4   Test of significance for $H_0 : b_{Y*X} = 0$

Source of variation	df	SS	MS	F_s
Explained due to linear regression	1	695713	695713	74.63
Unexplained, the error around the regression line	23	214403	9322	
Total	24	910116		

If the $MS_{explained}$ is significantly larger than the $MS_{unexplained}$, then there must be positive (Figure 11.9) or negative (Figure 11.10) covariance between the two variables. If this is the case, then there is necessarily a relationship between X and Y such that $b \neq 0$.

To complete the statistical analysis reflected in Table 11.4, we will use $\alpha = .05$. The degrees of freedom for the unexplained error equals $n - 2$. We lose one degree of freedom for each variable. The degrees of freedom for the explained variation is $a - 1$, which is always 1. The critical value for the F-distribution is again found in Appendix C. In this example, the critical value is $F_{[.05,1,23]} = 4.28$, which is considerably less than the test's F-value of 74.63. We reject the null hypothesis, and conclude that in fact the number of inhabitants affects the number of rooms in a settlement. Thus Dohm's proposition is supported in the historical record.

Another common way to present the significance of the result is to present the explained sums of squares as a proportion of the total sums of squares. This will identify the proportion of the total variation in the dependent variable that is explained by the independent variable. The value is called the *coefficient of determination*, and is symbolized as r^2 (verbalized as r-squared). r^2 is calculated using Equation (11.2).

The coefficient of determination

$$r^2 = \frac{SS_{explained}}{SS_{total}} \tag{11.2}$$

Here, $r^2 = \dfrac{SS_{explained}}{SS_{total}} = \dfrac{695,713}{910,116} = .76$. This means that 76% of the total variation in the number of rooms is explained by the influence of population size. As we would expect given that r^2 is a proportion, it can range from zero to one, with larger values reflecting a higher proportion of the total variation explained by the independent variable.

Archaeologists love using r^2, but it is a notoriously difficult statistic to interpret. At its base, r^2 reflects the *goodness of fit* between the regression line and the observed data. If all of the data points fall directly on the regression line such that the expected and observed values are identical, then the goodness of fit between the data and regression line is perfect and $r^2 = 1$. In contrast, we can have a significant linear relationship in which there is a great deal of variation around the regression line, causing r^2 to be quite low.

In this case, $b \neq 0$, but the goodness of fit between the data and the regression line indicates variables other than population size determine roughly 24% of the variation in the dependent variable. This in turn complicates the interpretation of the relationship between the dependent and independent variables. Does an r^2 value of .76 indicate that enough variation in total room count is explainable by total population to allow it to be meaningfully predicted by archaeologists using the total population? Perhaps, but there really isn't a clear cutoff between a "good fit" and a "not so good fit." Determining what is "good" or "not" rests on your analytic structure, theoretical structure, and instincts; sometimes it is enough to show that there is some relationship, even if it isn't exceptionally strong whereas other times our analytic structure may require a demonstration that the relationship is so strong that no other variables significantly impact the dependent variable. An $r^2 = .76$ seems pretty good to us here, in that a single variable seems to account for 76% of the variation in room counts. Others might disagree. If an r^2 seems too low, then you may want to consider the influence of additional variables, or rethink whether all of the data included in the analysis reflect a coherent group or whether the variable chosen as the independent variable is the ideal variable. In regards to formally presenting the test's results, you can't go wrong presenting both a regression table of the sort illustrated in Table 11.4 and the r^2 value, but the r^2 by itself *does not* constitute a meaningful statistical evaluation of the strength of the regression relationship.

Using Regression Analysis to Predict Values

Archaeologists will find it useful to know that there is a relationship between population size and the total number of rooms in a pueblo, especially since the regression analysis allows them to predict the size of a pueblo necessary to house a given population. For example, various models of community size and organization have been proposed for various parts of the American Southwest (Ruscavage-Barz, 1999). Areas such as Mesa Verde in southern Colorado have excellent chronological control allowing archaeologists to have outstanding estimates of the numbers of rooms inhabited through time. Based on the variation in settlement size and location, it has been suggested that local communities of about 600 people aggregated together when the environment allowed excellent agricultural returns, but then dispersed during environmental contexts that were less favorable (Cordell, 1997). An archaeologist using the regression analysis we presented could determine an estimate for the total number of rooms likely required to house a community of 600. In this case $\hat{Y} = 43 + .41(X) = 43 + .41(600) = 289$ rooms. This number can then be compared to the number of rooms in aggregated communities to determine if a community size of 600 is plausible. Of course, our knowledge of probability tells us that it is unlikely that a community of 600 will have *exactly* 289 rooms, even if the functional relationship specified in the regression analysis holds. The use of the regression's predicted value to evaluate the hypothesis of interest thus would be strengthened if there was a means of placing a confidence interval around the predicted value, such that we could evaluate whether a specific observed number was "close enough" to 289 rooms to support the proposition. Given that $\hat{Y}$ is actually a mean for Y_i at X_i, such confidence intervals are easily derived, but they take a slightly different form than might initially be suspected. We illustrate how to compute various confidence intervals in the following sections.

Placing confidence intervals around the regression coefficient

Given that we have only a single Y_i value from each X_i value in Table 11.2, and no actual observed Y_i values for the X_i value that interests us in this case ($X = 600$ people), our confidence intervals around each $\hat{Y}$ will not be based on the standard deviation of a directly observed sample of individual Y_i values that surround $X = 600$ people. Instead, the confidence interval will be based on the variation of the observed dependent values around the whole of the regression line. This probably makes intuitive sense to you, but it complicates the creation of confidence intervals around the regression line for a couple of reasons. First, the regression coefficient (b) is a statistic that estimates the true population parameter. This means that we can (and indeed should) compute confidence intervals around this measure to quantify the likely range containing the parameter. However, as we mentioned previously the regression line must go through the point $(\overline{X}, \overline{Y})$. This means that the slope of the regression line will pivot on this point, making the confidence intervals around the slope larger at the regression line's ends and smaller near $(\overline{X}, \overline{Y})$ (Figure 11.16). The upshot of this is that we are more confident about the accuracy of $\hat{Y}$ as an estimate of the true μ_{Y_i} as X_i approaches $\overline{X}$, but are less confident for more distant X_i values. Our confidence intervals must consequently account for: (i) the variation in the distribution of Y_i around $\hat{Y}$ and (ii) the variation in the placement of $\hat{Y}$ resulting from variation in the regression coefficient. A useful first step, then, is to quantify the potential variation in the placement of $\hat{Y}$ resulting from the variation in the regression coefficient.

So how do we go about calculating confidence intervals around b? Given that the regression coefficient reflects the placement of means, the appropriate measure

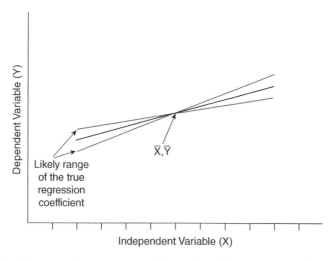

Figure 11.16 The range of regression coefficients that likely encompass the true regression coefficient

of dispersion will be the standard error, which is calculated using Equation (11.3). s_{Y*X}^2 is derived using Equation (11.4), which corresponds with the $MS_{\text{unexplained}}$ from the ANOVA analysis illustrated in Table 11.4.

Standard error for a regression coefficient

$$S_b = \sqrt{\frac{s_{Y*X}^2}{\sum x^2}} \tag{11.3}$$

Calculation of the variance of Y at X

$$s_{Y*X}^2 = \frac{\sum d_{Y*X}^2}{n-2} \tag{11.4}$$

Using Dohm's data, the standard error for b is calculated as

$$S_b = \sqrt{\frac{s_{Y*X}^2}{\sum x^2}} = \sqrt{\frac{9322}{4146532}} = .047$$

Once we have the standard error of the regression coefficient, we can build confidence limits using the t-distribution where the degrees of freedom equal $n - 2$. (We lose one degree of freedom for both the X and the Y variables.) For 95% confidence intervals, the limits are:

$$t_{.05[23]}S_b = 2.069\,(.047) = .098$$

$$L_1 = b - t_{.05[23]}S_b = .410 - .098 = .312$$

$$L_2 = b + t_{.05[23]}S_b = .410 + .098 = .508$$

Thus, we are 95% certain that the true regression coefficient lies somewhere between $b = .312$ and $b = .508$ (Figure 11.17). This information can be useful in several contexts. First, it provides another way to formally test the null hypothesis $H_0 : b = 0$. If the null hypothesis is true, then the confidence interval should encompass zero. Although this approach does not provide as much information as the ANOVA-style regression analysis illustrated in Table 11.4, it might be useful in some contexts. Second, it can be used to compare the slopes of two or more functional relationships $(H_0 : b_1 = b_2)$. Consider for example a researcher who wishes to evaluate if the regression coefficient determined using Dohm's data also characterized other groups who lived in above-ground, contiguous structures. A regression analysis could be performed for the other groups, and then compared to the slope determined here. If one or both of the confidence intervals encompassed the other distribution's regression coefficient, then the null hypothesis cannot be rejected; it is possible that the relationship holds across different cultures. If neither of the con-

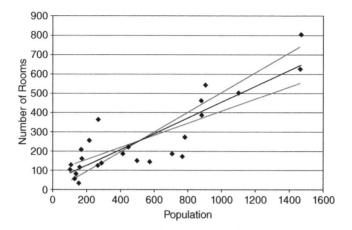

Figure 11.17   Confidence intervals for the regression coefficient illustrated in Figure 11.15

fidence intervals encompass the other regression coefficient, then the null hypothesis can be rejected, causing us to conclude that the relationship in Dohm's data may not be widely applicable to other groups. Be sure to use the Bonferroni method to control for the problem of compounding alpha errors when comparing more than two slopes. Also, be sure that you understand that the slopes of two functional relationships may be the same, yet the regression lines may not overlap with one another, if the intercepts are different (i.e., differing intercepts will produce parallel lines when the slopes are identical). Whether differences in the intercepts are important depends on your analytic problem and theoretical structure.

For completeness, we wish to present one final means of evaluating the null hypothesis $H_0 : b = 0$ sometimes used in archaeology. It is based on the t-distribution, using a modified version of the t-score (Equation (8.6)). The equation is changed by substituting the regression coefficient (b, formally symbolized here as b_{X*Y}) for $\bar{Y}$, 0 for μ (based on the fact that we specify μ is 0 in the null hypothesis), and the standard error of the regression coefficient for the standard error. This produces Equation (11.5), where the degrees of freedom are $n - 2$.

t-test evaluating $H_0 : b = 0$

$$t_s = \frac{b_{X*Y} - 0}{s_b} \tag{11.5}$$

For this example:

$$t_s = \frac{.410}{.047} = 8.641$$

Using an alpha value of .05, $t_{[.05,23]} = 2.096$. Given that 8.641 is larger than 2.096, we reject the null hypothesis, and for the third time conclude that there is a significant functional relationship between total population and the number of rooms in a pueblo. Which of the three methods presented for evaluating $H_0 : b = 0$ is the best? In our opinion, the analysis of variance illustrated in Table 11.4 provides the most information given that it quantifies the amount of explained and unexplained variation in the dependent variable. We encourage you to use this approach. However, the other two methods (confidence intervals around the regression coefficient and the t-test presented in Equation (11.5)) are perfectly acceptable, if you wish to use them instead.

Confidence Limits around $\hat{Y}$ for a Given X_i

Let's return to the issue that we originally wished to address, which is placing confidence limits around $\hat{Y}$ to determine if a particular observation is different than expected (i.e., $H_0: Y_i = \hat{Y}$). Evaluating this null hypothesis will allow us to determine if a specific variate likely does not reflect the same functional relationship underlying a significant regression relationship. In cases where the null hypothesis is rejected, the analyst might find it useful to consider alternate explanatory factors. Considering this null hypothesis might even be useful when we do not have a specific Y_i value to compare to $\hat{Y}$, in that it may be useful to know the range of likely values represented by the line. Depending on the research design, analytic structure, and theoretical system, our conclusions might be very different if there is a wide range of likely variation instead of a very narrow range.

Confidence limits are most easily calculated around the sample mean $\bar{Y}$, which, as previously mentioned, is the anchor point through which the least squares regression line must pass. Calculating the confidence limits requires a standard error for $\hat{Y}$, which should not be confused with the standard error for b (S_b; Equation (11.3)). In the special case of the sample mean $\bar{Y}$, we can calculate the standard error by simply modifying the general standard error formula (Equation (8.1)) using the $MS_{\text{unexplained}}$ (Equation (11.4)) for the variance. In this case, the standard error of the predicted value at $\bar{Y}$ is:

$$s_{\bar{Y}} = \sqrt{\frac{s^2_{Y*X}}{n}} = \sqrt{\frac{9322}{25}} = 19.31 \text{ rooms}$$

95% confidence limits for the mean are thus:

$$t_{[.05,23]} = 2.069$$

$$L_1 = \bar{Y} - t_{[.05,23]}(s_{\bar{Y}}) = 257.56 - (2.069 \times 19.31) = 217.61 \text{ rooms}$$

$$L_2 = \bar{Y} + t_{[.05,23]}(s_{\bar{Y}}) = 257.56 + (2.069 \times 19.31) = 297.51 \text{ rooms}$$

Thus, the average size of settlements with 523 people is between 218 and 298 rooms 95% of the time, according to the regression analysis. Great, except what we really want are confidence limits around $\hat{Y} = 289$ rooms at $X = 600$ people. As previously mentioned, calculating confidence intervals around this and any other $\hat{Y}$ value that is not exactly $\overline{Y}$ is complicated by the uncertainty associated with our estimate of the regression coefficient. The standard error must take this additional variation into account, which is accomplished through changes in the formula for the standard error for $\hat{Y}$ (Equation (11.6)).

The standard error of $\hat{Y}$ for a given X_i

$$s_{\hat{Y}} = \sqrt{s^2_{Y*X}\left(\frac{1}{n} + \frac{(X_i - \overline{X})^2}{\sum x^2}\right)} \qquad\qquad (11.6)$$

Notice that this value will increase exponentially as the distance between X_i and $\overline{X}$ increases. For $X_i = 600$:

$$s_{\hat{Y}} = \sqrt{9322\left(\frac{1}{25} + \frac{(600 - 523)^2}{4146532}\right)} = 19.65\,\text{rooms}$$

Now this value is ideal for placing confidence limits around $\hat{Y}_i$ for $X_i = 600$, but it is a standard error term. This, of course, means that it reflects the distribution of means, not variates. As a consequence, confidence intervals created using the t-distribution will reflect the likely range of the means, specifically the likely placement of μ_{Y_i}, the population parameter estimated by $\hat{Y}_i$. Knowing the likely placement of μ_{Y_i} could be useful, especially when evaluating whether empirically derived samples from X_i produce a mean consistent with the functional relationship specified in the regression analysis (e.g., an analyst considering the Ventana Cave point data presented earlier in this chapter might wish to evaluate whether five projectile points with the same maximum length produce an average weight consistent with the regression analysis as a means of testing whether the functional relationship identified using the sample of 10 points is generally applicable to the entire assemblage).

Here we are interested in the distribution of individual variates, so we must transform the standard error into a standard deviation term. This can be done using Equation (11.7).

The standard deviation of Y_i around $\hat{Y}_i$ at a given X_i

$$\hat{s}_Y = \sqrt{s^2_{Y*X}\left(1 + \frac{1}{n} + \frac{(X_i - \overline{X})^2}{\sum x^2}\right)} \qquad\qquad (11.7)$$

$\hat{s}_Y$ for $\hat{Y}_i$ at $X_i = 600$ is calculated as:

$$\hat{s}_Y = \sqrt{9322\left(1+\frac{1}{25}+\frac{(600-523)^2}{4146532}\right)} = 98.53 \text{ rooms}$$

Using this value, we can determine that the 95% confidence intervals for $\hat{Y}_i$ at $X_i = 600$ are:

$\hat{Y} = 298 \text{ rooms}$

$L_1 = \hat{Y} - t_{[.05,23]}\hat{s}_Y = 289 - (2.069 \times 98.53) = 85.14 \text{ rooms}$

$L_2 = \hat{Y} + t_{[.05,23]}\hat{s}_Y = 289 + (2.069 \times 98.53) = 492.86 \text{ rooms}$

Even a cursory glance at these limits suggests that there is a substantial range in the predicted values: pueblos housing 600 people are expected to differ by up to 408 rooms, a tremendously substantial number given the typical size and structure of Southwestern pueblos. The range of variation is directly attributable to the relatively large amount of unexplained variation within the regression analysis. There are many variables impacting the total number of rooms in addition to total population, which was the point of Dohm's (1990) original analysis. Our predictions for the size of pueblos will consequently be relatively poor until these factors are controlled. In cases where the explained variation comprises a greater proportion of the total variance (i.e., when r^2 is larger), the confidence limits for Y_i around $\hat{Y}_i$ will be substantially smaller.

Estimating *X* from *Y*

In most (experimental) sciences, the predictive power of regression is tremendously useful. It allows scientists to make predictions that can be compared to observations, which can in turn be used to evaluate competing ideas of how the world is structured. Archaeology can use this, but we are in the somewhat unusual position of often being more interested in reconstructing the independent variables, typically cultural or behavioral phenomena, using the dependent variables, which are some aspect of the archaeological record. It might be useful to predict the number of rooms based on hypothesized community sizes, but archaeologists would find it even more useful to predict population size based on variables such as room counts (Hassan, 1981).

The nature of the functional relationship makes it *seem* as if this can be easily done. After all, if a total population of 600 people corresponds with an estimated room total of 289, then shouldn't a pueblo with 289 rooms lead to a predicted population level of 600 people? Sadly, no. Using the *dependent* variable to estimate the *independent* variable violates two assumptions of regression:

1   The dependent variable is not measured without error. It reflects the operation of any additional independent variables that affect *Y*, and is consequently only an estimate of $\mu_{\bar{Y}}$ at X_i. As an analytic fact, it has error relative to variable *X* even when measured with accurate and precise methods.

2   The independent variable is not normally distributed or homoscedastic (or necessarily independent) at any given $\mu_{\bar{Y}}$. To the contrary, the independent variable isn't distributed in *any way* with respect to the dependent variable. It is the dependent variable that varies with respect to the independent variable.

The result of predictions of a dependent variable obtained by simply putting a value for *Y* into the equation formula to determine *X* is quite poor, and reflects little more than "an educated guess".

Despite the difficulties, though, it is indeed possible to estimate the independent variable using the dependent variable by means of confidence limits through a process called *inverse prediction*. It begins with Equation (11.8) to create $\hat{X}$ (pronounced X-hat), an initial estimate of the independent variable using the dependent variable. This formula is simply derived from the functional equation of $\hat{Y} = a + b_{Y*X}(X_i)$

The estimate of $\hat{X}$ at Y_i

$$\hat{X}_i = \frac{(Y_i - a)}{b_{Y*X}}$$ (11.8)

This estimate is not particularly reliable for the reasons outlined above, but it serves as an anchor point for establishing a possible range that likely does include the true value for X_i. Creating a confidence interval around this prediction is more complicated than establishing the previously illustrated confidence intervals. It requires the calculation of two terms: *D* (Equation (11.9)) and *H* (Equation (11.10)).

Calculation of D, *one of the terms necessary to estimate* $\hat{X}$

$$D = b_{Y*X}^2 - t_{[\alpha,n-2]}^2 s_b^2$$ (11.9)

Calculation of H, *one of the terms necessary to estimate* $\hat{X}$

$$H = \frac{t_{[\alpha,n-2]}}{D} \sqrt{s_{Y*X}^2 \left[D\left(1 + \frac{1}{n}\right) + \frac{(Y_i - \bar{Y})^2}{\sum x^2} \right]}$$ (11.10)

Applying these equations to $Y_i = 289$ rooms for $\alpha = .05$, produces the following results:

$$\hat{X}_i = \frac{(289 - 43.15)}{.41} = 600 \text{ people}$$

$$D = (.41)^2 - (2.069)^2 (.047)^2 = .159$$

$$H = \frac{2.069}{.159} \sqrt{9322 \left[.159 \left(1 + \frac{1}{25} \right) + \frac{(289 - 257.56)^2}{4146532} \right]} = 511.8 \text{ people}$$

These values can then be used to calculate confidence limits using Equations (11.11) and (11.12).

Lower confidence interval for estimate of $\hat{X}$ at Y_i

$$L_1 = \bar{X} + \frac{b_{Y*X}(Y_i - \bar{Y})}{D} - H \qquad (11.11)$$

Upper confidence interval for estimate of $\hat{X}$ at Y_i

$$L_2 = \bar{X} + \frac{b_{Y*X}(Y_i - \bar{Y})}{D} + H \qquad (11.12)$$

Note that the value specified in Equations (11.11) and (11.12) is $\bar{X}$, not $\hat{X}$. This is counterintuitive, but is necessary given the problems with the regression assumptions associated in inverse prediction. Again, applying these equations to estimating $\hat{X}$ for $Y_i = 289$ rooms produces the following results:

$$L_1 = 523.44 + \frac{.41(289 - 257.56)}{.159} - 511.8 = 92.85 \text{ people}$$

$$L_2 = 523.44 + \frac{.41(289 - 257.56)}{.159} + 511.8 = 1116.53 \text{ people}$$

These confidence limits do encompass $\hat{X}$, but are not quite symmetrical around it. We are 95% confident that the true value for X_i falls within these limits. Of course, this is a very expansive confidence interval, because of the uncertainty caused by the confounding influence of variables other than X on the dependent variable. A stronger relationship between variables X and Y would result in narrower confidence intervals. However, this approach provides a much more meaningful reflection of the true likely location of X_i than simply estimating it at 600 people. Simply using $\hat{X}$ would leave the underlying uncertainty hidden and possibly lead us to erroneous conclusions. We know that these formulas are not particularly intuitive, but using them to place confidence intervals around $\hat{X}$ is far preferable than simply using $\hat{X}$ by itself.

The Analysis of Residuals

There is an additional set of tools that is useful for archaeologists using regression analysis: the analysis of residuals. One of the things archaeologists frequently try to do is identify "odd" cases that deviate from the "typical" pattern. These cases are often behaviorally significant. They are the excessively large "ceremonial rooms" used for community integration, elite burials that show the development of social complexity through the abundance of grave goods, the expansive aggregated sites that serve as regional capitals, the enormous pots made by specialists for community feasting, etc. Part of the reason why archaeologists use a comparative approach is to identify the extraordinary. Being able to identify what is extraordinary (or ordinary for that matter), and where/when it occurs can be invaluable to understanding past behavior.

Regression provides a formal means of identifying individuals (however defined) that are atypical in a data set by comparing the difference (called the residual) between each actual observation (Y_i) and its predicted value ($\hat{Y}$) for each X_i. This is of course the same residual value used to calculate the unexplained sum of squares ($Y_i - \hat{Y}$) (see Tables 11.1 and 11.3), but in this case, we are using them individually instead of squaring and summing them.

Atypical variates may reflect the exceptional, the important, the individuals central to describing and explaining the archaeological record. They could also reflect analytic blunders or inconsistencies in measurement protocol (e.g., a pot whose volume was entered into the database incorrectly, differing definitions of "maximum flake length"). Solving these methodological problems may considerably improve the strength of the regression and other statistical analyses.

Residuals for Dohm's pueblo room example are presented in Table 11.5 and illustrated in Figure 11.18. The residual plot (Figure 11.18) is quite easy to make, with the x-axis reflecting the value of the independent variable, and the y-axis reflecting the magnitude of the residual. Put another way, the y-axis reflects the variation around the regression line (which is, after all, just the line formed by all of the expected values).

The residual plot can be informative in and of itself. For example, it can be used to ensure that the assumption of homoscedasticity is met in the data. Consistently

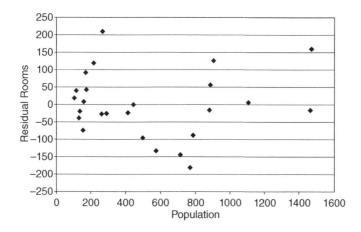

Figure 11.18   Regression residuals for the pueblo room data

Table 11.5   Residuals $d_{Y \cdot X}$ calculated as $Y - \hat{Y}$

Pueblo	Date	Observed value	Predicted Y	Residuals
Acoma	1948	387	403	−16
Cochiti	1952	225	225	0
Isleta	1948	804	645	159
Jemez	1948	459	405	54
Laguna	1948	189	334	−145
Nambe	1948	34	107	−73
Picuris	1948	59	96	−37
San Felipe	1948	276	364	−88
San Ildefonso	1948	207	113	94
San Ildefonso	1973	189	212	−23
San Juan	1948	176	358	−182
Sandia	1948	80	100	−20
Santa Ana	1948	136	161	−25
Santa Ana	1975	152	247	−95
Santa Clara	1948	144	278	−134
Santo Domingo	1948	500	496	4
Shipaulovi	1882	131	89	42
Shongopavi	1882	253	132	121
Sichomovi	1882	105	86	19
Taos	1948	543	415	128
Taos	1973	627	642	−15
Tesuque	1948	116	109	7
Tewa Village	1882	158	115	43
Walpi	1882	363	154	209
Zia	1948	126	153	−27

increasing or decreasing distance from zero as the value of the independent variable increases indicates unequal variances (or heteroscedasticity), which is a violation of the assumptions of regression as discussed at the start of this chapter (Figure 11.19). Similarly, residuals also reflect problems with the assumption of linearity (Figure 11.20). Curvilinear relationships such as Figure 11.21 will be reflected by residual plots similar to Figure 11.22. Runs of individuals on either side of the line indicate that the assumption of the linear model is not met. In these cases, a curvilinear regression model will be more appropriate than the linear regression model described here.

As useful as graphs of such *raw residuals* are, residuals can be made even more useful by calculating two variants: *leverage coefficients* and *standardized residuals*. Leverage coefficients provide a measure of the relative impact that each individual variate has on the regression line. In some cases, the placement of even two variates may produce a "significant relationship" even when there isn't one. Consider Figure 11.23. This distribution will produce a regression coefficient that is not equal to zero, but that in truth reflects the relationship between two points, one on each end of the distribution. Without these variates, $b = 0$, given that the remaining variates

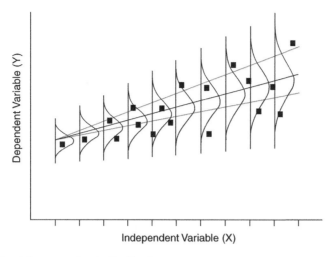

Figure 11.19 A heteroscedastic distribution

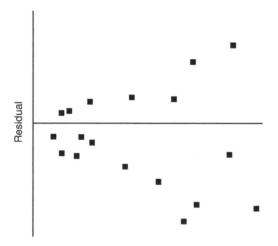

Figure 11.20 Residual pattern indicative of heteroscedasticity with increasing variance as X_i increases

are scattered randomly around $\bar{Y}$. A prudent analyst looking at the graph might suggest that these data do not satisfy the assumption of a linear relationship, but then again, perhaps not. We have certainly seen presentations at the Society for American Archaeology Annual Meetings and other professional contexts in which people conducted regression analysis of data distributed very much like this. It is questionable that a relationship based on these two "outliers" should be used to describe the relationships among the remaining data. Even in less extreme cases where the assumption of a linear relationship is more clearly met, it would be worthwhile to know whether the regression coefficient is reflecting a general pattern in the data, or is primarily the result of a limited number of variates. Leverage coefficients are the means of determining this.

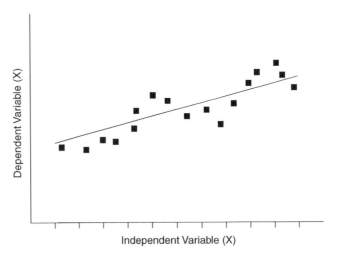

Figure 11.21   A curvilinear, as opposed to linear, distribution

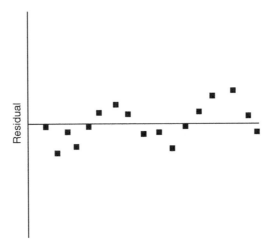

Figure 11.22   A residual pattern indicative of a curvilinear distribution

Leverage coefficients are calculated using Equation (11.13). The values range from 0 to 1, and reflect the "leverage" that a given X_i has on the regression coefficient. In effect, h_i is a measure of the amount of variation a particular X_i contributes to the squared deviations of $X_i - \overline{X}$. As illustrated in Figure 11.23, the farther an X_i is from $\overline{X}$, the greater its leverage is. Evaluating the leverage coefficients allows us to determine whether there is comparatively good representation of many data points in these outlying edges, or if there are only a handful of points that are controlling the entire regression analysis. If there are a few dominant points, the analyst should seriously consider whether these points are representative of the underlying distributions, and if the subsequently identified relationship is applicable to the entire data set in general.

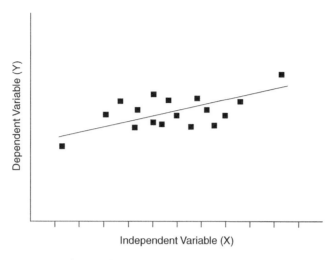

Figure 11.23   A scatter plot in which the two end points contribute disproportionately to the regression relationship

Calculation of the leverage coefficient

$$h_i = \frac{1}{n} + \frac{\left(X_i - \bar{X}\right)^2}{\sum x^2} \tag{11.13}$$

Table 11.6 and Figure 11.24 report the leverage coefficients for the pueblo data in Table 11.2. The smallest leverage coefficient is .04, not surprisingly associated with X_i values near $\bar{X}$, and the largest coefficients are .25 and .26, associated with the comparatively large pueblos of Isleta and Taos. The results are encouraging in that most of the variates do contribute something to the regression analysis, but the largest variates do overwhelm the others by a large magnitude. They are roughly three times larger than the average leverage coefficient, which is .08. We don't see anything in these leverage coefficients that cause us to suggest that the regression analysis characterizes the relationship among just these few points, but the larger contribution of the large pueblos might cause a researcher to review these data to ensure that they are analytically and theoretically comparable with the remaining data. As a rule of thumb, you should always be leery of using regression analysis when there are a small number of excessively distant outliers for the independent variable (i.e., a few individuals with large h_i values). These outliers will by mathematical necessity largely dictate the regression relationship.

Standardized residuals allow us to identify those observations that are statistically outside of the range expected for a given $\hat{Y}_i$. This enables us to statistically differentiate which values are exceptional at a given alpha level relative to the other variates. The standardized residual is calculated using Equation (11.14), which creates a t-score for each Y_i. These can then be compared to a critical value from the t-distribution for a given α and $n - 2$ degrees of freedom.

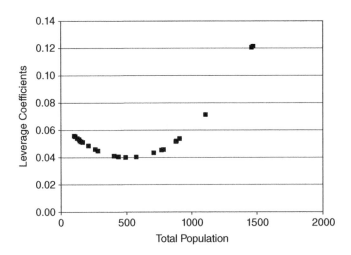

Figure 11.24 The leverage coefficients reported in Table 11.6

Table 11.6 Leverage coefficients for the data reported in Table 11.2

Pueblo	Map date	Total population	h_i
Acoma	1948	879	0.07
Cochiti	1952	444	0.04
Isleta	1948	1470	0.26
Jemez	1948	883	0.07
Laguna	1948	711	0.05
Nambe	1948	155	0.07
Picuris	1948	130	0.08
San Felipe	1948	784	0.06
San Ildefonso	1948	170	0.07
San Ildefonso	1973	413	0.04
San Juan	1948	768	0.05
Sandia	1948	139	0.08
Santa Ana	1948	288	0.05
Santa Ana	1975	498	0.04
Santa Clara	1948	573	0.04
Santo Domingo	1948	1106	0.12
Shipaulovi	1882	113	0.08
Shongopavi	1882	216	0.06
Sichomovi	1882	104	0.08
Taos	1948	907	0.08
Taos	1973	1463	0.25
Tesuque	1948	160	0.07
Tewa Village	1882	175	0.07
Walpi	1882	270	0.06
Zia	1948	267	0.06

The standardized regression residual

$$SR = \frac{d_{Y*X}}{s_{Y*X}\sqrt{(1-h_i)}} \tag{11.14}$$

d_{Y*X} is simply the raw residuals calculated as $Y_i - \hat{Y}$ we described at the beginning of this section (Table 11.5; see also Table 11.3). s_{Y*X} is the square root of the value calculated using Equation (11.4), which is also the $MS_{Unexplained}$ from Table 11.4. h_i is the leverage coefficient we just calculated. The standardized residuals for the pueblo data are presented in Table 11.7 and Figure 11.25. The critical value for $t_{[.05,23]}$ is 2.069. Thus, any standardized residual larger than 2.069 or smaller than −2.069 reflects a statistically significant departure from the regression's functional relationship. Only one standardized residual exceeds the critical value, the measurement for Walpi, which had 209 more rooms than expected given its population of 270 people (Table 11.5). The remaining deviations fell within the expected range of the regression line,

Table 11.7   Standardized residuals for the pueblo data

Pueblo	Map date	Total population	Total rooms	d_{Y*X}	h_i	Standardized residual
Acoma	1948	879	387	−16	0.07	−0.17
Cochiti	1952	444	225	0	0.04	0.00
Isleta	1948	1470	804	159	0.26	1.91
Jemez	1948	883	459	54	0.07	0.58
Laguna	1948	711	189	−145	0.05	−1.54
Nambe	1948	155	34	−73	0.07	−0.78
Picuris	1948	130	59	−37	0.08	−0.40
San Felipe	1948	784	276	−88	0.06	−0.94
San Ildefonso	1948	170	207	94	0.07	1.01
San Ildefonso	1973	413	189	−23	0.04	−0.24
San Juan	1948	768	176	−182	0.05	−1.94
Sandia	1948	139	80	−20	0.08	−0.22
Santa Ana	1948	288	136	−25	0.05	−0.27
Santa Ana	1975	498	152	−95	0.04	−1.00
Santa Clara	1948	573	144	−134	0.04	−1.42
Santo Domingo	1948	1106	500	4	0.12	0.04
Shipaulovi	1882	113	131	42	0.08	0.45
Shongopavi	1882	216	253	121	0.06	1.30
Sichomovi	1882	104	105	19	0.08	0.20
Taos	1948	907	543	128	0.08	1.38
Taos	1973	1463	627	−15	0.25	−0.18
Tesuque	1948	160	116	7	0.07	0.08
Tewa Village	1882	175	158	43	0.07	0.46
Walpi	1882	270	363	209	0.06	2.23
Zia	1948	267	126	−27	0.06	−0.29

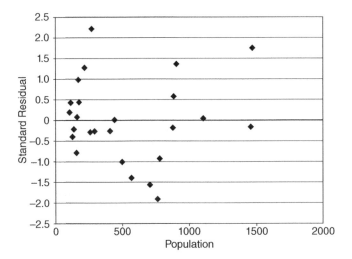

Figure 11.25   Standardized residuals for the residuals reported in Table 11.5

suggesting that the functional relationship it specified fits these data well. An interested researcher could now begin to explore the behavioral, historical, cultural, or analytical factors contributing to Walpi's excessive number of rooms.

In archaeology, it typically is not a "problem" when there are significant standardized residuals, given that these specific Y_i values are likely to be interesting in many research contexts. Although not the case here, the exception is when values with exceptionally high leverage coefficients are also significant. These values likely impact the regression analysis such that it does not reflect the functional relationship among most of the data effectively. (Walpi had a leverage coefficient of only .06, so it is not exerting undue influence on the regression analysis.) The analyst will want to give special consideration to the analytical implications of these individuals and address whether they truly should be included with the other data. Perhaps these significant outliers are not truly part of the population containing the other variates (e.g., including a more recent "pueblo-like" dorm complex built by the Federal Government to house students attending a reservation school would almost certainly produce an outlier, because of the substantial differences in the use and organization of space). In such cases, refining the sample would likely improve the overall analysis. If the analyst is satisfied that the sample truly is homogeneous, then the outlier might reflect some behaviorally or taphonomically significant distinction among the variates included in the sample. Our theoretical and analytical frameworks will then help provide meaning to the observed differences.

Some Final Thoughts about Regression

Now that you understand the basics of regression, let us take a moment to address a few miscellaneous issues that are likely to arise as you use (or evaluate other's use of) regression analysis.

Selecting the right regression model

There are several different varieties of regression analysis. The *least squares linear regression* (LSLR) that we presented here is by far the most commonly used method in science, but other methods are available. LSLR is typically preferred because it maximizes the explained sums of squares using the regression equation, which is an ideal strategy when using regression to evaluate the strength of the relationship between the dependent and independent variables or when predicting values for the dependent variable. However, it suffers from the "drawbacks" that it may not actually pass through any of the observed points (e.g., Figure 11.4) and is excessively impacted by outliers. Other methods (e.g., nonparametric regression, regression minimizing absolute deviations, quartile regression) will seek to create regression equations based on other criteria, such as creating a line that maximizes the number of observed points through which it passes. This method will be less prone to being pulled by outliers, but it does not maximize the amount of variation explained using the regression coefficient. We won't outline these alternate approaches, but be aware that there are alternatives to the least squares linear regression method described here.

Do not extrapolate beyond the boundaries of the observed data

Regression is useful for predicting Y_i for a given X_i, because of the presence of the functional relationship $Y_i = a + b(X_i)$. If the assumptions outlined at the start of this chapter are met, then the use of regression for prediction is justified. However, it cannot be assumed that the functional relationship identified during the regression analysis extends outside of the observed data range. It is always possible that there is some sort of threshold phenomenon that occurs as a variable becomes increasingly large or small. The functional relationship between the dependent and independent variable can then shift, making the previous regression analysis a very poor predictor of the dependent variable. For example, increasing political complexity associated with urbanism and populations larger than 5,000 people might lead to a very different regression coefficient describing the relationship between population size and the total number of rooms than calculated using the pueblo data. As a result, you should not predict values for the dependent variable for X_i outside of the observed range.

Use the right methods when creating reverse predictions

Archaeologists love to use reverse prediction, but they seldom consider the inherent difficulties this entails. This is an unfortunate mistake. Using the dependent variable to predict the independent variable leads to a very poor prediction that can be improved using the methods outlined above. Building arguments on $\hat{X}$ alone, however, provides a false sense of accuracy and can lead to preventable mistakes, as it is built upon a faulty analytic foundation. When using reverse prediction, take the time to consider the range that likely includes the true value of X_i. It will strengthen your analytic foundation while clarifying the implications of your statistical analysis.

Be aware of the assumptions for regression analysis

To serve as a handy reference, we relist the four assumptions of regression analysis here:

1   X is measured without error.
2   The expected (mean) values for Y fall on a straight (as opposed to curvilinear) line.
3   For any given X_i, each Y value is independent of other Y values.
4   For each X_i, the distribution of Y_i is (i) normally distributed and (ii) homoscedastic.

If your data do not meet these requirements, regression is not an appropriate quantitative tool. More significantly, it is incumbent on the researcher to demonstrate that these assumptions are met. It is always wise to use the analysis of residuals to look for problems with these assumptions.

You may be able to transform your data to create a linear relationship from a curvilinear relationship

Curvilinear functional relationships are fairly common in the natural world and in archaeological contexts, but they cannot be characterized using LSLR because they violate Assumption 2 above. One possible strategy for analyzing such distributions is to transform them using a function, often *logarithmic transformation*. This is appropriate when percentage changes in the dependent variable vary directly with changes in the independent variable to create an exponential-based relationship (e.g., Figure 11.26). Logarithmically transforming these data produces the relationship in Figure 11.27. While LSLR is obviously inappropriate for the data in Figure 11.26, the transformed data in Figure 11.27 do meet the requisite assumption. When faced with a curvilinear relationship, you may transform either or both of the independent and dependent variables. Be sure to include the transformation in your regression equation (e.g., $\hat{Y} = \log(a) + b(\log[X_i])$) when the independent variable is transformed) and transform your data back into the original units if you use the regression for prediction.

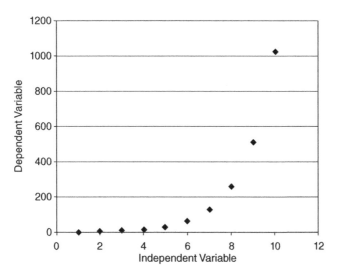

Figure 11.26   A curvilinear relationship based on exponential growth

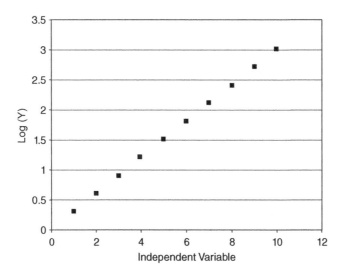

Figure 11.27   The logarithmical transformed relationship illustrated in Figure 11.26

Use the right confidence limits

Confidence intervals are tremendously useful for linear regression, in large part because the line itself reflects the distribution of means. We presented a plethora of confidence limits used to estimate various attributes that were each associated with their own measure of dispersion. It is easy to confuse the various confidence limits with each other. We suspect that at least some readers will join us in finding it hard to keep them all straight. However, mixing them up and using the wrong confidence limits will invalidate your analysis. For your convenience, we present a summary of the confidence limits and their error terms in Table 11.8.

We have now covered the basics of regression. When we wish to examine the nature of a relationship between two continuously measured variables where an argument of cause cannot be made, we turn to correlation, the subject of the next chapter.

Table 11.8 Summary of confidence limits and measures of dispersion used in regression analysis

Purpose	Measure of dispersion	Basic formula
Confidence intervals around the regression coefficient (b_{Y*X})	Standard error for regression coefficient (Equation (11.3)) $$s_b = \sqrt{\frac{s_{Y*X}^2}{\sum x^2}}$$	$L_1 = b - t_{[\alpha,n-2]}s_b$ $L_2 = b + t_{[\alpha,n-2]}s_b$
Confidence intervals for means around $\bar{Y}$ (useful for determining the likely range of μ_Y at $\bar{X}$)	Standard error of the regression coefficient at $\bar{Y}$ $$s_{\bar{Y}} = \sqrt{\frac{s_{Y*X}^2}{n}}$$	$L_1 = \bar{Y} - t_{[\alpha,n-2]}(s_{\bar{Y}})$ $L_2 = \bar{Y} + t_{[\alpha,n-2]}(s_{\bar{Y}})$
Confidence intervals for means around $\hat{Y}$ when it is not $\bar{Y}$ (useful for determining the likely range of μ_Y at X_i)	Standard error of $\hat{Y}$ for a given X_i (Equation (11.6)) $$s_{\bar{Y}} = \sqrt{s_{Y*X}^2 \left(\frac{1}{n} + \frac{(X_i - \bar{X})^2}{\sum x^2} \right)}$$	$L_1 = \hat{Y} - t_{[\alpha,n-2]}(s_{\bar{Y}})$ $L_2 = \hat{Y} + t_{[\alpha,n-2]}(s_{\bar{Y}})$
Confidence intervals for variates around $\hat{Y}$ (useful for determining the likely range of Y_i at X_i)	The standard deviation of Y_i around $\hat{Y}_i$ at a given X_i (Equation (11.7)) $$\hat{s}_Y = \sqrt{s_{Y*X}^2 \left(1 + \frac{1}{n} + \frac{(X_i - \bar{X})^2}{\sum x^2} \right)}$$	$L_1 = \hat{Y} - t_{[\alpha,n-2]}(\hat{s}_Y)$ $L_2 = \hat{Y} + t_{[\alpha,n-2]}(\hat{s}_Y)$
Confidence intervals for X_i around $\bar{X}$ (useful for reverse prediction of independent variable using the dependent variable)	Two necessary terms (Equations (11.9) and (11.10)) $D = b_{Y*X}^2 - t_{[\alpha,n-2]}^2 s_b^2$ $$H = \frac{t_{[\alpha,n-2]}}{D} \sqrt{s_{Y*X}^2 \left[D\left(1 + \frac{1}{n}\right) + \frac{(Y_i - \bar{Y})^2}{\sum x^2} \right]}$$	$L_1 = \bar{X} + \frac{b_{Y*X}(Y_i - \bar{Y})}{D} - H$ $L_2 = \bar{X} + \frac{b_{Y*X}(Y_i - \bar{Y})}{D} + H$

Practice Exercises

1 What are the assumptions of regression analysis? Explain why each is necessary.

2 In an experimental study evaluating the potential use of gourds as floats for fishing nets, Hart *et al.* (2004: 145) present the following data concerning gourd diameter and the weight supported by the float. Presumably there is a causal relationship between these two variables, such that larger gourds will support more weight.

 (a) Create a scatter plot of the data and evaluate whether the assumptions for regression analysis appear to be met.

 (b) Perform a regression analysis ($\alpha = .05$). Is there a significant relationship between gourd diameter and the weight the float can support?

 (c) Calculate the regression coefficient and the coefficient of determination. How much of the variation in the weight supported can be explained by changes in the gourd diameter?

 (d) Draw the regression line on the scatter plot you created in part (a) of this question.

 (e) Predict the weight that a gourd 6.0 cm in diameter could support. Place confidence intervals around this prediction.

Float	Maximum diameter (cm)	Weight supported by float (g)
Gourd 1	6.1	90
Gourd 2	5.8	90
Gourd 3	6.2	120
Gourd 4	6.4	120
Gourd 5	6.5	150

3 Use the following data to test the hypothesis that there is no linear relationship between the height of a ceramic vessel (independent variable) and the average thickness of the vessel wall (dependent variable). Predict the average wall thickness that you would expect for a pot 19 cm tall. Place 95% confidence intervals for Y_i around this predicted mean. Use reverse prediction to estimate the expected pot height associated with a sherd that is 2.1 cm thick.

Pot height (cm)	Average thickness (cm)
25	3.2
10	1.0
14	1.7
20	2.9
9	1.2
9	.09
18	2.0
17	2.2
12	1.5
14	1.8

4   Foster and Cohen (2007: 42) present the following data from the Hite Bowl site in the southeastern US reflecting the depth of palynological samples and their excess ^{210}Pb concentrations, which can be used to date the samples and determine sedimentation rates. Presumably depth (time) controls the amount of ^{210}Pb.

(a)   Prepare a scatter plot reflecting these data. Do they appear to meet the requirements for regression analysis?

(b)   Perform a regression analysis on the data ($\alpha = .05$). Is there a significant relationship? If so, is it positive or negative?

(c)   Draw the regression line on the scatter plot created in part (a) of this exercise.

(d)   Conduct an analysis of the regression residuals. Draw a plot of the residuals. What does the plot indicate about the appropriateness of the assumption of a linear relationship?

Depth (cm)	Excessive ^{210}Pb
1	13.66
3	13.30
5	13.17
7	6.54
9	4.19
11	3.56
13	2.53
15	1.45
17	1.27
19	0.73

12

Correlation

Correlation is so similar to regression that they are occasionally confused with one another, but there is a very significant difference – regression is used to explore the relationship between an *independent* variable and a *dependent* variable, whereas correlation is used to consider a relationship between two *dependent* variables. To illustrate this difference, consider the relationships between stature, femur length, and humerus length of 10 men from the University of New Mexico's documented skeletal collection (Table 12.1). Given that stature is, at least in part, a function of femur length, regression can be used to explore the relationship between femur length (the independent X variable) and stature (the dependent variable Y). Figure 12.1 presents this relationship graphically, as well as the regression equation and r^2.

What about the relationship between stature and humerus length? Figure 12.2 seems to indicate that there is some sort of relationship between the two variables such that stature increases as humerus length increases, but can this relationship be modeled using regression? The answer is no, because the assumption of a causal relationship required for regression is not met; stature as it is typically defined is not a function of the length of an individual's upper arm. Still, it seems reasonable to suggest that tall people might have longer arms than shorter people, and that stature might *correlate* with humerus length, even if humerus length does directly control stature. The two variables may correspond with each other as a result of the influence of the same genetic predispositions or environmental factors such as nutrition. That is to say, both stature and humerus length may be dependent variables that are at least in part controlled by the same independent variables. The more important the shared independent variables are in determining the two dependent variables, the greater the correspondence between them will be.

Quantitative Analysis in Archaeology, Todd L. VanPool and Robert D. Leonard
© 2011 Todd L. VanPool and Robert D. Leonard

Table 12.1   Stature and long bone lengths of 10 males from the UNM documented skeletal collection

Stature (cm)	Femur length (cm)	Humerus length (cm)
176.8	47.9	34.6
172.7	45.6	31.4
158.8	43.7	31.4
166.4	46.9	34
182.9	48.7	35
167.6	43.8	31.3
170.2	46.1	33.9
167.6	44	32.3
171.5	46.2	33.6
175.3	47.5	33.5

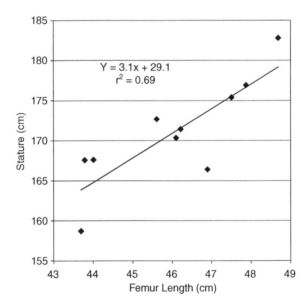

Figure 12.1   The regression relationship between femur length and stature for 10 males from the University of New Mexico's skeletal collection

Still, why can't we go ahead and apply regression analysis when there is a strong correspondences between variables, even though there may be no direct causal link between them? The reason is that neither of the variables is measured without error. Remember that regression requires that the independent variable be measured without error. This means more than simply using a tool that is adequately precise and accurate to take a good measurement, but is instead the assumption that the measurement reflected in the independent variable represents a specific value (that is, a point) as opposed to an estimate of a specific value (that is, a distribution

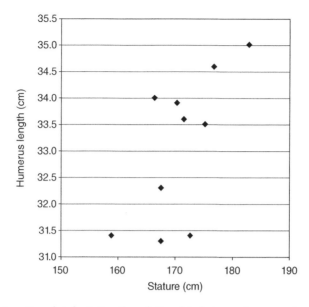

Figure 12.2 A scatter plot depicting the relationship between humerus length and stature for 10 males from the University of New Mexico's skeletal collection

around a mean) relative to the dependent variable. In this case, the values for both variables reflect the influence of all of the underlying independent variables, some or all of which may be shared. This in turn means that the values for both variables are estimates of the average value expected given the various states of the independent variables and their interaction (i.e., both variables reflect a relationship based on $Y_i = a + bX_i + e_i$, which explicitly does include an error term). Hence, neither variable is measured "without error". The result of this codependence is that any model of the "linear relationship" in Figure 12.2 must account for variation in both variables (Figure 12.3), which is not the case for Figure 12.1. Variation in both variables makes it impossible to mathematically fix a predictive, linear line to the relationship, as is done with regression (compare Figures 11.6 and 12.3).

To deal with variables linked to shared independent variables, archaeologists need a mathematical and analytic model that will allow them to determine the degree to which the shared independent variable(s) influence the dependent variables. This would allow the variation in seemingly distinct variables to be linked to the same underlying factors, and would be a very useful tool for archaeologists who frequently use multiple lines of evidence to argue for the presence or absence of some underlying environmental or behavioral attribute. Fortunately for archaeologists, correlation is useful in these situations. Unlike regression, correlation *does not* allow for the direct prediction of one variable from the other or measure how much of the variation in one attribute is caused by variation in the other. It instead estimates the degree to which both variables are the product of one or more shared independent variables. The most common correlation method used for interval or ratio scale data is the Pearson's product-moment correlation coefficient.

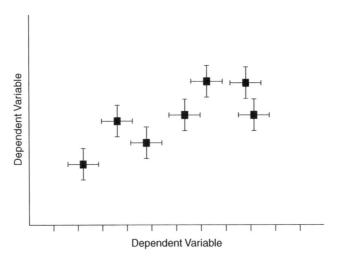

Figure 12.3 95% confidence intervals illustrating the variation in both dependent variables

Pearson's Product-Moment Correlation Coefficient

The best way to start any analysis using Pearson's product-moment correlation coefficient (which is commonly shortened to Pearson's correlation coefficient or even the correlation coefficient) is to prepare a scatter plot such as Figure 12.2. Here we arbitrarily place stature on the horizontal axis, but it doesn't matter which variable is selected for the *x*-axis or *y*-axis given that they are both dependent variables. We will discuss a couple of important things to look for in the graph after we have explored the assumptions of correlation analysis a bit, but Pearson's correlation coefficient may be appropriate if there appears to be a linear, more or less continuous distribution of points. It should not be applied to cases where there are extreme outliers, a curvilinear relationship, or tight clusters of data. Spearman's rank order coefficient, which we also outline below, is superior in these contexts.

Pearson's correlation coefficient is determined using Equation (12.1), which is most easily calculated using the following nine quantitative steps (Table 12.2).

Quantity 1: $\sum Y_1 = 1709.8 \, \text{cm}$

Quantity 2: $\sum Y_1^2 = 292732.2 \, \text{cm}^2$

Quantity 3: $\sum Y_2 = 331 \, \text{cm}$

Quantity 4: $\sum Y_2^2 = 10973.48 \, \text{cm}^2$

Quantity 5: $\sum Y_1 Y_2 = 56649.57 \, \text{cm}^2$

Table 12.2 Calculation of sums used to compute a Pearson's correlation coefficient

Specimen number	Stature (cm)		Humerus length (cm)		Y_1Y_2
	Y_1	Y_1^2	Y_2	Y_2^2	
1	176.8	31258.24	34.6	1197.16	6117.28
2	172.7	29825.29	31.4	985.96	5422.78
3	158.8	25217.44	31.4	985.96	4986.32
4	166.4	27688.96	34.0	1156.00	5657.60
5	182.9	33452.41	35.0	1225.00	6401.50
6	167.6	28089.76	31.3	979.69	5245.88
7	170.2	28968.04	33.9	1149.21	5769.78
8	167.6	28089.76	32.3	1043.29	5413.48
9	171.5	29412.25	33.6	1128.96	5762.40
10	175.3	30730.09	33.5	1122.25	5872.55
Σ	1709.8	292732.20	331.0	10973.48	56649.57

Quantity 6, the sum of squares of Y_1: $\sum y_1^2 = \sum Y_1^2 - \dfrac{\left(\sum Y_1\right)^2}{n} = $ Quantity 2 $-$

$$\dfrac{\left(\text{Quantity 1}\right)^2}{n} = 292732.3 - \dfrac{\left(1709.8\right)^2}{10} = 390.60 \text{ cm}^2$$

Quantity 7, the sum of squares of Y_2: $\sum y_2^2 = \sum Y_2^2 - \dfrac{\left(\sum Y_1\right)^2}{n} = $ Quantity 4 $-$

$$\dfrac{\left(\text{Quantity 3}\right)^2}{n} = 10973.48 - \dfrac{\left(311\right)^2}{10} = 17.38 \text{ cm}^2$$

Quantity 8, the sum of products: $\sum y_1 y_2 = \sum Y_1 Y_2 - \dfrac{\left(\sum Y_1\right)\left(\sum Y_2\right)}{n} = $ Quantity 5 $-$

$$\dfrac{\left(\text{Quantity 1}\right)\left(\text{Quantity 3}\right)}{n} = 56649.57 - \dfrac{\left(1709.8\right)\left(331\right)}{10} = 55.19 \text{ cm}^2$$

Pearson's product-moment correlation coefficient

$$r_{1,2} = \dfrac{\sum y_1 y_2}{\sqrt{\sum y_1^2 \sum y_2^2}} = \dfrac{\text{Quantity 8}}{\sqrt{(\text{Quantity 6})(\text{Quantity 7})}} = \dfrac{55.19}{\sqrt{(390.596)(17.38)}} = .67 \quad (12.1)$$

The notation for these quantities should be familiar from the least squares regression method introduced in Chapter 11, because Pearson's correlation coefficient also uses the covariance (see Quantity 8) to measure the degree of shared variance in the two variables. Also, like regression analysis, the correlation coefficient uses a term (r, as opposed to regression's r^2) to reflect the amount of covariance. These two coefficients (r and r^2) are even connected such that the square of Pearson's correlation coefficient (r) equals regression's coefficient of determination (r^2) calculated for the same data. However, Pearson's correlation coefficient can range from -1 (a perfect negative correlation in which one variable decreases as the other increases) to 1 (a perfect positive correlation), whereas a regression coefficient of determination (a squared value) must be a positive value varying between 0 and 1. Also as with regression's coefficient of determination, Pearson's correlation coefficients close to zero indicate that there is no correspondence between the two variables of interest.

But how close must Pearson's r be to 0 to conclude that there isn't a significant correlation between the variables? The most common significance test for Pearson's correlation coefficient uses a t-test (Equation (12.2)) to evaluate the null hypothesis $H_0 : \rho = 0$, where ρ (vocalized as rho or roe) is the population parameter estimated by the correlation coefficient, r. If r is close to 0, then the null hypothesis that $\rho = 0$ cannot be rejected, and it is possible that the two dependent variables are not significantly determined by the same underlying variables. In contrast, if r is significantly different from 0, then we can reject the null hypothesis and conclude that there is a significant correlation.

The t-test's formula (Equation (12.2)) is derived from Equation (8.6) by replacing μ with 0 (which is the specific value of ρ we wish to evaluate), $\bar{Y}$ with r, and the standard error $s_{\bar{y}}$ with the correlation coefficient's standard error s_r, which is calculated using Equation (12.3).

The t-test for evaluating the significance of Pearson's correlation coefficient

$$t = \frac{r-0}{s_r} \tag{12.2}$$

The standard error of the Pearson's correlation coefficient

$$s_r = \sqrt{\frac{1-r^2}{n-2}} \tag{12.3}$$

Applying these equations to the example above produces:

$$t = \frac{r-0}{\sqrt{\frac{1-r^2}{n-2}}} = r\sqrt{\frac{n-2}{1-r^2}} = .67\sqrt{\frac{10-2}{1-.45}} = 2.56$$

Comparing the t-score to the critical value of $t_{.05[8]} = 2.306$, we reject H_0 and conclude that there is a significant correlation between stature and humerus length, just as we would suspect.

Table 12.3 ANOVA table for Pearson's correlation coefficient

Source of variation	SS	df	MS	F_S
Explained	$(r)^2 * \sum y_1^2$	$a - 1$	$\dfrac{(r)^2 * \sum y_1^2}{1}$	$\dfrac{MS_{explained}}{MS_{unexplained}}$
Unexplained	$\left(1-[r]^2\right) * \sum y_1^2$	$n - 2$	$\dfrac{\left(1-[r]^2\right) * \sum y_1^2}{n-2}$	
Total	$\sum y_1^2$	$n - 1$	$\dfrac{\sum y_1^2}{n-1}$	

It is also possible to complete an ANOVA analysis to evaluate the null hypothesis $H_0 : \rho = 0$ by calculating an explained and unexplained sums of squares using Equations (12.4) and (12.5). It doesn't matter which variable is selected to be Y_1. The results will be the same. An ANOVA table (Table 12.3) can even be created, although for whatever reason this is rarely done when using Pearson's correlation coefficient. If you don't wish to bother with the ANOVA table, simply divide $SS_{explained}$ and $SS_{unexplained}$ by their degrees of freedom to create the explained and unexplained means sums of squares, and then divide $MS_{explained}$ by the $MS_{unexplained}$. Note that both $SS_{explained}$ and $SS_{unexplained}$ contain $\sum y_1^2$, the sums of squares for the selected variable (compare Equations (12.4) and (12.5)). When dividing $MS_{explained}$ by $MS_{unexplained}$, this term will cancel out, reducing the formula for the F-score to Equation (12.6).

Explained sums of squares for Pearson's product-moment correlation coefficient

$$SS_{exp} = (r)^2 \times \sum y_1^2 \tag{12.4}$$

Unexplained sums of squares for Pearson's product-moment correlation coefficient

$$SS_{Unex.} = \left(1-[r]^2\right) \times \sum y_1^2 \tag{12.5}$$

F-score comparing the explained and unexplained sums of squares for Pearson's product-moment correlation coefficient

$$F = \dfrac{(r)^2}{\dfrac{\left(1-[r]^2\right)}{(n-2)}} = \dfrac{(r)^2 \times (n-2)}{\left(1-[r]^2\right)} \tag{12.6}$$

Illustrating the application of these equations using humerus length as Y_1 produces the following results:

Table 12.4 ANOVA analysis of the correlation between the humerus length and stature of 10 skeletons from UNM's skeletal collection

Source of variation	SS	df	MS	F_S
Explained	175.34	1	175.34	6.52
Unexplained	215.26	8	26.91	
Total	390.60	9	43.40	

$$SS_{exp} = (.67)^2 \times 390.60 = 175.34 \text{ cm}^2$$

$$SS_{Unex.} = \left(1 - [.67]^2\right) \times 390.60 = 215.26 \text{ cm}^2$$

$$F = \frac{(.67)^2 \times (10-2)}{1-(.67)^2} = 6.52$$

These same results are reflected in Table 12.4. Using an alpha of .05, the critical value for $H_0 : \rho = 0$ is $F_{.05[1,8]} = 5.32$. Given that 6.52 is greater than the critical value, we again reject the null hypothesis and conclude that there is a significant correlation between humerus length and total stature.

The assumptions of Pearson's product-moment correlation coefficient

As with all quantitative methods, there are assumptions that must be met for Pearson's correlation coefficient to be effectively used. These are:

1 There should be no known causal relationship between the two variables such that one is dependent on the other. If there is, use regression. The correlation analysis won't be wrong *per se* if this assumption is violated, but it won't properly model the strength of the causal relationship. Although the correlation coefficient and regression coefficient of determination are mathematically linked such that $(r)^2$ equals r^2, the two measures actually reflect different population parameters. Using a correlation coefficient when regression analysis is warranted will not properly reflect the strength of the causal relationship between the variables, while using the regression analysis when a correlation analysis is warranted will underplay the influence shared among the two dependent variables. Using the right tool for the right job will help archaeologists better understand the relationships between the variables being considered.

2 The measurements must be independent of one another such that one measurement doesn't determine another. For example, the correlation of wall lengths of opposing walls in rectangular rooms is not a valid use for Pearson's correlation coefficient, because the lengths are interdependent.

3   The distributions of both variables should be normally distributed, and their bivariate (combined) distribution should be normal as well. We can evaluate this assumption by using graphical methods. Simple histograms can help determine if the distributions of both variables are normally distributed. A scatter plot such as illustrated in Figure 12.2 will help determine if the combined distribution in normally distributed. If it is, the points will be spread relatively uniformly throughout the distribution with variates becoming more frequent towards the center of the distribution (Figure 12.4). Extreme outliers should be absent. When extreme outliers are present, a nonparametric correlation analysis should be used. In cases where there are obvious gaps in the distribution (Figure 12.5), other quantitative methods such as cluster analysis will be more useful for understanding the relationship between the two dependent variables and the underlying independent variable(s).

Archaeological data often meet the first two assumptions. The archaeological record is replete with cases in which two or more mechanically distinct attributes are determined by the same independent variable. For example, wall thickness and rim thickness of ceramic pots may correlate with each other, despite the mechanical independence of these two variables, because they are both controlled in part by vessel size. Likewise, original cobble size may impact both the length of lithic debitage and the size of the resulting tool in certain flaked stone technologies. The third assumption requiring normality is frequently more problematic given that archaeological assemblages often produce skewed data distributions. The outliers in skewed distributions will exert an inordinate influence on the measures of covariance used to determine Pearson's correlation coefficient, possibly producing spurious results. When there is any question about whether one or both of the distributions are too

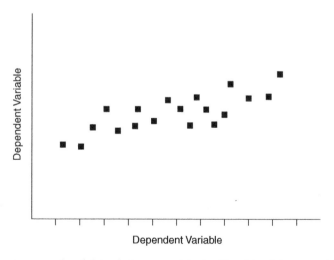

Figure 12.4   An example of data that are consistent with a bivariate normal distribution without significant outliers

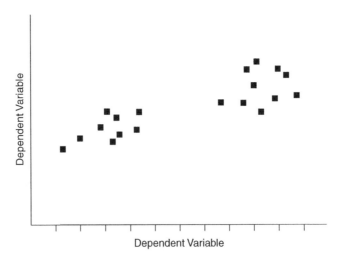

Figure 12.5   An example of clustered data for which Pearson's correlation coefficient is inappropriate

heavily skewed to be normally distributed, use Spearman's rank order correlation coefficient instead.

Spearman's Rank Order Correlation Coefficient

Spearman's correlation coefficient is a powerful nonparametric test that can be used when ratio or interval data are too heavily skewed to allow for the Pearson's correlation coefficient, or for ordinal data. It is tremendously robust, and can also be applied to any data that Pearson's correlation coefficient can. Unlike Pearson's correlation coefficient, it is not based on the measurement of covariance but is instead derived by comparing the ranking of variates between the two variables. The underlying idea is quite straightforward. If a positive correlation is present, the largest/highest ranked individual will produce the largest variates for both variables, the next largest individual will produce the second largest variates for both variables, etc. If a negative correlation is present, then the individual producing the largest variate for one variable will produce the smallest variate for the other, the individual producing the second largest variate for the first variable will produce the second smallest variate for the other, etc. If there is no correlation between the variables, the ranks will be independent, in that highly ranked variates of one distribution will not consistently correspond with high or low ranked members of the other distribution. The variation in the rank order of the two variables can consequently be measured to determine if the rankings of the two variables are significantly correlated.

Table 12.5 The ranking of fish species by their abundance at the Newbridge and Carlin settlements

Taxon	Newbridge (%)	Newbridge rank (R_1)	Carlin (%)	Carlin rank (R_2)	$(R_1 - R_2)$	$(R_1 - R_2)^2$
Bullhead	33.96	1	11.27	3	−2	4
Bowfin	21.05	2	16.67	2	0	0
Buffalo	12.90	3	10.29	4.5	−1.5	2.25
River Catfish	10.53	4	19.12	1	3	9
Bass	6.11	5	10.29	4.5	.5	.25
Sunfish	4.41	6	0.98	9	−3	9
Pike	3.40	7	3.92	8	−1	1
Redhorse	3.06	8	5.88	7	1	1
Freshwater Drum	2.38	9	9.80	6	3	9
Crappie	.51	10	0.49	10	0	0
Total						35.50

Consider the data presented in Table 12.5. These are rank order fish bone abundances from two sites in the lower Illinois Valley (Styles, 1985: 42). These data reflect the percentage of the NISP, the number of identifiable specimen, of the total assemblage composed of each fish species, which is a notoriously problematic measure of abundance. Different sizes of fish, different processing strategies, and differences in bone preservation, among other factors, make it difficult to tie NISP to the actual number of individual animals represented in an assemblage. This means that Pearson's correlation coefficient cannot be applied to these data, because they are not in reality measured at a ratio or interval scale (see Lyman, 2008). However, we might be willing to accept NISP as a relative measure of abundance, such that a higher proportion of the NISP reflects the presence of more fish than a lower percentage of the NISP, even if we don't claim to know *exactly* how many fish are represented by either percentage. We can consequently use NISP to create an ordinal ranking of the abundance of each fish species, even if we don't know exactly how many fish were consumed. Such a ranking is all that Spearman's correlation coefficient requires.

As is typical of faunal remains, understanding fish choice could provide information about subsistence patterns, seasonality, past environments, and a host of other archaeologically important attributes. A researcher might wish to compare the two sites to determine if their inhabitants were consuming the same fish (reflecting similarities in underlying attributes such as settlement patterns and environmental setting) or were consuming different fish (reflecting differences in the underlying variables). To be clear, the question is not whether the inhabitants consumed the same *amounts* of fish, but rather, did the inhabitants utilize the same kinds of fish?

To evaluate this hypothesis of interest, we will evaluate the null hypothesis $H_0 : r_s = 0$, where r_s reflects Spearman's rank order correlation coefficient (Equation (12.7)). As with Pearson's correlation coefficient, values of Spearman's r range from 1 (perfect positive correlation) to −1 (prefect negative correlation) with values close to 0 indicating that there is no correlation between the two variables.

Spearman's rank order correlation coefficient

$$r_s = 1 - \frac{6\sum (R_1 - R_2)^2}{n(n^2 - 1)} \tag{12.7}$$

In Equation (12.7), $R_1 - R_2$ is simply the difference between the rankings for each variate, and n is the sample size. Don't worry about the origin of the "6". It is a constant in the equation that never changes. It also doesn't matter which variable is selected for R_1 or R_2, or if the ranking is organized from smallest to largest or vice versa. The results won't change. In cases where there are two or more variates with the same value, their ranking is determined by averaging the ranks they would otherwise occupy (e.g., if two variates with the same value are in line to be ranked 7 and 8, they both are ranked 7.5; see the Carlin site rankings for Bass and Buffalo in Table 12.5 for another example). For samples larger than 10, the significance of Spearman's r can then be evaluated using Equation (12.2), the t-test presented above for evaluating Pearson's correlation coefficient. Critical values for smaller samples are listed in Table 12.6.

We demonstrated the ranking process and the calculation of $R_1 - R_2$ in Table 12.5. Applying these data to Equation (12.7) produces the following result:

Table 12.6 Critical values for Spearman's r when sample size is equal to or smaller than 10

n	Significance level (one-tailed test)	
	.05	*.01*
4	1.000	
5	.900	1.000
6	.829	.943
7	.714	.893
8	.643	.833
9	.600	.783
10	.564	.746

$$r_s = 1 - \frac{6 \sum (R_1 - R_2)^2}{n(n^2 - 1)} = 1 - \frac{6(35.5)}{10(100 - 1)} = 1 - \frac{213}{990} = .78$$

Given that our sample size is 10, we use Table 12.6 to determine that the critical value for $\alpha = .05$ is .56, which is considerably smaller than .78. We reject the null hypothesis and conclude that the ranking of the fish species is correlated between the Newbridge and Carlin sites.

For fun, let's apply the same process to compare the ethnobotanical remains from the Newbridge site and a third site, the Weitzer site (Table 12.7; Styles, 1985: 52). The null hypothesis is again set at $H_0 : r = 0$ and the level of rejection (α) is set at .05. Table 12.7 lists the rankings of various types of plant resources, as well as $R_1 - R_2$. Spearman's r is:

$$r_s = 1 - \frac{6 \sum (R_1 - R_2)^2}{n(n^2 - 1)} = 1 - \frac{6(21)}{7(49 - 1)} = 1 - \frac{126}{336} = .375$$

This Spearman's r is much closer to 0 than was the r_s for the Newbridge and Carlin fish assemblages. Comparing it to the critical value of .714 listed in Table 12.6 for $n = 8$ indicates that we cannot reject the null hypothesis. The ethnobotanical remains between the two settlements differ with oily cultivated plants ranking much higher at the Newbridge settlement that at the Weitzer settlement. Such differences could reflect variation in subsistence strategies, settlement location, environmental conditions, or various other factors. With a proper theoretical and analytical framework, this (and similar) differences may become key insights to quantifying and explaining differences in the archaeological record.

Table 12.7 Paleobotanical information from the Newbridge and Weitzer sites

Seed group	Newbridge (%)	Newbridge rank (R_1)	Weitzer (%)	Weitzer rank (R_2)	($R_1 - R_2$)	($R_1 - R_2$)²
Starchy cultivated (?)	84.82	1	26.67	2	−1	1
Miscellaneous	10.75	2	48.33	1	1	1
Oily cultivated	2.80	3	0	6.5	−3.5	12.25
Starchy non-cultivated	1.41	4	6.67	4.5	−.5	.25
Sumac	.09	5	11.67	3	2	4
Fleshy fruits	.08	6	6.67	4.5	1.5	2.25
Weed seeds	.07	7	0	6.5	.5	.25
Number of seeds	15009		2868			
Total						21

Some Final Thoughts (and Warnings) about Correlation

It is an oft-stated truism in the philosophy of science, statistics, and even as a cliché in daily life, that correlation does not equal causation. We encourage you to keep this in mind. A significant regression relationship does equal (an argument for) causation, but a significant correlation at most supports the proposition that there are underlying shared causal factors affecting both dependent variables. These underlying factors can be central to scientific explanations, but they can also be relatively trite. Statisticians talk about "nonsense correlations", a classic example of which is that the number of Baptist ministers and the total amount of liquor consumed correlate quite strongly in communities larger than 10,000 throughout the United States. Such correlations aren't really nonsense so much as they are analytically meaningless. They *do* reflect shared independent variables, but these are so broadly formed as to provide little insight. The number of Baptist ministers and the amount of alcohol consumed correlate well with each other because they are both the product of population size, not because Baptist ministers drink lots of alcohol. We are confident that the number of hardware stores, McDonald's restaurants, fire hydrants, and automobile dealerships will all correlate with the number of Baptist ministers (and alcohol consumed), for the exact same reason. Yet such a relationship doesn't provide significant insight into any of the correlated variables. The mere presence of a correlation can never be considered proof of a causal relationship between the two variables, and may not even be significant to the analysis at all. The archaeological record is necessarily replete with nonsense correlations based on attributes such as population size, the presence/absence of certain taphonomic processes, age of deposits, excavation methods, etc. It is not surprising for example that the frequency of iron artifacts increases in more recent historic settlements relative to older historic settlements. This simply reflects the reality that iron rusts. It would be a mistake to uncritically interpret this pattern as necessarily reflecting an increase in the use of iron through time.

Remember also that it is always possible to commit Type I and Type II errors when conducting any hypothesis testing. A Type I error in the context of a correlation would lead to the conclusion that a correlation is present when in fact the two variables are not the product of shared underlying independent variables. Given enough correlation analyses, any researcher is guaranteed to identify spurious correlations. Likewise, it is possible to commit Type II errors, and incorrectly conclude that there is not a significant correlation when in fact there is. This is especially common when two variables share some but not all of their underlying causal attributes. As ρ gets closer to 0, it will become increasingly likely to fail to identify weak correlations that really do exist. A useful, but not foolproof, way to help mitigate this is through large samples that will tend to more robustly reflect any correlation that is present. When you expect that two variables share some but not all underlying independent variables, be sure to keep in mind the likelihood of

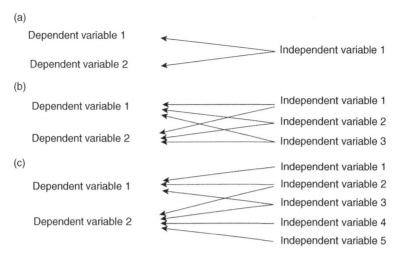

Figure 12.6 Examples of possible relationships that can result in significant correlations

committing a Type II error and maximize the sample size as best as is possible. Still, weak correlations will always be susceptible to Type II errors.

Finally, the presence of a correlation reflects that the two dependent variables share one or more underlying independent variables, but the relationships between these variables can be complex. Figure 12.6 reflects a couple of the innumerable possible relationships underlying significant correlations. In the simplest possible relationship, both variables could reflect the operation of one and only one shared variable (Figure 12.6(a)). More complex forms could include a number of shared independent variables that all contribute to both dependent variables (Figure 12.6(b)). In even more complex relationships, the dependent variables could be the product of some shared attributes but also reflect the operation of other unique variables (Figure 12.6(c)). Determining the relationship underlying correlations will require a clearly defined analytic and theoretical structure.

Practice Exercises

1 What are the assumptions of correlation analysis? How do these differ from regression analysis's assumptions?

2 Identify three examples where two or more archaeologically relevant dependent variables might be controlled by the same independent variable(s). Use your imagination to derive three possible "nonsense" correlations. Briefly discuss why the archaeologically relevant examples you selected are different than a nonsense correlation.

3   Following are the length and width measurements for various lithic tools from different periods of occupation at the White Paintings Rock Shelter in Botswana (Donahue *et al.*, 2002–2004: 158). Treat the entire data set as a single sample.

   (a)   Create a scatter plot reflecting the two variables. Intuitively, does there appear to be a relationship between them? If so, is it positive or negative?

   (b)   Perform a Pearson's product-moment correlation analysis to determine if the variables correspond with each other ($\alpha = .10$). If so, what is the strength of the shared variation? What is the underlying variable that is likely causing any correspondence you identify?

Culture	Artifact type	Length (cm)	Width (cm)
Late Stone Age	Double-backed Point	14	5
Middle Holocene Late Stone Age	Burin	21	10
Late Pleistocene Late Stone Age	Blade	19	14
Early Late Stone Age	Blade	63	24
Middle Stone Age	Point	28	25
	Blade	36	16
	Blade	40	19
	Point	30	20
	Multipurpose tool	27	18
	Point	34	28
	Point	21	16

4   Bustard (1996: 292–3) presented the flowing data for the numbers of above ground rooms and pit structures for small sites in Chaco Canyon, New Mexico. Which correlation method (Pearson's or Spearman's Correlation Coefficient) would be best to evaluate whether there is a relationship between these variables? Use the appropriate method to determine if there is a correlation between the number of above ground rooms and pit houses ($\alpha = .05$).

Site number	Above ground rooms	Pit structures
Bc 50	20	4
Bc 51	45	6
Bc 53	17	4
Bc 54	4	3
Bc 57	9	4
Bc 58	13	1
Bc 59	14	5
Bc 126	5	2
Bc 192	11	1
BC 236	9	1
Bc 362	18	2

5 Following are data related to the minimum number of individuals for various artifact assemblages grouped by use categories/associations from various historic households (Van Bueren and Wooten, 2009: 120). Use a Spearman's rank order correlation to determine if the frequencies of the items correlate with one another ($\alpha = .05$).

Type	All male households	Family dwellings
Male items	50	40
Health items	22	106
Alcohol	76	378
Tobacco	61	11
Other personal	366	189
Household	176	1000+
Hunting	131	68

13

Analysis of Frequencies

"Within recent years there appears to have been an increasing awareness on the part of archaeologists that certain statistical techniques offer economical methods of extracting information of cultural significance from archaeological data" (Spaulding, 1953). With this opening sentence, Albert Spaulding introduced a now commonly used statistical procedure to archaeology called chi-square analysis. Spaulding was dissatisfied by archaeological classification procedures, and sought to make typology rigorous, replicable, and scientific. At the time, archaeologists typically created archaeological types by laying artifacts out on a table, and then putting similar objects together in a pile. What constituted "similar" was up to the individual building the piles. Of course, what one individual determined to be similar differed from what another might see. This led (and continues to lead) to no end of problems (Whittaker *et al.*, 1998) – one of the most significant being that a select group of individuals who originally created types in a region became the "experts" on the types they created. As a consequence, if a student didn't learn types at the feet of the Master (or one of the Master's apprentices), his or her work was forever suspect.

Spaulding sought to put an end to this procedure, and offered numerical taxonomy (the use of statistical methods to create analytic units) as an alternative. He proposed that by demonstrating non-random associations between attributes of artifacts, archaeologists would be able to "discover" the meaningful emic types that the original makers had in mind and consistently produced. Although there were many alternatives to accomplish this task, the rather straightforward method that Spaulding suggested was to use the chi-square test to compare the frequencies of various associations. His idea was that artifact types would be characterized by the consistent co-occurrence of attributes as people consistently made the tools they

Quantitative Analysis in Archaeology, Todd L. VanPool and Robert D. Leonard
© 2011 Todd L. VanPool and Robert D. Leonard

desired. A test of abundance such as the chi-square test would allow archaeologists to see which associations were particularly common, and would thereby allow archaeologists to discover the underlying types that structured past human technology. Spaulding's (1953) proposal was critiqued at the time (Ford, 1954) and was never generally accepted by archaeologists, although some continue trying to develop methods of numerical taxonomy to fulfill Spaulding's original goal (e.g., Benfer and Benfer, 1981; Christenson and Read, 1977; Cowgill, 1990; Phagan, 1988). However, chi-square analysis continues to be a mainstay of archaeological quantification.

Let us consider the calculation and utility of the chi-square test using Spaulding's example. Table 13.1 presents two hypothetical variables that Spaulding sought to use to discover ceramic types based on sherd frequencies. These data create a matrix in which the correspondence of two or more variables is reflected. In this example there are an equal number of sherds reflected in each cell, which is the expected pattern if there is no association between temper type and surface treatment. Here, the two variables are distributed independently relative to each other such that a smooth surfaced vessel is as likely to have grit temper as shell temper and a grit tempered vessel is as like to have a stamped surface as a smooth surface. Contrast this with Table 13.2, which presents a perfect association between temper type and surface treatment. Here, grit temper is always associated with a stamped surface and shell temper with a smooth surface. To Spaulding, such non-random associations would constitute emic types that reflect the will of the potters involved in making the original ceramics.

The non-overlapping associations in Table 13.2 are quite unambiguous, and an archaeologist really wouldn't need to use any statistical test to determine that there

Table 13.1 Pottery sherds classified by their surface treatment and temper type

Surface texture	Grit temper	Shell temper	Total
Stamped surface	25	25	50
Smooth surface	25	25	50
Total	50	50	100

Table 13.2 Sherds classified by their surface treatment and temper type in which there is a strong association between the two variables

Surface texture	Grit temper	Shell temper	Total
Stamped surface	50	0	50
Smooth surface	0	50	50
Total	50	50	100

is a difference in the temper of smooth surfaced and stamped surface pottery. However, archaeologists also realize that the archaeological record rarely reflects such nice, neat associations (which was Ford's (1954) point in his debate with Spaulding). Archaeologists more commonly deal with data that looks something like Table 13.3. There may be a strong positive association between stamped surface pottery and shell temper, and between smooth surfaced pottery and grit temper, but are these associations beyond what would be expected due to chance if the variables varied independent of each other? Well, thinking about it, if there isn't an association, we would expect that an equal percentage of grit tempered ceramics would have stamped surfaces as have smooth surfaces. Grit tempered pottery makes up 59.6% of the total pottery assemblage (234 grit tempered sherds/392 total sherds), so 59.6% of the 146 stamped surface sherds (or 87.2 sherds) and the 256 smooth surface sherds (or 146.8 sherds) would be expected to have grit temper by chance (Table 13.4). Given that 40.4% of the assemblage has shell temper ($n = 158$ sherds), we would expect that 40.4% of the stamped pottery (58.8 sherds) and the smooth surfaced pottery (99.2 sherds) would have shell temper (Table 13.4). These values reflect the expected values if there is no relationship between pottery surface texture and temper type. Again, intuitively there seems to be a big difference between the observed values (Table 13.3) and the expected values (Table 13.4), but we would naturally expect some variation between the observed and expected values. Are the differences between the observed and expected values so large that they indicate a non-random association between smooth surface pottery and grit temper and between stamped pottery and shell temper? A chi-square test, represented using the symbol χ^2, is ideal for answering this question. It does so by directly comparing the observed frequencies with the expected frequencies using Equation (13.1), where O is the observed frequency and E is the expected frequency of an

Table 13.3 An additional set of hypothetical frequencies of sherds classified by their surface treatment and temper type

Surface texture	Grit temper	Shell temper	Total
Stamped surface	12	134	146
Smooth surface	222	24	246
Total	234	158	392

Table 13.4 Expected frequencies for the sherds in Table 13.3 assuming that there is no association between surface treatment and temper type

Surface texture	Grit temper	Shell temper	Total
Stamped surface	87.2	58.8	146
Smooth surface	146.8	99.2	246
Total	234.0	158.0	392

association. Note that a chi-square value is calculated for each cell in a table like Table 13.3, but that the chi-square test will reflect the difference between all of the observed and expected values in the matrix.

The chi-square test

$$\chi^2 = \sum \frac{(O_{ij} - E_{ij})^2}{E_{ij}} \tag{13.1}$$

Table 13.5 reports the chi-square test derived from the observed values (Table 13.3) and the expected values (Table 13.4) in our example. This χ^2 value can be compared to critical values from the chi-square distribution to determine if the differences are greater than expected by chance. The formal hypothesis being tested is $H_0 : O_{ij} = E_{ij}$, where O_{ij} reflects the observed values and E_{ij} reflects the expected values. The chi-square test uses the chi-square distribution, which is a unimodal distribution heavily skewed to the right (Appendix D). Like the t and the F distributions, the chi-square distribution changes shape depending on the degrees of freedom, but the degrees of freedom for this distribution are based on the number of character states for each variable, not actual sample size. This is best understood by thinking about the data in the form of a data matrix.

Table 13.3 is a matrix with four cells that each reflects the frequency of the co-occurrence of the variable states represented as rows (surface texture) and columns (temper type). Temper type has two character states (grit temper and shell temper), which results in two columns. Surface texture likewise has two character states (stamped and smooth surfaces), resulting in two rows. This creates a 2×2 data matrix, regardless of the actual number of sherds in each cell (e.g., Tables 13.2 and 13.3 both are 2×2 data matrixes despite the differences in total sherds reflected in the table). The degrees of freedom are calculated based on the size of the data matrix as illustrated in Equation (13.2), where R is the number of rows in a matrix and C is the number of columns. In the example above (Table 13.3), there are two columns and two rows. The degrees of freedom consequently are $(2 - 1) \times (2 - 1) = 1$. This is true for all 2×2 data matrixes, regardless of the number of observed values.

Table 13.5 Chi-square test of the data presented in Table 13.3

Texture/Temper	Observed	Expected	$(O - E)$	$(O - E)^2$	$(O - E)^2/E$
Stamped grit	12	87.2	−75.2	5647.52	64.80
Smooth grit	222	146.8	75.2	5647.52	38.46
Stamped shell	134	58.8	75.2	5647.52	95.96
Smooth shell	24	99.2	−75.2	5647.52	56.96
Total	392	392.0	0	22590.08	$\chi^2 = 256.18$

Degrees of freedom for the chi-square test

$$v = (R-1) \times (C-1) \tag{13.2}$$

Using the degrees of freedom, the critical value for the chi-square test can be determined from Appendix D. The critical value for $X^2_{.05[1]} = 3.84$, which is much smaller than the calculated chi-square value of 256.18. We therefore reject the null hypothesis and conclude that the observed sherd frequencies are different than those expected by chance if there was no association between surface texture and temper type. This of course means that certain combinations (smooth surfaced grit tempered sherds and stamped shell tempered sherds) are more frequent than expected, which is why the chi-square test is considered a "test of abundance". Once the analyst knows which cells have more than expected and which have fewer, the significance of these differences can be understood based on the analytic design and theoretical structure a researcher employs.

Let's take a moment to think about a couple of characteristics of the chi-square test. First, the chi-square test uses nominal scale data. The data are divided into categories, and the counts within each cell of the matrix are compared to expected values. There is no assumption about the differences between the various categories, beyond simply that they are mutually exclusive. Even when the data are measured at the ordinal scale (e.g., small, medium, large), the chi-square test does not assume any order or ranking in the categories. Switching the order of ordinal data in a matrix so that small follows large instead of medium will have no impact on the chi-square analysis. Further, chi-square analysis of data measured at the interval or ratio scales is only possible if these data are divided into distinct categories.

Second, the chi-square test does not test to determine if there are differences in the frequencies of two character states. Rather, it is a means of determining if the distributions of the various character states are random with respect to (i.e., independent of) the other variable. This is unfortunately an easily confused point, and it is easy to misstate what the results of a chi-square test indicate. To understand this point, consider that Table 13.6 has many more smooth surfaced sherds than stamped sherds, but that the distribution of grit tempered and shell tempered sherds is the same as reflected in Table 13.1 (i.e., 50% of the sherds from each texture category are grit tempered and 50% are shell tempered). Performing a chi-square test on Table 13.6 will not lead to the rejection of the null hypothesis $H_0 : O_{ij} = E_{ij}$,

Table 13.6 A matrix of sherd frequencies in which the number of smooth surfaced sherds greatly outnumber the stamped surfaced sherds

Surface texture	Grit temper	Shell temper	Total
Stamped surface	25	25	50
Smooth surface	250	250	500
Total	275	275	550

Table 13.7 Chi-square test of the data presented in Table 13.6

Texture/Temper	Observed	Expected	$(O - E)$	$(O - E)^2/E$
Stamped grit	25	25	0	0
Smooth grit	250	250	0	0
Stamped shell	25	25	0	0
Smooth shell	250	250	0	0
Total	550	550		$\chi^2 = 0$

despite the large differences in the number of smooth surfaced and stamped sherds (Table 13.7). The chi-square test is designed to take as a given that there will be differences in the frequencies of the various character states, but will determine if the various categories are distributed evenly given these differences.

Finally, it is possible to calculate the expected values as we intuitively did for Tables 13.5 and 13.6, but Equation (13.3) provides an easier way to do so, especially when dealing with large matrixes.

The expected values for the chi-square test

$$E = \frac{RT \times CT}{GT} \tag{13.3}$$

The equation provides the expected value for each cell of the matrix based on the cell's row total (RT) and column total (CT), and the matrix's grand total (GT). To illustrate this, consider the cell for stamped, grit tempered pottery in Table 13.3. The expected value for this cell can be calculated by taking the sum of all of the cells in the first row (the cells reflecting sherds with stamped surfaces, which equals 146), the sum of all of the cells in the first column (grit tempered sherds, which equals 234), and the sum of all of the sherds from all of the cells (total sample size, which is 392 sherds). These values are then placed into Equation (13.3) as follows:

$$E = \frac{RT \times CT}{GT} = \frac{146 \times 234}{392} = 87.2 \text{ sherds}$$

This is exactly the same value we determined in Table 13.4 using the intuitive method that the cell should hold 59.6% of the stamped surface sherds. Using Equation (13.3) is frequently easier, however, especially when analyzing larger data sets using statistical software or spreadsheets, because it allows the expected value for each cell to be calculated without first calculating the various percentages associated with each column. This eliminates a step when determining the expected values, thereby saving time and helping to eliminate simple calculation mistakes.

Determining the Source of Variation in a Chi-Square Matrix

As previously mentioned, the chi-square test reflects the combined total of all of the chi-square values created by comparing the observed and expected values. Rejecting the null hypothesis indicates that at least some of the expected and observed values are significantly different, but it doesn't specify exactly which ones these are. In Table 13.5 it is fairly obvious which cells have more and which have fewer sherds than expected, but this often is not the case, especially when dealing with large matrixes. Researchers will need to use a test of association to determine which cells contribute significantly to the significant χ^2 value when it isn't obvious. Consider Table 13.8, which presents the frequencies of flaked stone artifacts recovered from three room blocks and a plaza area at Galeana, a large pueblo-like settlement in northwestern Mexico, during our initial survey of the site. We are interested in determining whether there are any differences in the frequencies of flaked stone artifacts grouped by their raw material, which could in turn reflect possible differences in activities performed at each location.

A quick examination of the table indicates that there are differences in the frequencies of flaked stone artifacts. For example, 39 chalcedony artifacts were recovered from Room Block 2 but only 13 were recovered from the plaza, despite the fact that the plaza produced more flaked stone artifacts. Are this and other differences in frequencies greater than expected given chance and the general differences in flaked stone frequency from each provenience? We can evaluate this hypothesis of interest using the chi-square test as illustrated in Table 13.8. We determined the expected values using Equation (13.3) and the chi-square value using Equation (13.1). The level of significance is set at .05. The chi-square value from the test is 49.10, which exceeds the critical value of 21.03 for 12 degrees of freedom (Table 13.9).

We reject the null hypothesis and conclude that at least some of the raw materials are present in different relative frequencies at each provenience. But which ones? The matrix has 20 cells, any or all of which could be reflecting a significant difference between the expected and observed values. Is there more chert than expected at Room Block 2? Well, yes; 86 chert artifacts were recovered but only 80.7 were

Table 13.8 Frequencies of flaked stone artifacts grouped by provenience and raw material

Raw material	Room Block 2	Room Block 3	Room Block 4	Plaza	Total
Chert	86	21	38	97	242
Chalcedony	39	10	12	13	74
Obsidian	4	3	3	8	18
Quartzite	16	8	7	6	37
Igneous	217	63	81	353	714
Total	362	105	141	477	1085

Table 13.9 Chi-square test comparing the frequencies of lithic raw materials from four areas of Galeana, Chihuahua, Mexico

Provenience	Raw material	Observed	Expected	$(O - E)^2/E$
Room Block 2	Chert	86	80.7	0.34
	Chalcedony	39	24.7	8.29
	Obsidian	4	6.0	0.67
	Quartzite	16	12.3	1.08
	Igneous	217	238.2	1.89
Room Block 3	Chert	21	23.4	0.25
	Chalcedony	10	7.2	1.13
	Obsidian	3	1.7	0.91
	Quartzite	8	3.6	5.45
	Igneous	63	69.1	0.54
Room Block 4	Chert	38	31.4	1.36
	Chalcedony	12	9.6	0.59
	Obsidian	3	2.3	0.19
	Quartzite	7	4.8	1.00
	Igneous	81	92.8	1.50
Plaza	Chert	97	106.4	0.83
	Chalcedony	13	32.5	11.73
	Obsidian	8	7.9	0.00
	Quartzite	6	16.3	6.48
	Igneous	353	313.9	4.87

$$X^2 = 49.10$$
$$\chi^2_{.05[12]} = 21.03$$

expected. Does this difference reflect a statistically significant difference between the expected and observed values? We can't tell by looking at Table 13.9. All of the observed and expected values differ somewhat, some by very little and some by a bit more. Which of these differences are significant?

We could reason (correctly) that the largest differences are likely responsible for the statistical difference reflected in the χ^2 value. After all, the largest differences between the observed and expected values do produce the largest χ^2 values, which in turn result in a larger cumulative χ^2 value for the chi-square test. For example, the difference between the observed and expected frequencies of chert at Room Block 2 produced a χ^2 value of only .34, which seems small when compared to the χ^2 values associated with chalcedony artifacts from the plaza (11.73) or quartzite from Room Block 3 (5.45). However, we don't really know at what point the differences shift from being small to significant. We could guess (e.g., everything above 2 is considered significant), but that seems too arbitrary.

Instead of guessing, we would ideally like to be able to calculate some sort of standardized measure of variation for each cell's χ^2 value, which in turn would allow us to determine whether the observed and expected values differed significantly at some significance level. Fortunately, and as you have probably already surmised, there is such a measure. It is called the chi-square residual (Equation (13.4)).

The chi-square residual

$$e_{ij} = (O_{ij} - E_{ij})/\sqrt{E_{ij}}$$
(13.4)

The chi-square residual is the mathematical equivalent of the square root of the chi-square value with the exception that it can be either positive or negative. Positive values indicate that the observed frequency was greater than expected whereas negative values indicate there were fewer than expected. Once e_{ij} has been calculated, it can be compared to critical values from the standardized normal distribution (e.g., ± 1.96 for $\alpha = .05$; see Appendix A). A residual greater than the critical value (e.g., $e_{ij} > 1.96$) or less than the negative version of the critical value (e.g., $e_{ij} < -1.96$) indicates a significant difference. Unfortunately, the chi-square residual has been demonstrated to be biased such that it tends to underestimate the significance of differences for small samples (i.e., it is prone to Type II errors in which differences that are real aren't identified). As a result, it is best to calculate the adjusted residual using Equation (13.5). CT, RT, and GT again stand for the column total, row total, and grand total for a given value in the matrix.

The adjusted chi-square residual

$$d_{ij} = \frac{e_{ij}}{\sqrt{\left(1 - \dfrac{CT}{GT}\right)\left(1 - \dfrac{RT}{GT}\right)}}$$
(13.5)

To illustrate the calculation of the adjusted residual, consider Table 13.10, which presents the observed and expected values from the Galeana survey. The residual for the first cell ($i = 1, j = 1$) in the matrix is calculated as:

$$e_{1,1} = \frac{O_{1,1} - E_{1,1}}{\sqrt{E_{1,1}}} = \frac{86 - 80.7}{\sqrt{80.7}} = \frac{5.3}{8.98} = .59$$

The adjusted residual is calculated as:

$$d_{1,1} = \frac{e_{1,1}}{\sqrt{\left(1 - \dfrac{CT}{GT}\right)\left(1 - \dfrac{RT}{GT}\right)}} = \frac{.59}{\sqrt{\left(1 - \dfrac{362}{1085}\right)\left(1 - \dfrac{242}{1085}\right)}} = \frac{.59}{\sqrt{.66 \times .78}} = .81$$

Because .81 is not larger than 1.96 or smaller than -1.96 (the critical Z-values for $\alpha = .05$), we conclude that the difference between the observed and expected frequencies of chert artifacts from Room Block 2 is not significant.

Table 13.10 Observed and expected values for the Galeana data

Raw material	Room Block 2	Room Block 3	Room Block 4	Plaza	Total observed
Chert	$O = 86$	$O = 21$	$O = 38$	$O = 97$	242
	$E = 80.7$	$E = 23.4$	$E = 31.4$	$E = 106.4$	
Chalcedony	$O = 39$	$O = 10$	$O = 12$	$O = 13$	74
	$E = 24.7$	$E = 7.2$	$E = 9.6$	$E = 32.5$	
Obsidian	$O = 4$	$O = 3$	$O = 3$	$O = 8$	18
	$E = 6.0$	$E = 1.7$	$E = 2.3$	$E = 7.9$	
Quartzite	$O = 16$	$O = 8$	$O = 7$	$O = 6$	37
	$E = 12.3$	$E = 3.6$	$E = 4.8$	$E = 16.3$	
Igneous	$O = 217$	$O = 63$	$O = 81$	$O = 353$	714
	$E = 238.2$	$E = 69.1$	$E = 92.8$	$E = 313.9$	
Total observed	362	105	141	466	1085

The adjusted residuals for the rest of the cells in Table 13.10 can be calculated in the same way. Table 13.11 presents the chi-square test previously reported in Table 13.9 with the addition of the adjusted residuals. Bolded values indicate adjusted residuals corresponding with differences significant at $\alpha = .05$. Using the adjusted residuals, we now can conclude that more chalcedony artifacts were recovered from Room Block 2, more quartzite artifacts from Room Block 3, and more crystalline ingenious artifacts from the plaza than expected. We likewise collected fewer igneous artifacts from Room Blocks 2 and 4, and fewer chalcedony and quartzite artifacts from the plaza than expected. Based on the differences among the raw materials, we can further conclude that the materials used in the plaza tended to be more crystalline (the igneous material), which produce more durable tools with dull edges when compared to the other raw materials that are more common in the room blocks. Furthermore, differences between the room blocks might provide additional information about changes in raw material selection through time, differences in the use of raw materials, and differences in access to raw materials among the inhabitants of Galeana. These differences then could (and did) help frame our research questions and field research (VanPool *et al.*, 2000).

We note that although the adjusted residual is very useful when the chi-square test results in the rejection of the null hypothesis, it should not be used when a significant difference is not present (i.e., when the null hypothesis is not rejected). When χ^2 is not significant, we can conclude that *none* of the differences between the expected and observed values in the table were significantly larger than expected by chance. Using adjusted residuals to try to identify significant differences when they aren't present will lead to Type I errors and the heartbreak of faulty conclusions.

Table 13.11 Adjusted residuals for the chi-square analysis of flaked stone raw materials recovered from Galeana, Mexico

Provenience	Raw material	Observed	Expected	Chi-square values	Adjusted residuals
Room Block 2	Chert	86	80.7	0.34	0.81
	Chalcedony	39	24.7	8.29	**3.66**
	Obsidian	4	6.0	0.67	−1.01
	Quartzite	16	12.3	1.08	1.30
	Igneous	217	238.2	1.89	**−2.88**
Room Block 3	Chert	21	23.4	0.25	−0.60
	Chalcedony	10	7.2	1.13	1.16
	Obsidian	3	1.7	0.91	1.01
	Quartzite	8	3.6	5.45	**2.50**
	Igneous	63	69.1	0.54	−1.32
Room Block 4	Chert	38	31.4	1.36	1.42
	Chalcedony	12	9.6	0.59	0.85
	Obsidian	3	2.3	0.19	0.47
	Quartzite	7	4.8	1.00	1.09
	Igneous	81	92.8	1.50	**−2.24**
Plaza	Chert	97	106.4	0.83	−1.38
	Chalcedony	13	32.5	11.73	**−4.74**
	Obsidian	8	7.9	0.00	0.04
	Quartzite	6	16.3	6.48	**−3.46**
	Igneous	353	313.9	4.87	**5.04**

$$X^2 = 49.10$$
$$\chi^2_{.05[12]} = 21.03$$

Assumptions of Chi-Square Analysis

The chi-square test is extremely powerful and can be very useful to archaeologists, who frequently deal with counts of artifacts, features, and other materials. It does have assumptions that must be met before it can be used, however. Two of these assumptions are obvious. First, the chi-square test requires that data be organized at the nominal or ordinal scale. Although ordinal and interval level data can be transformed into nominal scale data, it is seldom worthwhile to do so. Very powerful statistical tools such as the t-test, ANOVA, and correlation are available for use with these data. Second, the data must be independent of each other. This is true for the categories within each variable and for the variables themselves. For example, it would be improper in the Galeana example presented above to compare the frequency of crystalline, cryptocrystalline, and noncrystalline artifacts by raw material. The coarseness of flaked stone artifacts is a direct product of the stone's raw material. These variables are not independent, and are therefore inappropriate for the chi-square test.

The third assumption reflects the nature of the chi-square distribution, upon which the chi-square test is built. The shape of the distribution changes according to the degrees of freedom, but it can also change a bit with sample size. The distribution is stable (i.e., doesn't change much) for large samples, but statisticians realize that with a small enough sample size, the chi-square distribution will lose its characteristic shape and would begin to resemble a normal distribution. When this happens, the values from the χ^2 distribution will no longer accurately measure the probability space, and the results of a chi-square test will be spurious. The point at which this happened was unknown when the chi-square test came into common usage but a commonly accepted convention, called the rule of five, was developed that guaranteed the validity of the chi-square test.

The rule of five holds that at least 80% of the expected frequencies must be five or more before chi-square analysis is appropriate. This rule was agreed upon as a "safe bet" based on the fact that statisticians were sure that the chi-square test would be applicable under these conditions (Cochran 1954). It was widely accepted in the 1950s and has been enshrined in statistics books ever since. However, it is unnecessarily conservative. More recent research such as Larntz (1978), Lewontin and Felsenstein (1965), Roscoe and Byars (1971), Slakter (1966), and Yarnold (1970) have demonstrated that the chi-square test is generally applicable even if a significant proportion of the expected values are less than five (see also Everitt, 1992: 13–14, 39). Lewontin and Felsenstein (1965: 31) in fact argued based on the results of a simulation consisting of 1,500 chi-square values derived using randomly created data that the chi-square statistic will be correctly distributed as long as all of the expected values are one or greater, and, by extension, that the chi-square test will produce valid statistical results with these small samples (see also Roscoe and Byars, 1971: 758). However, traditions die hard. Most archaeologists (and others) continue to apply the rule of five, and look very skeptically at chi-square tests applied to samples with small expected values.

It is reasonable to question whether small samples do in fact accurately reflect the parent archaeological population, but the research noted above indicates that the chi-square test is as appropriate for these small samples as are the other statistical methods such as Fisher's exact test that are proposed as alternatives. We therefore note that you will wish to seriously consider whether a small sample resulting in many expected values less than five accurately reflects the parent population. If you decide that it does and if the expected values are all above one, then the chi-square test is likely as good as any of the commonly used alternatives. However, we present the Fisher's exact test and the Yate's continuity correction below, which are the most common alternatives used for small samples in archaeological contexts. You may use either of these instead.

If you don't feel comfortable with your small expected frequencies, there are several ways that you may be able to improve them. The first and perhaps easiest way is to eliminate categories represented by small samples. For example, when analyzing pottery frequencies of sites where a particular trade ware occurs infrequently, it might be worthwhile to eliminate it from the analysis so long as this doesn't undermine the hypothesis of interest. A second possibility is to collapse categories. For example, instead of having several different trade wares that occur in small quantities, it might be better to collapse them into a single category. This would increase their frequencies and would thereby increase the expected values. Again, care must be taken to ensure that this doesn't undermine the hypothesis of interest. Finally, increasing the sample size through additional observations will increase the frequencies of even rare categories. It may not always be possible to increase the sample size given the realities of the archaeological record and funding constraints, but this is often the best solution. After all, as sample size increases, so does our certainty that our sample accurately reflects the population we are trying to study. Increasing sample size thus has the benefit of improving both the strength of the statistical test and our confidence in the conclusions derived through the test.

The Analysis of Small Samples Using Fisher's Exact Test and Yate's Continuity Correction

The most commonly used alternative for the chi-square test when dealing with small sample sizes is Fisher's exact test. It is not based on the chi-square distribution, and is in truth one of the least intuitive tests we discuss in this text. We won't bother trying to explain the mathematics underlying it, but it is calculated using Equation (13.6) when dealing with a 2 × 2 table. Although statistical packages such as SAS can calculate a Fisher's exact probability for large matrixes, the equation becomes prohibitively complex to calculate by hand (Everitt, 1992: 39).

Fisher's exact probability for 2 × 2 tables

$$P = \frac{(a+b)!(a+c)!(c+d)!(b+d)!}{a!b!c!d!N!} \qquad (13.6)$$

The terms a, b, c, and d reflect the counts in each of the four cells when they are arranged as illustrated in Table 13.12. The numerator of Equation (13.6) can be more simply understood as the factorials for the two row totals and the two column totals. N in the denominator stands for the sum of all cells, which is of course the total sample size.

Table 13.12 Matrix explaining the symbolism for Fisher's exact probability test

	Character state A	Character state B	Total
Character state 1	a	b	$a + b$
Character state 2	c	d	$c + d$
Total	$a + c$	$b + d$	N

Let us provide an archaeological example to illustrate the application of Fisher's exact probability test. Alzualde *et al.* (2007) provide an analysis of status and ancestry reflected in the Aldaieta cemetery, which was used in the Basque region of Spain during the sixth and seventh century AD. They are interested in understanding the impact that the larger and more complex groups in post-Roman Europe had on this genetically distinct population. Part of their analysis focuses on the difference in grave goods between males and females. Table 13.13 reports the frequency of males ($n = 31$) and females ($n = 12$) associated with weapons and domestic utensils. Do males and females differ in their association with these goods? A Fisher's exact probability can be used to evaluate this hypothesis of interest.

Unlike most of the other methods presented here, the Fisher's exact probability test does not provide a value that is matched with an underlying distribution to determine a probability. Instead, it provides an actual probability value that can be directly compared to α. A consequence of this is that the null hypothesis can be directly stated in terms of the hypothesis of interest, without referencing a specific statistical term. The null hypothesis is thus H_0 : men = women. Applying Equation (13.6) to the data from Table 13.13 produces the following result.

$$P = \frac{39! \times 27! \times 9! \times 21!}{24! \times 15! \times 3! \times 6! \times 48!} = .09$$

Given that .09 is greater than .05, we fail to reject the null hypothesis, and conclude that despite the absolute differences in the quantities of grave goods, weapons and utilitarian utensils such as serving dishes are not differentially distributed according to gender in this sample. A fewer number of women were buried with grave goods, but the proportions of weapons and utilitarian goods among those with grave goods was similar to the burials of men. (The same conclusion would have been reached using a chi-square test, even though the expected values violate the rule of five discussed above.)

There are a couple of issues to keep in mind when using the Fisher's exact probability test. First, unless you are working with a 2 × 2 table, you will need to rely on a statistical package to calculate the result. Second, Fisher's exact probability has been demonstrated to be an overly conservative test that does not measure probabil-

Table 13.13 Frequency of burials from Aldaieta cemetery, Spain, associated with weapons and utilitarian utensils

	Buried with weapons	Buried with domestic, utilitarian utensils	Total
Men	24	15	39
Women	3	6	9
Total	27	21	48

ity at the stated α level and is prone to Type II errors when dealing with small sample sizes (which is of course when archaeologists are most likely to use it). This means that the test will consistently fail to identify statistically significant differences at a given α level, when in truth they are present. Everitt (1992:18–19) reports that even a moderately large difference of 30% in the probability of having a given trait with a sample size of 20 in each category has a power of only .53. Thus, given repeated samples of 20 men and 20 women in which men were 30% more likely to have weapons than women, we would still fail to detect the real difference in burial good associations nearly half of the time using a Fisher's exact probability test. To be honest, we suspect that we likely committed a Type II error when evaluating the data in Table 13.13. Roughly 75% of the men but only 25% of the women are associated with weapons, whereas both 50% of the men and 50% of the women are associated with utilitarian utensils. This difference strikes us as likely reflecting a differential association between gender and the presence of weapons, even though it is not statistically significant. Increased sample size or comparisons with other, similar burial sets would be needed to determine if the failure to reject the null hypothesis really is an error. Third, the Fisher's exact probability as presented above is a one-tailed test that reflects the probability of having differences equal to or greater than those observed among the cells. It is possible to use a two-tailed version would include the probability of having excessive uniformity in the cells as well. In such cases, the area of rejection will be split evenly between both ends of the distribution (e.g., the regions of rejection for $a = .05$ will be defined as <.025, which marks cases where the differences among the categories are greater than expected by chance, and >.975, which marks cases where the differences are less than expected by chance). Although potentially useful in some contexts, this is incidental to its typical use to evaluate archaeological hypotheses.

While some argue that the Fisher's exact probability test is better for small sample sizes than the chi-square test, it is an imperfect tool at best (Everitt, 1992: 18–19). Further, it is difficult to imagine very many archaeological contexts in which small samples really provide a good understanding of the underlying phenomenon. Ultimately, there is nothing wrong with applying the Fisher's exact probability test (or the chi-square test for that matter) to relatively small samples, but we encourage you to carefully consider whether you really should conduct any statistical analysis at all if the sample is that small. If you do perform the analysis, keep in mind the high probability of incorrectly failing to reject the null hypothesis.

You can also choose to use Yate's continuity correction (also called Yate's correction for continuity) for the chi-square test when evaluating 2×2 matrixes with small sample sizes. Yate's continuity correction modifies the chi-square test slightly in order to correct for a problem caused by using the continuous chi-square distribution to approximate the probability of the discrete number of observed frequencies (Everitt, 1992: 13; see also Sokal and Rahlf, 1995: 695 for a discussion of the underlying disjuncture between the "chi-square test" and the actual chi-square distribution). This creates a slight difference in the actual and calculated probabilities that is generally inconsequential for large samples and large matrixes, but can be significant for small samples in a 2×2 matrix. This difference can be corrected by subtracting .5 from positive values for $O_{ij} - E_{ij}$ and adding .5 to negative values for $O_{ij} - E_{ij}$. This is actually easier than it sounds, in that one simply uses Equation (13.7) instead of Equation (13.1) to determine the chi-square value for each cell of the matrix. The "brackets" around $|O_{ij} - E_{ij}|$ stand for "the absolute value", which means to change the value into a positive value even if it would otherwise be negative (e.g., $|2 - 4| = 2$, not -2).

Yate's continuity correction

$$\chi^2 = \sum \frac{(|O_{ij} - E_{ij}| - 0.5)^2}{E_{ij}} \tag{13.7}$$

Statisticians are mixed about the utility of Yate's continuity correction (e.g., Everitt, 1992: 13–14, 17 recommends it strongly whereas Sokal and Rohlf, 1995: 737 suggest it is nearly always unnecessary). What is true is that the Yate's continuity correction frequently causes the results of the chi-square test with excessively small samples to be virtually identical to Fisher's exact probability when it is modified for a two-tailed test as Everitt (1992: 17) describes. This also means that Yate's continuity correction produces an overly conservative result that misstates the true α level and is prone to Type II errors.

Again, comparing frequencies with small samples is problematic, and you should carefully weigh your options. No one will raise an eyebrow if you use Fisher's exact probability test or Yate's continuity correction, but it is unclear that these are substantially superior to the standard chi-square test in many cases. We recommend that you carefully evaluate the implications of making Type I and Type II errors in your analysis. If making Type I errors is a more significant problem, then the Fisher's exact probability test or Yate's continuity correction are ideal. (Yate's continuity correction is typically applied to a 2×2 matrix whereas Fisher's exact probability can be calculated for any sized matrix, with the help of a good statistical package.) If making a Type II error is more significant, then using the chi-square test when all of the expected values are greater than one is the more defensible decision.

The Median Test

In Chapter 10, we introduced ANOVA, a very powerful method for comparing the central tendencies of multiple data sets. While useful for archaeologists, it does have the severe drawback of being a *parametric* test, which requires each sample to be normally distributed with roughly equal variances. (We will discuss the differences between parametric and nonparametric tests in detail in the next chapter.) Yet archaeological data such as measurements of length, width, thickness, and weight are frequently skewed, because of the presence of a few large (or small) artifacts, features, elements, or whatever we are measuring. Also sometimes archaeological samples are too small to be certain of the shape of the underlying distribution (e.g., it may be hard to assess whether a sample of the volume of 10 ceramic vessels is normally distributed). Using the chi-square distribution, it is possible to compute a nonparametric ANOVA utilizing the median values as measures of central tendency for distributions of ratio or interval data. The test is also often called the median test (e.g., Conover, 1980: 171), which is the term we adopt here. The null hypothesis for the comparison is $H_0 : M_1 = M_2 = ... = M_i$, where M (the Greek letter) is the population parameter for the median of each group. Unlike the mean, which is heavily impacted by outliers, the median is a robust measure of a distribution's central tendencies in skewed distributions, because it is impervious to the absolute size of outliers. The comparison of medians will therefore still reflect a reliable assessment of the similarity of the central tendencies within skewed distributions, whereas the comparison of means may not.

The medians of two or more groups can be compared using the chi-square test is a straightforward way. Start by determining the median of all of the data pooled together. This is the grand median. If the medians of each sample are the same, then the grand median will be close to or the same as the medians of the groups, causing roughly half of the variates in each group to be larger than and half to be smaller than the grand median. If the null hypothesis is not true, then one or more of the samples must contain values that are consistently larger or smaller than the grand median. Using a chi-square test, the presence of groups that tend to be smaller or larger than the grand median can be detected.

Consider the data in Table 13.14, which are the length measurements of four classes of arrow points from the Hohokam period occupation of Ventana Cave. We might be interested in determining if any of the classes tend to be longer than the others. We could accomplish this by comparing the means of each sample of points.

Table 13.14 Length measurements (cm) for four classes of arrowheads from Ventana Cave, Arizona

Side-notched	Corner-notched	Straight stemmed	Triangular
1.98	2.71	2.10	2.31
2.20	2.08	3.07	2.01
2.39	2.55	2.25	2.57
2.61	2.45	2.40	2.80
2.40	2.25	2.81	2.40
2.59	2.98	2.27	3.00
3.28	2.61	2.72	2.40
3.41		3.02	2.89
3.81		3.72	2.82
2.81			3.32
2.55			

Is the parametric form of ANOVA introduced in Chapter 10 appropriate for analyzing these data? A quick glance at a frequency table derived from the data (Table 13.15) indicates that it probably is not. At least two and perhaps three of the distributions appear to be skewed (side-notched, corner-notched, and, perhaps, triangular-shaped points). Though ANOVA is probably not appropriate for these data, the median remains an excellent measure of the distributions' central tendencies. The median test is consequently a better choice to evaluate the hypothesis of interest.

To evaluate the hypothesis, we define the null hypothesis as $H_0 : M_{sn} = M_{cn} = M_{stemmed} = M_{tri}$, and set the level of rejection at .05. The first step in calculating the median test is to determine the grand median of all 37 variates in Table 13.14, which is 2.59 cm, the 19th largest value. The next step is to determine how many variates in each group are larger and how many are smaller than 2.59 cm. The magnitude of the difference is unimportant; only the direction matters. Table 13.16 provides this information. Note that the number of plus and minus signs must be (roughly) equal through the entire matrix, because there must be an equal number of variates above and below the grand median by definition. The variate(s) that directly correspond with the median are excluded.

Table 13.17 presents the counts for each class. If the null hypothesis is true and the medians are equal, than each class should have an equal number of variates greater than and less than the group median. This proposition can be evaluated with a chi-square test (Table 13.18). The chi-square value of .025 is not greater than the critical value of 7.81 for $\alpha = .05$ and three degrees of freedom. As a result we cannot reject the null hypothesis and we conclude that there are no significant differences in the median point lengths among the four groups. We would have reached the same conclusion using an ANOVA analysis, but we are certain that we are not violating any critical assumptions by using the median test here.

The median test does not make any assumptions about the shape of the underlying distribution, but it does require interval or ratio level data, which are necessary

Table 13.15 Frequency distributions of the Ventana Cave projectile points

Class mark (cm)	Side-notched	Corner-notched	Straight stemmed	Triangular
2.0	I	I		I
2.2	I	I	III	
2.4	II	I	I	III
2.6	III	II		I
2.8		I	II	III
3.0		I	II	I
3.2	I			
3.4	I			I
3.6				
3.8	I		I	

Table 13.16 The direction of difference between each variate listed in Table 13.14 and the grand median ("+" indicates that the variate is greater than the grand median, whereas "−" indicates it is smaller)

Side-notched	Corner-notched	Straight stemmed	Triangular
−	+	−	−
−	−	+	−
−	−	−	−
+	−	−	+
−	−	+	−
NA	+	−	+
+	+	+	−
+		+	+
+		+	+
+			+
−			

Table 13.17 Number of variates greater than and less than the median

	Side-notched	Corner-notched	Straight stemmed	Triangular
Great than median	5	3	5	5
Less than median	5	4	4	5

Table 13.18 Chi-square test comparing the frequencies of observed and expected values of arrow points that are greater than and less than the grand median

Direction of difference	Point class	Observed	Expected	Chi-square value
Larger	Side-notched	5	5	0.00
	Corner-notched	3	3.5	0.07
	Straight stemmed	5	4.5	0.06
	Triangular	5	5	0.00
Smaller	Side-notched	5	5	0.00
	Corner-notched	4	3.5	0.07
	Straight stemmed	4	4.5	0.06
	Triangular	5	5	0.00

$$X^2 = 0.25$$
$$\chi^2_{.05[3]} = 7.81$$

to calculate medians. When dealing with ordinal scale data, the chi-square test can be used to compare the frequencies of the size categories in each group, but this is not a test of $H_0 : M_1 = M_2 = M_3 = M_i$ and is not a median test. It is more properly considered a special application of the chi-square test.

Let's evaluate another projectile point assemblage, this time from Paleolithic sites in Europe. Knecht (1991) analyzes bone and antler projectile points dating to the Upper Paleolithic occupation of France, Belgium, and Germany to study technological innovation. Table 13.19 presents the maximum length of a sample of 11 points from four projectile types. As was the case for the Ventana Cave projectile points, the distributions of at least two of these types (split based points and Gravettian single-beveled points) appear to be skewed (Table 13.20). As a result, the median test is ideal for comparing the central tendencies of the distributions. The null hypothesis is $H_0 : M_{split} = M_{simple} = M_{single-beveled} = M_{lateral-beveled}$, and $\alpha = .05$.

The grand median for Table 13.19 is 91.5 mm. The frequency of variates greater than and less than the grand median is presented in Table 13.21. The chi-square test comparing these frequencies is presented in Table 13.22. Unlike the Ventana Cave example, the chi-square value of 10.55 exceeds the critical value of 7.81. We

Table 13.19 Maximum length (mm) of Upper Paleolithic bone points

Split-based points	Losange-shaped simple-based points	Gravettian single-beveled points	Gravettian lateral-beveled points
45	99	116	110
45	101	138	116
52	37	59	96
130	106	51	130
39	96	135	91
125	117	58	98
91	92	108	106
59	148	86	68
76	137	90	66
29	69	67	36
56	159	33	161

Table 13.20 Frequency distribution of the maximum lengths of the Upper Paleolithic bone points

Class marks (mm)	Split-based points	Losange-shaped simple-based points	Gravettian single-beveled points	Gravettian lateral-beveled points
30	II	I	I	I
50	IIIII		III	
70	I	I	I	II
90	I	III	II	III
110		III	II	III
130	II	I	II	I
150		II		
170				I

Table 13.21   The number of Upper Paleolithic points greater than and less than the grand median

	Split-based points	Losange-shaped simple-based points	Gravettian single-beveled points	Gravettian lateral-beveled points
Greater than median	2	9	4	7
Less than median	9	2	7	4

Table 13.22   Chi-square test comparing the frequencies of observed and expected values of Upper Paleolithic bone points that are greater than and less than the grand median

Direction of difference	Point type	Observed	Expected	Chi-square Value	Adj. residuals
Greater than median	Split-based points	2	5.5	2.23	**−2.44**
	Simple-based points	9	5.5	2.23	**2.44**
	Single-beveled points	4	5.5	0.41	−1.04
	Lateral-beveled points	7	5.5	0.41	1.04
Less than median	Split-based points	9	5.5	2.23	**2.44**
	Simple-based points	2	5.5	2.23	**−2.44**
	Single-beveled points	7	5.5	0.41	1.04
	Lateral-beveled points	4	5.5	0.41	−1.04
				$X^2 = 10.55$	
				$\chi^2_{.05[3]} = 7.81$	

therefore reject the null hypothesis, and conclude that there are significant differences among the medians. The adjusted residuals indicate that simple based points tend to be larger than the grand median and the split based points tend to be smaller than the grand median. This of course indicates that the medians for these groups are different. The significance of this finding can be derived using an appropriate analytic and theoretical framework.

Now that you have added the analysis of frequencies to your quantitative tool kit, you have mastered the basics of archaeological quantification. There are a wide variety of additional quantitative methods that are used on occasion, but most of them are some variant or sibling of the methods outlined here and in the preceding 12 chapters. With your knowledge, you should be able to master them easily. In the next chapter, we will introduce some of these related methods that are (or should be) commonly used in archaeological contexts.

Practice Exercises

1   What are the assumptions of the chi-square test? What are various solutions and their relative advantages/disadvantages to potential problems associated with small expected frequencies? What statistical method(s) should archaeologists use when encountering a 2×2 data matrix with a small sample size?

2   Abbott *et al.* (2006) report the following data concerning temper and pottery types from various settlements. Perform a chi-square analysis to determine if there is an association between the two variables ($\alpha = .10$). If so, use the adjusted residuals to determine which associations deviate significantly from the expected frequencies.

Assemblage	Temper type		
	South Mt. Granodiorite	Estrella Gneiss	Mica Schist
Las Canopas Plain ware	244	19	5
Pueblo Viejo Plain ware	429	14	17
Pueblo Viejo Red-slipped ware	171	9	8
Farmstead Plain Ware	62	64	2

3   Below is the sex distribution of two samples of adult burials of the Reigh site collection from the Great Lakes Region studied by researchers during different studies (Pfeiffer, 1977: 105). Subsequent researchers evaluating and comparing the results might wish to determine if the differences in the proportions of each sex are statistically significant, which might impact their comparability if there are sex-based differences.

   (a)   Use the chi-square test to evaluate whether there is a significant difference in the sex distribution considered by the two researchers ($\alpha = .01$).

   (b)   Use the Yate's continuity correction for the chi-square test. Did it significantly impact the results?

(c) What are the advantages and drawbacks of using or not using the Yate's continuity correction?

Researcher	Males	Females
Hsu	26	7
Pfeiffer	13	16

4 Use the Fisher's exact probability test to evaluate whether there are significant differences in hand placement between the Early and Middle Period burials at St. Mary's City Cemetery (Riordan, 2009: 90).

Period	Pelvis	Side
Early	6	5
Middle	8	1

5 Data regarding the number of dog burials oriented in different directions is presented below.

(a) Evaluate whether there are significant differences between the observed and expected frequencies using a chi-square analysis ($\alpha = .05$).

(b) Reevaluate the null hypothesis using Yate's continuity correction for the chi-square analysis. Did your conclusion change?

(c) Use a Fisher's exact probability test to test for differences. Did your conclusion change?

(d) Which of the three tests is/are preferable in this situation? Why? Defend your answer to the best of your ability.

Site	Oriented N–S	Oriented E–W
Perry site	3	7
Ferb site	5	3

6 Following is data reporting the diameters (cm) for postholes from three structures, one with two occupations. Perform a median test to evaluate whether the medians of the four samples are different ($\alpha = .01$). If so, which are different?

House 1A	House 1B	House 2	House 3
12.3	11.4	12.9	12.5
10.5	12.1	14.6	9.8
11.0	10.8	17.4	13.1
12.6	12.2	13.3	11.1
10.9	10.7	15.7	10.3
10.8	9.9	15.2	10.7
11.3	12.4	12.6	11.0
12.0	10.1	14.5	10.5
10.2	11.3	13.4	10.8
11.2	10.6	16.9	12.1

14

An Abbreviated Introduction to Nonparametric and Multivariate Analysis

Our previous discussion of the Spearman's correlation coefficient and median test introduced the idea of *nonparametric tests*, which are also called *distribution-free methods* (e.g., Gibbons, 1985: 3). These techniques do not require the data to conform to a given distribution, but instead are useful with a wide range of differently shaped distributions. Most of these approaches focus on rankings or some other method of indicating the relationship between variates instead of characterizing distributions using measures of central tendency and dispersion to estimate population parameters. As a result, these methods make fewer or no assumptions about the structure of the underlying data. Many of these nonparametric techniques are also relatively easy to compute, which may lead one to wonder why archaeologists and other researchers even bother with the parametric tests. The reason is that when the assumptions are met, parametric tests tend to provide more powerful tests in the sense of accurately measuring α and helping to reduce Type II errors (Sokal and Rohlf, 1995: 423). Further, most of the parametric tests are relatively robust to departures from the expected distribution (most commonly the normal distribution) so long as the shape is approximately right (distributions are symmetrical and unimodal without major outliers). Although skewed distributions are common in archaeology, so are normal distributions, especially when dealing with means because of the central-limit theorem. As a result, the parametric tests are often the better, and more familiar, analytic tools. Still, nonparametric methods are very useful when the assumptions for parametric methods clearly are not met, or when dealing with ordinal data, which cannot be analyzed using parametric methods. Here we introduce two commonly used nonparametric tests: the Wilcoxon two-sample test, which is a nonparametric alternative to the t-test; and the Kruskal–

Wallis nonparametric ANOVA analysis. Many other nonparametric tests exist (see Conover, 1980 and Gibbons, 1985 for detailed descriptions of many of them), but these two tests are commonly used by archaeologists, have proven to be robust and useful quantitative tools, and are comparatively easy to compute and interpret.

Nonparametric Tests Comparing Groups

Wilcoxon two-sample test

The Wilcoxon two-sample test is a nonparametric test for comparing two samples of ordinal, interval, or ratio scale data. Because it is nonparametric, it is used in archaeological contexts to compare samples when people are uncomfortable with the assumptions of a t-test because of a sample's skewedness or size. It can also be used when an order is known, but the exact difference among the variates is not (i.e., with ordinal data). The test's null hypothesis is formally $H_0 : R_1 = R_2$, where R is a sum of ranks for each sample. This null hypothesis is not intuitively meaningful to many people, causing some to state that the Wilcoxon two-sample test compares the medians of the two samples. This is unfortunately incorrect in that the Wilcoxon two-sample test does not employ *any* measure of central tendency. Instead, the sums of the rankings within each group are being compared.

While the hypothesis may be easily misunderstood, the idea underlying the Wilcoxon two-sample test is quite intuitive. Imagine you ranked the data in two samples from the smallest variate (ranked #1) to the largest variate (ranked #N) and then summed the rankings within each sample. Obviously, if one of the distributions is consistently larger than the other, then the sums of the ranks of the larger distribution will be much larger than the other, given that it will contain the largest variates with the highest ranks. Tremendously different sums in the ranks will indicate, then, that the two distributions likely do not represent the same parent population. If two distributions do reflect the same parent population, then neither of them should be consistently larger or smaller than the other, and the sum of their rankings should be roughly identical. Comparing the rankings between the distributions will therefore provide a means of determining if the distributions are more or less similar.

Consider the following data that Cory Hudson and Matt Boulanger collected on the friction coefficient of pottery with various surface treatments (Table 14.1). Archaeologists from around the world have proposed explanations for the presence and nature of surface treatment on pottery, including that the rougher surfaces of stamped or corrugated pottery help their users hold onto vessels without dropping them. Hudson and Boulanger (2007) wished to evaluate the plausibility of this hypothesis by determining if there is a meaningful difference in vessel wall friction caused by the surface treatments. To complete their analysis, they produced pottery test tiles with various surface treatments and then measured the friction coefficients

Table 14.1 Friction coefficients of test tiles with differing surface treatments

Smoothed	Rectangular dentate stamped
0.69	0.96
0.72	0.97
0.84	1.04
0.85	1.07
0.88	1.08
0.95	1.09
0.95	1.10
1.04	1.13
1.09	1.18
1.12	1.18

Table 14.2 Frequency distribution for the friction coefficients of pottery test tiles with different surface treatments

Class mark	Smoothed	Rectangular dentate stamped
0.65	I	
0.75	I	
0.85	III	
0.95	II	II
1.05	II	IIII
1.15	I	IIII

for the pottery. The data in Table 14.1 reflect the resulting friction coefficients for the smoothed surface pottery (the control sample) and rectangular dentate pottery, which has a rough surface caused by scoring with a sharp stick to form a series of close parallel lines. Here, our hypothesis of interest is whether there is a significant difference between the friction coefficients of the two surface treatments. This would be a perfect case for using a t-test, but the frequency distributions of the data presented in Table 14.2 suggest that rectangular dentate pottery might not be normally distributed. The distribution isn't too far from the expected shape, but the small sample size prevents us from being completely certain that it is normal. As a result, the Wilcoxon two-sample test is a reasonable choice to evaluate the hypothesis of interest.

If there is no difference in the friction coefficient for the two samples of test tiles, then the sums of their rankings should be roughly equal. If the surface treatment does produce a significant difference, then the variates of one of the distributions will be consistently larger, meaning that the sum of the rankings will be larger. The

U value that is calculated using the Wilcoxon two-sample test can be used to determine if the difference between the rankings of the two samples is significantly different.

The first step in calculating the Wilcoxon U value is to rank the data. An easy way to accomplish this is illustrated in Table 14.3, which is an ordered array of the data from Table 14.1 with the ranking of each variate indicated in parentheses. As before, ties are dealt with by averaging the ranks of the tied variates (e.g., two tied variates that would be ranked sixth and seventh are both ranked 6.5). Now sum the ranks in each column to determine R (Table 14.3). These values reflect the differences in rankings between each sample. For "Smoothed" surface treatments $R = 70$, and for "Rectangular dentate stamped" surface treatments $R = 140$. The Wilcoxon U value is calculated by taking the *smaller* of the R values and calculating U using Equation (14.1), where n is the sample size of the sample corresponding with the smaller R value. (We can compute a U value for both samples (i.e., calculate U_{Smoothed} and $U_{\text{Rectangular Dentate}}$), but only the smaller U statistic is important for evaluating the null hypothesis.)

Table 14.3 Ranking of the pottery test tile data

	Smoothed	Rectangular dentate
	0.69 (1)	
	0.72 (2)	
	0.84 (3)	
	0.85 (4)	
	0.88 (5)	
	0.95 (6.5)	
	0.95 (6.5)	
		0.96 (8)
		0.97 (9)
	1.04 (10.5)	
		1.04 (10.5)
		1.07 (12)
		1.08 (13)
	1.09 (14.5)	
		1.09 (14.5)
		1.10 (16)
	1.12 (17)	
		1.13 (18)
		1.18 (19.5)
		1.18 (19.5)
Summed rankings	70	140

The Wilcoxon two-sample test

$$U = R - \frac{n(n+1)}{2}$$ (14.1)

In this case, $n = 10$ and $R = 70$. Solving for U produces the following result:

$$U_{smoothed} = 70 - \frac{10(10+1)}{2} = 15$$

The U value literally represents the difference between the observed sum of rankings and the lowest possible sum of rankings for that variable. In other words, the lowest possible sum of rankings for the smoothed surface sherds is $(1 + 2 + 3 + \ldots + 10) = 55$, which is 15 less than 70. If there is no difference between the distributions, the smaller of the two R values should still be somewhat larger than the smallest possible R value (i.e., U should be comparatively large). If the distributions are different, then the smaller R should be pretty close to the lower limit of R, necessitating that U is a value somewhat close to 0.

The question then is whether $U = 15$ is close enough to 0 to indicate that the distribution of smoothed surfaced test tiles is different than the stamped test tiles. Unlike most quantitative methods where a null hypothesis is rejected if the computed value exceeds the critical value, *the null hypothesis for the U test is rejected if the value is smaller than the critical value.* Critical values for the U test when both samples sizes are 20 or less are presented in Appendix E. Because this is a nonparametric test, there are no degrees of freedom *per se*, but the probabilities do change according to the size of the samples. Here $U_{smoothed} = 15$ is *smaller* than the critical value of 23 associated with $n_1 = 10$ and $n_2 = 10$ for $\alpha = .05$, prompting us to **reject** the null hypothesis. By convention, n_1 is considered the sample used for determining the pertinent U value. We therefore conclude that the two samples are different with the rough surfaced rectangular dentate test tiles producing a higher friction coefficient. This in turn lends support to the suggestion that the pottery with a stamped surface was easier for people to hold without dropping, which may help explain the independent development of texturing around the necks and bodies of jars and bowls in early pottery traditions from around the world.

The values in Appendix E will be slightly conservative if there are ties in data set, meaning that the actual α level of the test will be slightly less than stated. Fortunately the impact of ties tends to be analytically unimportant unless there are a great many of them, which is unlikely if both sample sizes are equal to or less than 20. Still, if nearly all of your data reflect ties, then you might consider whether the data are measured with adequate precision and accuracy to identify any differences that might be present.

If either of the samples is larger than 20 (or if you don't have ready access to Appendix E), you can test the significance of the U test by converting U to a Z-score using Equation (14.2). You can then use Appendix A to directly determine a probability or to identify a critical value.

Approximating U *using the normal distribution*

$$Z_U = \frac{U - \left(\frac{n_1 n_2}{2}\right)}{\sqrt{\frac{n_1 n_2 (n_1 + n_2 + 1)}{12}}}$$

(14.2)

For this example,

$$Z_U = \frac{15 - \left(\frac{(10 \times 10)}{2}\right)}{\sqrt{\frac{(10 \times 10)(10 + 10 + 1)}{12}}} = -2.64$$

The value of −2.64 is smaller than the critical Z-score for $\alpha = .05$, which is of course ±1.96. As a result, we again reject the null hypothesis, and conclude that the rectangular dentate produces higher friction coefficients than does a smoothed surface. Again, the presence of ties has a small but generally negligible impact on the Z-score. The four ties in Table 14.1 produce a difference of less than .0001 in the actual probability associated with the Z-score determined above (i.e., instead of being .05, α truly equaled slightly more than .0499). This is a trivial difference in most analyses.

We wish to emphasize that the Wilcoxon two-sample test can be used for ordinal data, which means it is unnecessary to quantify distributions compared with the Wilcoxon two-sample test beyond their ranking. Given that archaeologists can often derive sequences even when they don't have detailed knowledge of the differences among their data, this can be invaluable. For example, if an archaeologist is able to order the occupation of sites in a river valley and an adjacent upland area based on the presence of diagnostic artifacts, radiocarbon dates, stratigraphic associations, and all of the other absolute and relative dating methods at our disposal, then he or she can perform a Wilcoxon two-sample analysis to see if either of these groups of sites tend to predate the other, even in the absence of secure absolute dates. This can provide useful information about settlement patterns and land use, even if the archaeologist has no idea of the actual magnitude of the differences in the dates among the sites. Building sequences, grave lot values, demographic variables, and a whole host of other data can be evaluated the same way, so long as the variates can be ranked relative to each other. Of course, archaeologists often have more than two groups they wish to compare. The Kruskal–Wallace nonparametric ANOVA is an excellent tool in such cases.

Kruskal–Wallis nonparametric ANOVA

Like the Wilcoxon two-sample test, the Kruskal–Wallis is a ranking-based test. Its calculation begins by ranking the pooled variates from all of the groups from the smallest to the largest. Tied variates are again averaged according to the ranks they would otherwise occupy. The ranks are then substituted for the variates in the groups. Once this is done, the average ranking for each group and for the combined groups is calculated and used to solve Equation (14.3).

The Kruskal–Wallis test

$$H = \left[\frac{12}{N(N+1)} \sum^{a} \frac{\left(\sum R_i\right)^2}{n_i} \right] - 3(N+1) \tag{14.3}$$

Equation (14.3) is a rather daunting looking formula, but is actually easier to calculate than it looks. N is the combined sample size of all of the groups. $\sum^{a} \dfrac{\left(\sum R_i\right)^2}{n_i}$ are instructions to divide the squared sum of ranks for each group by the group's sample size, and then add these values together. As with the Wilcoxon two-sample test's U value, H reflects whether the rankings are similar between groups. The null hypothesis is consequently $H_0 : R_1 = R_2 = R_3 = R_a$, where R again reflects the sum of the rankings in each group. We won't try to explain the derivation of the equation, but its resulting H is distributed according to the chi-square distribution as $\chi^2_{[a-1]}$, where a is the number of groups being compared. In this case, H must exceed the critical value to reject the null hypothesis.

To illustrate the use of the Kruskal–Wallis test, consider the data for the ratio of ash of burned bone compared to the bone's total weight before burning presented in Table 14.4. These data were collected by Miller Wieberg (2006: 28) during an experimental archaeological study of the breakage patterns of green bone. She buried fresh pig bones, and then retrieved samples at 28-day increments for a period of 141 days. The recovered bones were then broken, and differences in breakage patterns were recorded. The goal of the study was to determine if bones broken

Table 14.4   Ratio of ash to original bone weight for burned bone

6/19/2005	7/17/2005	8/14/2005	9/11/2005	10/9/2005	11/6/2005
14.67	36.16	30.01	41.96	39.32	29.28
15.35	37.07	30.31	43.49	42.39	34.00
15.41	38.13	30.44	44.28	44.75	37.45
15.64	38.20	37.38	44.45	45.12	38.14
15.75	38.99	38.48	44.77	45.66	42.21
16.61	39.89	38.50	45.67	47.21	44.61
16.88	40.37	42.59	50.10	48.46	44.98
18.19	41.70	45.15	50.16	52.41	46.24
19.94	44.71	45.25	50.17	52.93	47.36
21.26	45.79	45.39	53.71	55.02	49.16

immediately before or at death could be differentiated from those broken somewhat shortly after death as a result of burial practices, scavengers, and other taphonomic processes. The ash to total weight ratio reflects the moisture in the bone, and is therefore an excellent measure of how "green" (or fresh) the bone is.

It is of course known that bone dries over time, but 141 days is not a particularly long period. Looking at the frequency distributions for the data (Table 14.5), the ratios clearly change after the first 28 days (ending 6/19/2005), but it is less clear that they change substantially after that. As a result, an interested researcher might wish to determine if there is a significant difference in the ash to total weight ratio for bones collected between July 17 and November 6 in 2005.

Table 14.6 presents the rankings for the period from July 17 through November 6, 2005. The null hypothesis is that $H_0 : R_{7/17} = R_{8/14} = R_{9/11} = R_{10/9} = R_{11/6}$, where each

Table 14.5   Frequency tables for the ash to total weight ratio for the various time periods

Ash to total weight ratio	6/19/2005	7/17/2005	8/14/2005	9/11/2005	10/9/2005	11/6/2005
Less than 22.00	IIIIIIIII					
30.5			III			I
33.5						I
36.5		II	I			I
39.5		IIIII	II		I	I
42.5		I	I	II	I	I
45.5		II	III	IIII	III	III
48.5					II	II
51.5				III	II	
54.5				I	I	

Table 14.6   Rankings for the ash to total weight ratios

	7/17/2005	8/14/2005	9/11/2005	10/9/2005	11/6/2005
	6	2	20	16	1
	7	3	24	22	5
	10	4	25	29	9
	12	8	26	32	11
	15	13	30	36	21
	17	14	37	40	27
	18	23	44	42	31
	19	33	45	47	39
	28	34	46	48	41
	38	35	49	50	43
ΣR	170	169	346	362	228

subscript reflects the date the bone was collected. Using these data to solve for Equation (14.3) yields the following results:

$$H = \left[\frac{12}{50(50+1)} \sum^a \left(\frac{170^2}{10} + \frac{169^2}{10} + \frac{346^2}{10} + \frac{362^2}{10} + \frac{228^2}{10} \right) \right] - 3(50+1) = 16.51$$

The critical value for $\chi^2_{.05[4]} = 9.49$. Given that 16.51 is greater than 9.49, we reject the null hypothesis and conclude that the rankings are different. Looking at the cumulative rankings, the ash to total bone weight 9/11/2005 and 10/9/2005 are indeed larger than the earlier periods, although the final period ending at 11/6/2005 has a lower cumulative ranking. Miller Wieberg (2006) suggests rain the day before the bone was collected produced the decrease associated with 11/6/2005. Ultimately, the results indicate that the bones did dry appreciably, if not uniformly, during the period. Miller Wieberg (2006) was able to tie these patterns to the differences she identified in her breakage study, in which she found that she could begin to identify breaks as postmortem only after they had weathered about four months after death, a finding that may be important to archaeologists looking at variables such as cause of death or the likelihood of violent death.

Now that we have considered some nonparametric methods for comparing distributions, let's turn to some powerful parametric methods for multivariate analysis and the comparisons of groups.

Multivariate Analysis and the Comparison of Means

Given the complexity of human-related phenomena, archaeologists can reasonably expect that the causal factors influencing the archaeological record reflect complex interactions of many different variables. For example, both the length of the growing season and the average rainfall might impact farming and settlement strategies in an area. The length of lithic debitage may reflect the original size of the core, the raw material, the reduction technology, the stage or intensity of reduction, and the morphology of the desired tool. Quantitative methods such as ANOVA and the t-test are excellent for comparing variation resulting from any of these variables (e.g., comparing the length of debitage of different raw materials), but they are fundamentally limited to examining the operation of these variables one at a time. This isn't necessarily a bad thing, given that we could compare the various attributes that have been demonstrated to reflect lithic reduction technology and use among flaked stone assemblages to gain insight into the assemblage (e.g., Andrefsky, 2005; Shott, 1994). This "one-dimensional" approach is useful, in that at the end of all of the independent analyses, the archaeologist will know a lot more about the archaeological record than when he or she started.

However, sometimes archaeologists need to understand the interaction of the variables above and beyond what can be learned by looking at them in isolation. We have already explored this somewhat in Chapters 11 and 12 when discussing regression and correlation, which are means for evaluating the relationship *between* variables. The same thing is true for the chi-square test discussed in Chapter 13, which is a formal means of determining if two variables vary independently of each other. These tests are *multivariate* tests, in the sense that they consider the action and interaction of more than one variable. Similar multivariate analyses are available for comparing the means of groups. These take the form of more complex ANOVAs. Given the importance of the mean in characterizing and comparing distributions in archaeological analyses, these methods hold a special potential for the clever archaeologist. We begin this discussion by reviewing several important conceptual issues related to ANOVA that will be central to understanding the methods we describe, and then introduce the two-way ANOVA and the nested ANOVA.

A review of pertinent conceptual issues

ANOVA allows us to examine the sources of variation in a data set. In a standard ANOVA, these are broken down as follows:

$\bar{Y} - \bar{\bar{Y}}$ *among groups*: the variation attributed to the differences from the group means to the grand mean. This can reflect fixed effects (treatment effects) in Type I ANOVA or random effects in Type II ANOVA.

$Y - \bar{Y}$ *within groups*: the variation attributed to the differences of each Y_i from its mean. This is assumed to be random in ANOVA analysis, and is the result of the influence of other unspecified variables, random chance, and measurement error.

$Y - \bar{\bar{Y}}$ *total*: the variation attributed to the differences of each Y_i from the grand mean. The total variation is equal to the variation among groups and within groups, i.e., $\left(Y_{ij} - \bar{\bar{Y}}\right) = \left(\bar{Y}_j - \bar{\bar{Y}}\right) + \left(Y_{ij} - \bar{Y}_j\right)$. It therefore reflects the variation among groups attributable to random or fixed effects and the variation within groups caused by the action of unspecified variables and measurement error.

The relationship among the various sources of variation can be expressed in a slightly different manner as $Y_{ij} = \mu + a\bar{Y}_j + \varepsilon_{ij}$ where a reflects a specific treatment effect impacting $\bar{Y}_j$ when dealing a Type I ANOVA, or $Y_{ij} = \mu + A\bar{Y}_j + \varepsilon_{ij}$ where A reflects a random effect for Type II ANOVA. The impact of $A\bar{Y}_j$ or $a\bar{Y}_j$ will be reflected in the variation measured among groups and ε_{ij} will be reflected in the variation within groups. There can be more than a single variable impacting variation among groups, though. Imagine for example a case in which a researcher wanted to compare the average stature of two groups of humans who live under very different ecological conditions to determine if there are adaptive differences. The height of an individual reflects many factors including but certainly not limited to age, sex, nutrition, general health, and ancestry. Further, these variables might interact with one another such that the ecological conditions might impact the degree of sexual dimorphism through factors such as nutrition, parasite infections, and disease. Because of these confounding issues, the conceptual model of the "one-way" ANOVA we previously introduced isn't complete, in that it doesn't account for the interaction of multiple variables *at the same time*. Two-way ANOVA does.

Two-way ANOVA

The two-way ANOVA partitions the variation resulting from two variables and their interaction in a mathematically and conceptually elegant manner to evaluate the null hypothesis $H_0 : \bar{Y}_1 = \bar{Y}_2 = \bar{Y}_3 = \bar{Y}_a$. Let's say that the researcher interested in comparing the stature of various groups limits his or her study to adults, thereby eliminating one confounding variable (age), and wishes to control for the influence of sexual dimorphism and geography/ecology on stature. A more satisfactory conceptual model would then be $Y_{ij} = \mu + A\bar{Y}_j + B\bar{Y}_j + (AB)\bar{Y}_j + \varepsilon_{ij}$, where A reflects the impacts of sexual dimorphism, B reflects ecological conditions, and AB reflects the interaction between the two variables, both of which are random effects since they aren't under the researcher's control. This model reflects a Model II two-way ANOVA. If both variables were under the control of the researcher, then the conceptual model would be $Y_{ij} = \mu + a\bar{Y}_j + b\bar{Y}_j + (ab)\bar{Y}_j + \varepsilon_{ij}$, where a and b reflect two different treatment effects and ab reflects their interaction (a Model I two-way ANOVA). This model can of course be extended to a "mixed effects" ANOVA in which one variable is fixed and the other is random (i.e., $Y_{ij} = \mu + a\bar{Y}_j + B\bar{Y}_j + (aB)\bar{Y}_j + \varepsilon_{ij}$). In each of the models, the variation in the variates is conceived as reflecting the population mean, plus the effects of two fixed or random effects *and* their interaction, plus random error caused by the influence of other variables, measurement error, etc. This conceptual model is much more likely to fit many analytic situations that archaeologists encounter than any model specifying the action of a single variable in isolation. It will also provide a much better understanding of the variation in Y_{ij}, given that the variation attributable to the second fixed/random effect and its interaction with the first fixed/random effect would otherwise be lumped into the error term (resulting in larger within-group variation). Being able to control this variation will consequently reduce the amount of unexplained variation in the archaeologist's study, and allow the identification of significant relationships that might otherwise be obscured.

To illustrate the calculation of the two-way ANOVA, consider the cranial lengths for two samples of males and females from geographically distinct cultures (Table 14.7). We wish to evaluate the null hypothesis $H_0 : \bar{Y}_{Nrs.male} = \bar{Y}_{Nrs.female} = \bar{Y}_{And.male} = \bar{Y}_{And.female}$ at $\alpha = .05$. Table 14.8 provides the summary statistics for the four samples, and the following quantities allow us to calculate a one-way ANOVA (Table 14.9) as discussed in Chapter 10.

$$Q1 = \sum Y_{ij} = \text{sum of all observations} = 1,872 + 1,768 + 1,676 + 1,584 = 6,900 \, \text{mm}$$

Table 14.7 Cranial length for samples of Norse and Andaman Islander populations (mm)

| | Norse | | Andaman | |
	Male	Female	Male	Female
1	189	182	168	151
2	182	170	161	159
3	191	180	165	161
4	191	177	163	159
5	178	180	169	164
6	194	180	169	159
7	186	176	171	159
8	186	172	178	158
9	189	180	165	156
10	186	171	167	158

Table 14.8 Summary statistics for the data presented in Table 14.7

Group	N	$\overline{Y}$	s^2	$s_{\overline{Y}}$
Norse males	10	187.2	22.0	1.5
Norse females	10	176.8	19.1	1.4
Andaman males	10	167.6	22.5	1.5
Andaman females	10	158.4	11.2	1.1

Table 14.9 ANOVA analysis comparing the cranial lengths of Norse and Andaman Islander populations

Source of variation	SS	df	MS	F_S
Between groups	4574	3	1524.67	81.68
Within groups	672	36	18.67	
Total	5246	39		

$$Q2 = \sum (Y_{ij})^2 = \text{sum of squared observations} = 1,195,496$$

$$Q3 = \frac{\overset{j}{\sum}\left(\sum Y_i\right)^2}{n} = \text{sum of each sample's total squared divided by the sample size}$$

$$= \frac{\sum (1872)^2 + (1768)^2 + (1676)^2 + (1584)^2}{10} = 1,194,824$$

$$Q4 = \text{Correction Term} = \frac{(Q1)^2}{\sum n_j} = \frac{(6,900)^2}{40} = 1,190,250$$

$Q5 = SS_{Total} = $ Sums of Squares Total $ = Q2 - Q4 = 5{,}246$
$Q6 = SS_{subgroup} = Q3 - Q4 = 4{,}574$
$Q7 = SS_{within} = Q5 - Q6 = 672$

In this case, $F_s = 81.7$, which exceeds the critical value of 2.9 for an alpha level of .05. We consequently reject the null hypothesis, and conclude that there are differences between the means. But what is causing these differences? Is it because males and females are different because of sexual dimorphism? Or is it because Andaman and Norse populations are different because of their ancestry and somatic adaptation to different ecological settings? Or is it a product of some combination of the two? A two-way ANOVA will allow us to answer these questions.

To demonstrate the rest of the computation, we reorganize Table 14.7 as illustrated in Table 14.10 so that it is now analogous to a two-by-two matrix with sex as the variable listed in the rows and culture/ecology listed in the columns. This

Table 14.10   The data in Table 14.7 reorganized to more clearly differentiate between the variables sex and culture area

	Norse	Andaman
Female	182	151
	170	159
	180	161
	177	159
	180	164
	180	159
	176	159
	172	158
	180	156
	171	158
Male	189	168
	182	161
	191	165
	191	163
	178	169
	194	169
	186	171
	186	178
	189	165
	186	167

reorganization is then carried further in Table 14.11, which reflects the sums of the variates in each group. Computation of the two-way ANOVA continues using the following steps.

$Q8 = SS_{rows} = \sum^r cn(\bar{R} - \bar{\bar{Y}})^2$. Multiply the squared difference between the row mean and grand mean by the sample size and the number of columns, and sum these values for all of the rows. Here, SS_{rows} is the SS due to sex:

$$\sum^r CN(\bar{R} - \bar{\bar{Y}})^2 = (10*2*[167.6-172.5]^2) + (10*2*[177.4-172.5])^2 = 960.4$$

$Q9 = SS_{columns} = \sum^C rn(\bar{C} - \bar{\bar{Y}})^2$. Multiply the squared difference between the column mean and grand mean by the sample size and the number of rows, and sum these values for all of the columns. Here $SS_{columns}$ is SS due to culture area/ecological setting: $\sum^C rn(\bar{C} - \bar{\bar{Y}})^2 = (10*2*[182-172.5]^2) + (10*2*[163-172.5]^2) = 3610.0$

$Q10 = SS_{interaction} = $ SS reflecting the interaction between sex and culture area $= Q6 - Q8 - Q9 = 4574 - 960.4 - 3610.0 = 3.6$

The two-way ANOVA table is then constructed as illustrated in Table 14.12. Applying this to the quantities calculated above produces Table 14.13. The critical value for the F statistics is identified using Appendix C according to α and the

Table 14.11 Two-by-two matrix reflecting the sums of the data presented in Table 14.10

	Norse	Andaman
Female	1768	1584
Male	1872	1676

Table 14.12 The structure of a two-way ANOVA table

Source of variation	df	SS	MS	F
$\bar{Y}_a - \bar{\bar{Y}}$ Between rows	$r - 1$	SS_{rows} (Q8)	SS_{rows}/df	MS_r/MS_w
$\bar{Y}_b - \bar{\bar{Y}}$ Between columns	$c - 1$	$SS_{columns}$ (Q9)	$SS_{columns}/df$	MS_c/MS_w
$\bar{Y}_{ab} - \bar{Y}_a - \bar{Y}_b + \bar{\bar{Y}}$ Interaction	$(r - 1)(c - 1)$	$SS_{interaction}$ (Q10)	$SS_{interaction}/df$	MS_i/MS_w
$Y_{ij} - \bar{\bar{Y}}$ Within (error)	$rc(n - 1)$	SS_{within} (Q7)	SS_{within}/df	

Table 14.13   Two-way ANOVA comparing the mean values for Norse and Andaman Islander males and females

Source of variation	df	SS	MS	F
$\overline{Y}_a - \overline{\overline{Y}}$ Between rows	$r - 1 = 1$	960.4	960.4	51.45
$\overline{Y}_b - \overline{\overline{Y}}$ Between columns	$c - 1 = 1$	3610.0	3610.0	193.39
$\overline{Y}_{ab} - \overline{Y}_a - \overline{Y}_b + \overline{\overline{Y}}$ Interaction	$(r - 1)(c - 1) = 1$	3.6	3.6	0.19 (not significant)
$Y_{ij} - \overline{\overline{Y}}$ Within (error)	$rc\,(n - 1) = 2 \times 2 \times$ $(10 - 1) = 36$	672.0	18.7	

degrees of freedom for the terms, as outlined in Chapter 10. Using $\alpha = .05$, the critical values for all three F statistics are $F_{.05[1,36]} = 4.11$. We reject the null hypothesis for the variation between rows (sex) and between columns (culture area), but not for the variation attributable to the interaction of the rows and columns. From this we conclude that the average cranial length of males and females differ (presumably due to sexual dimorphism) and that the Andaman Islanders and Norse groups differ (presumably due to ancestry and ecological conditions), but that there is no interaction between these variables causing sexual dimorphism to vary differentially according to the culture area. What is also very significant is that a two-way ANOVA allows us to directly determine the proportion to the total variation attributable to the interaction of each variable. Using the sums of squares from Table 14.13 and the total sums of squares from Table 14.9, we can tell that sexual dimorphism accounts for 18% of the total variation (960.4/5246), ancestry and ecological conditions accounts for 69% of the variation (3610/5246), the interaction of these two variables accounts for less than 1% of the total variation (3.6/5246), and 13% of the variation remains unexplained within the groups resulting from measurement error and the operation of other variables that do not affect all of the members in a sample equally (672/5246). Thus, differences in ancestry and ecological conditions control a majority of the variation between the means, differences in sex controls some, their interaction controls effectively none, and there is still a bit, about 13% of the variation, that is attributable to causes we haven't identified (e.g., class differences within the culture, variation in subsistence strategies within the culture, variation in ecological conditions within each geographic area, etc.). Understanding how much of the total variation is explained by each variable and their interactions, and how much of the variation is attributable to the operation of other, unspecified variables and other sources of error, can be tremendously important to archaeologists grappling with the operations and significance of various variables. The two-way ANOVA gives us a means of knowing what portion of the variation we can explain, and what proportion of the variation remains to be explored.

Obviously, the conceptual models for two-way ANOVA can be expanded to include any number of additional variables (e.g., a three-way ANOVA, a five-way ANOVA, a ten-way ANOVA). The general mathematical process of calculating ANOVAs can likewise be expanded (Sokal and Rohlf, 1995: 369–88), although the computational structures quickly exceed most archaeologists' patience to directly compute. Various statistical software packages are available that will allow such comparisons. Their interpretation proceeds in exactly the same manner as illustrated here. We believe that once you begin to use these quantitative tools, you will find them invaluable. Let's now turn to a related method that directly builds on the desire of researchers to understand the variation that may be caused by unknown or hidden variables: the nested ANOVA.

Nested ANOVA

The two-way ANOVA is ideal for situations where an archaeologist has measured two variables that he or she suspects impact some aspect of the archaeological record. There will be times, however, when an archaeologist is concerned that some (possibly unspecified) variables that aren't consistent across a data set might contribute to variation within samples. This is especially common when using multiple samples of data derived at different times, by different researchers, or from different places. For example, an archaeologist studying variation in point morphology might want to compare the average length of several types of projectile points to see if they differ. This seems like the perfect case for a one-way ANOVA, which it is. But what if the assemblages of points come from different sites? Archaeologists know that the length of points can reflect the initial size of the parent stone nodule, the degree of core reduction (which in turn can reflect factors such as the distance from the raw material's source), the presence and degree of repair and reworking, isochrestic variation reflecting local traditions, and so on. These variables can and often do vary from site to site based on access to raw material sources, site use, and cultural differences in isochrestic or emblemic expressions. Do projectile points of the same type from different sites really reflect a homogeneous population? Are there other random (with respect to the point type) factors that contribute significantly to variation within types? And if other factors are contributing to the variation, what is their significance to the variation among the projectile point types? The archaeologist might chose to simply perform a one-way ANOVA regardless of the potential difficulties, and hope that any intersite variation within the types is limited when compared to the variation between types, but this "compute and hope you're right" approach to statistical analysis can (and has) led to some unfortunate errors. The archaeologist is effectively ignoring the potential of committing a Type II error, which weakens any analysis as discussed in Chapter 9. A better way to address this problem of intersample variation is with a nested ANOVA.

People frequently confuse nested ANOVAs with two-way ANOVAs, which is an understandable but preventable mistake. Two-way ANOVAs evaluate the variation resulting from two variables and their interaction across samples of data such that the variables are held constant for the entire data set. In the cranial length example used above, each skeleton was assigned to the classes of Female/Male, and Andaman/Norse without exception. The two-way ANOVA thus allows the evaluation of the relationship between these variables. The variation within these variables can be random effects (Model II), or fixed effects (Model I), depending on whether they are controlled by the archaeologist or not. In contrast, the nested ANOVA is focused on variation that is not consistent across the entire data set. To illustrate this, let's modify the preceding example such that we are comparing the average cranial capacity of multiple samples of Norse males and females from different burial contexts as reflected in Table 14.14. This could be approached as a one-way ANOVA or even a t-test in which the samples are combined into the classes of females and males, regardless of their provenience. However, the researcher might legitimately be concerned whether the samples themselves are internally homogeneous. Perhaps differences in time, location, status, or measurement methods have contributed to variation among samples of the same sex. Collapsing the samples together into "female" and "male" samples would subsume this variation, and might lead to a larger within-group variation than necessary that reflects the operation of unspecified difference between the samples. The archaeologist would legitimately want to know that these differences exist, and explore the data to determine their source. Perhaps the researcher will even choose to exclude one or more samples because they are not analytically comparable with the others, and thereby improve the power of the test to evaluate the hypothesis of interest.

A nested ANOVA is designed to help researchers faced with such cases where they are concerned that variation within the groups may be artificially elevated

Table 14.14   The cranial lengths for three samples each of Norse males and females (mm)

	Female			Male	
Sample 1	Sample 2	Sample 3	Sample 1	Sample 2	Sample 3
172	182	183	189	189	180
170	170	177	182	189	192
181	180	173	191	191	193
177	177	183	191	187	185
181	180	181	178	185	193
175	180	174	194	194	193
175	176	175	186	183	194
178	172	182	186	185	184
178	180	178	189	193	194
171	171	180	186	180	193

because of differences among the samples. It is effectively a means of evaluating the statistical power of the test to help determine the likelihood of committing a Type II error – failing to identify differences when in fact they are present. Unlike the two-way ANOVA, the nested ANOVA organizes variation hierarchically, such that it examines the differences within samples of groups that are then compared against one another. The model can be conceptually presented as $Y_{ijk} = \mu + A_j\overline{Y}_j + B_{jk}\overline{Y}_{jk} + \varepsilon_{ijk}$, where $A_j\overline{Y}_j$ reflects a random effect between groups (that is, the highest level of classification, which in this case is sex), $B_{jk}\overline{Y}_{jk}$ reflects random effects acting upon samples within the groups, and ε_{ijk} reflects the random variation of variates within the samples caused by measurement error, chance, and the operation of other variables. Of course, the effects affecting the groups could be fixed effects under the control of an archaeologist conducting some sort of experiment. This would produce the model $Y_{ijk} = \mu + a_j\overline{Y}_j + B_{jk}\overline{Y}_{jk} + \varepsilon_{ijk}$. However, the effects operating at the lower hierarchical level must *always* be random. Otherwise, a researcher should complete a standard one-way ANOVA to examine the effects of the treatments. The assumption here, then, is that there are no random effects operating below the level of the group, and that samples *within* the same group should be statistically identical, even if the groups differ from one another. Nested ANOVA is a means of determining if this is the case (i.e., if B in the term $B_{jk}\overline{Y}_{jk}$ equals 0).

The conceptual model for nested ANOVA superficially looks like the two-way ANOVA, except it is missing the interaction term AB_{ij}. The interaction term is unnecessary because B_{jk} is subsumed *within* each individual A_j and presumably differs across groups, in that the random effects impacting samples in one group need not be the same random effects impacting samples in other groups. For our examples, the random effects creating differences among samples of females may be different than the random effects creating differences among male samples, and the random effects impacting projectile point morphology may differ among projectile point types. These random effects are therefore not consistent across the entire data set, and there structurally cannot be consistent interactions between the random effects affecting individual samples and the random or fixed effects impacting the higher level groups. Note also that the nested ANOVA can be extended for any number of hierarchical levels (i.e., samples within samples within samples, etc.) so long as we can reasonably assume that the samples at each level are normally distributed. Here we will illustrate the nested ANOVA using the simplest possible structure of two levels.

Like the one-way ANOVA, nested ANOVA evaluates the null hypothesis $H_0 : \overline{Y}_1 = \overline{Y}_2 = \overline{Y}_3 = \overline{Y}_a$. To understand the calculation procedure, keep in mind that "a" refers to the highest level (in this case, sex) and "b" refers to the lower level (in this case, the unknown variables that might be impacting the samples within each sex). To facilitate the computation, Table 14.15 presents the sums and averages for each sample presented in Table 14.14. Computing the nested ANOVA begins by calculating the following four quantities:

Table 14.15　Summary statistics of samples of Norse females and males presented in Table 14.14

(Level a)	Female			Male		
(Level b)	Sample 1	Sample 2	Sample 3	Sample 1	Sample 2	Sample 3
$\sum Y_{ij}$	1758	1768	1786	1872	1876	1901
$\bar{Y}_j$	175.8	176.8	178.6	187.2	187.6	190.1

Q1.　$\displaystyle \bar{\bar{\bar{Y}}} = \frac{\overset{a}{\sum}\overset{b}{\sum}\overset{n}{\sum}Y_i}{abn} = \text{Grand\ mean: } \bar{\bar{\bar{Y}}} = \frac{1758+1768+1786+1872+1876+1901}{2\times3\times10}$

$= 182.7$

Q2.　$SS_{among} = nb\overset{a}{\sum}\left(\bar{Y}_j - \bar{\bar{Y}}\right)^2$. Multiply the sum of the squared differences between each group (defined at the highest hierarchical level) and the grand mean by the number of samples in each group and the sample size for each sample. In this case, this means multiplying the sum of the squared differences between the two sexes and the grand mean by the number of samples within each sex (3) and the number in each sample (10): $SS_{among} = 10\times3\times\sum\left([177.0-182.7]^2 + [188.3-182.7]^2\right) = 1892.8$.

Q3.　$SS_{subgr} = n\overset{a}{\sum}\overset{b}{\sum}\left(\bar{Y}_b - \bar{Y}_a\right)^2$. The sum of squares subgroup is a new term that refers to the variation within each of the groups. In this case, it is the variation of the samples within the group "Females" and within the group "Males". The term is calculated by multiplying the number in each sample by the sums of the summed squared difference between the means of each sample and the associated group mean:

$SS_{subgr} = 10\overset{a}{\sum}\overset{b}{\sum}\left([175.8-177.1]^2 + [176.8-177.1]^2 + \cdots + [190.1-188.3]^2\right) = 89.7$

Q4.　$SS_{within} = \overset{a}{\sum}\overset{b}{\sum}\overset{n}{\sum}\left(Y_i - \bar{Y}_b\right)^2$. Here, the sum of squares within refers to the variation within each sample of males and females. It is calculated by summing the squared differences between each Y_{ijk} and its sample mean:
$SS_{within} = \sum(137.6+171.6+126.4+197.6+178.4+232.9) = 1044.5$

The ANOVA table is then constructed using these quantities and their degrees of freedom as illustrated in Tables 14.16 and 14.17.

　　Interpreting the results of the ANOVA table starts with considering the variation among the samples (i.e., the lowest level in the hierarchy) and then moving upward. Significant variation between samples suggests that the variation within groups is excessively large because of one or more random effects differentially impacting the samples. If there is significant subgroup variation, then the variation among groups must be evaluated with care, because the variation within groups will be excessively

Table 14.16   The construction of a nested ANOVA table

	Source of variation	df	SS	MS	F_S
$\bar{Y}_A - \bar{\bar{Y}}$	Among groups	$a - 1$	Q2	$\dfrac{Q2}{a-1}$	$\dfrac{MS_{among}}{MS_{subgr}}$
$\bar{Y}_B - \bar{Y}_A$	Among subgroups within groups	$a(b - 1)$	Q3	$\dfrac{Q3}{a(b-1)}$	$\dfrac{MS_{subgr}}{MS_{within}}$
$Y_i - \bar{Y}_B$	Within subgroups	$ab(n - 1)$	Q4	$\dfrac{Q4}{ab(n-1)}$	
$Y_i - \bar{\bar{Y}}$	Total	$abn - 1$	Q2 + Q3 + Q4		

Table 14.17   Results of the nested ANOVA analysis comparing the cranial lengths of samples of Norse males and females

Source of variation	df	SS	MS	F_S
Among groups	$2 - 1$	1892.8	1892.8	84.5
Among subgroups within groups	$2(3 - 1)$	89.7	22.4	1.2 (ns)
Within subgroups	$2 \times 3(10 - 1)$	1044.5	19.3	
Total	$(2 \times 3 \times 10) - 1$	3027.0		

large. This means that real differences among the groups might be obscured by the action of random effects creating differences between the samples within the groups. This in turn leads to a high probability of committing a Type II error.

Looking at the results of the nested ANOVA (Table 14.17), we find that the variation within subgroups (i.e., the variation among the samples of the groups) is not significant at $\alpha = .05$. F_S is equal to 1.2, which is smaller than the critical value of $F_{.05[4,54]} = 2.54$. From this we conclude that the samples *within the groups* are consistent with each other such that they reflect the same population. This in turn indicates that the comparison between the males and females is not complicated by inconsistent samples within them. We can be assured that the comparison of means among the groups will be as powerful as the size of the samples and the strength of any differences among the groups allows (i.e., we are not prone to committing an excessive number of Type II errors, relative to the inherent likelihood of these errors given the structure of the underlying data). Given this, we can feel comfortable continuing to the evaluation of the null hypothesis itself. The F value of 84.5 for the comparison among groups (Table 14.17) does exceed the critical value of $F_{.05[1,54]} = 4.02$. As a result, we reject the null hypothesis and conclude that the cranial length of the samples of Norse females and males differ.

What if, unlike this example, the subgroup variation was significant? A t-test or other test of association can be used to determine exactly where the differences lie. The archaeologists can then modify the analytic design to identify and account for the random effect(s), to eliminate the inconsistent samples, or define new groups that better divide the samples into meaningful units. Defining new analytic groups and eliminating inconsistent samples must be done with care after considering the impact of the changes given the hypothesis of interest, the research design, and the theoretical foundation for the research, however. Otherwise, the archaeologist runs the danger of "cooking the data" to produce a specific outcome by choosing the data that "fits" and eliminating contrary samples that might reflect real differences (e.g., perhaps poorly defined pottery types do not reflect a homogeneous population).

The power of quantitative analyses for examining the influence and interaction of multiple variables on archaeological data should now be clear. In the case of two-way ANOVA, the variables are specifically defined, but, as illustrated by the nested ANOVA, it is possible to control for sources of differences that are unknown and/or inconsistent throughout the samples of the groups. A multitude of multidimensional methods are available to the interested researcher, although their mathematical complexity may make them a bit more daunting to many archaeologists. In the next chapter, we turn to a more general (and less mathematically intensive) consideration of factor analysis and principal component analysis, other multivariate techniques that have proven useful in archaeological analyses, especially for exploratory data analysis.

Practice Exercises

1   Compare and contrast the conceptual models of the one-way ANOVA (discussed in Chapter 10), the two-way ANOVA, and the nested ANOVA. How are they similar? How do they differ from one another? Under what circumstances would you select each of them? Provide an example in which a researcher might use each of them to evaluate an archaeologically relevant issue.

2   In an experimental study focused on bronze casting, Ottaway and Wang (2004:18) examine the microhardness of the resulting metal using the Vickers scale (HV). The analysts control the way that the bronze cast is cooled (air-cooled and water-quenched) and the bronze's composition (presence or absence of lead). Perform a two-way ANOVA to evaluate whether there are differences and interactions between the average hardness resulting from the two variables ($\alpha = .10$). Is this an example of a Model I, Model II, or Mixed Model two-way ANOVA?

Air-cooled		Water-quenched	
Added lead	*No lead*	*Added lead*	*No lead*
58.5	62.9	57.9	64.1
54.3	77.7	57.1	74.8
54.5	99.9	55.2	104.0
74.5	126.8	84.2	153.0
69.6	249.4	68.0	263.6

3 Weights (g) for samples of two types of Greek coins are presented below (data gleaned from Ireland 2000). Given that each type of coin was minted at a single mint, the weights of the coins should be consistent within each type, but possibly different between the types. Use a nested ANOVA to evaluate whether the average weight is different between the coin types ($\alpha = .05$). What is the significance of the variation within subgroups on your conclusion?

Helmeted head of Athena with Perseus holding Medusa head on the reverse side		Gorgon's head with Nike holding palm on the reverse side	
Sample 1	*Sample 2*	*Sample 1*	*Sample 2*
18.12	18.69	7.05	6.52
18.55	17.83	7.67	7.36
19.14	17.75	7.42	6.83
18.61	19.13	6.89	6.87
19.53	18.46	7.81	6.46
18.84	19.04	9.64	7.61
19.01	19.01	5.28	6.86
19.32	18.62	7.25	6.06

4 Following is the maximum width (mm) of two "types" of Benton points from the Middle Archaic period Ryan site in Alabama (Baker and Hazel, 2007: 47). Use the Wilcoxon two-sample test to evaluate if the assemblages differ ($\alpha = .01$). Reevaluate the null hypothesis using a t-test to compare the means. Do the results differ? Which test is more appropriate? Defend your answer.

Benton knife/spear burial offerings	Benton projectile points/darts
41.7	25.6
42.0	25.7
48.4	30.4
31.0	21.9
32.4	

5   Di Peso *et al.* (1974: 5) present wall lengths (m) for rooms in several architectural clusters. Use a Kruskal–Wallis nonparametric ANOVA to evaluate whether there are significant differences in the wall lengths ($\alpha = .05$).

Unit 11	Unit 12	Unit 14	Unit 16	Unit 22
7.28	4.22	3.10	5.58	2.65
14.35	4.26	4.99	4.51	1.90
11.50	4.28	6.35	5.38	7.08
10.60	5.00	6.08	4.65	1.95
7.85	5.00	6.78	5.49	1.98
7.20	4.08	4.75	4.82	2.75
7.40	4.08	4.28	4.70	3.54

15

Factor Analysis and Principal Component Analysis

In all of the previous chapters, we present both the interpretive and mathematical structure of the methods we discuss. The importance of both aspects are self-evident; without understanding the interpretative side of a given statistical method, the analyst can't properly interpret the results, but without understanding the mathematical side of the method, the analyst cannot understand the assumptions, structure, and limitations of a given approach. A good analyst needs to understand both of these aspects to use any particular quantitative method to its fullest. However, we are going to eschew this approach in this chapter because of the computational complexity of most multivariate techniques. We promised to hold our discussion to a level that requires no greater background than simple algebra. We also doubt that most archaeologists really are interested enough in matrix algebra and the other mathematical structures that underlie these methods to bother reading a detailed discussion of their computation. We will instead focus only on the conceptual underpinnings and proper interpretation of methods such as factor analysis, but encourage the interested reader to consult texts such as Jolliffe (2002), which will give more detail concerning the computational procedures.

Using regression analysis, the archaeologist can demonstrate the degree to which one variable controls the form of a second variable. This is useful because it establishes the strength of a causal relationship. Correlation is likewise useful because it reflects the degree to which two dependent variables are controlled by one or more shared independent variables. Imagine a case where three variables were all dependent on a single independent variable (i.e., three variables correlated strongly with each other and are characterized by a regression relationship with a fourth variable). By controlling the single independent variable, the archaeologist could explain much of the variation in the three dependent variables. If the relationship was

Quantitative Analysis in Archaeology, Todd L. VanPool and Robert D. Leonard
© 2011 Todd L. VanPool and Robert D. Leonard

strong enough, the three dependent variables could even be collapsed into the single independent variable, allowing the archaeologist to shift from trying to measure and account for the variation in all four variables to working with a single independent variable whose effects on other variables is known. This process is called *dimensional scaling*, in that the variation in multiple dimensions (variables) is reduced to a smaller number of independent variables that control the remainder of the variation. This is useful because it clarifies causal relationships, reduces the amount of data the analyst needs to manage, and allows the interaction of causal relationships to be more easily studied.

Two types of dimensional analysis are commonly used in archaeology: principal component analysis and the closely related factor analysis. Principal component analysis and factor analysis are generic names given to a class of multivariate approaches used to quantify the structure underlying data matrices. They both seek to define a set of common underlying dimensions that structure the data. These methods are effectively exploratory data analyses, in that they do not explicitly evaluate previously defined null hypotheses. Instead, they look at shared variation among a set of variables that can be mathematically modeled. They then produce a measure of the amount of shared variation that can be tied to hypotheses of interest given an appropriate analytic and theoretical structure, but that are not formal statistical tests in and of themselves.

The conceptual structure of the methods is really quite easily understood. Imagine that an analyst has measured various attributes of prehistoric pottery including pot volume, pot height, and pot width. These three variables are likely interconnected, such that a large pot will tend to be taller and wider than a small pot. A scatter plot of the three variables would likely indicate a reasonably tight correspondence, indicating considerable redundancy in the variation (i.e., the variation in the three variables is linked). It intuitively seems obvious that we could define a mathematical model, rather like a regression line, that could capture the "essence" of this linked variation. This single *new* variable (as opposed to previously defined variable), which could be called "pot size", could then effectively summarize and even replace the other three variables in the analyst's efforts to explain the variation in the data, communicate the structure of the data to others, and perform additional analyses. Of course, the analyst didn't measure this newly defined variable directly, but it captures the relationship between the variables he or she measured in a way that is explanatorily useful.

Factor analysis and principal component analysis are closely related means of identifying "new" dimensions that capture the "essence" of the correspondence among the original variables. These new dimensions, which are called "common factors" or just "factors" in factor analysis and "components" in principal component analysis, allow the analyst to determine how much of the variation in each variable is shared with the other variables and to postulate the mechanisms that link the variables together. This information can then be used to summarize the data such that a much smaller number of dimensions can be used to characterize the associations underlying the data when compared to the original list of variables.

In effect, a larger number of linked variables can be replaced by a single factor (in factor analysis) or component (in principal component analysis) that reflects the relationship between the various independent and dependent variables. We can even go so far as to substitute the scores derived from each factor or principal component for the original variate data to simplify the data set (e.g., get rid of different units of measurement and number magnitudes) and facilitate further mathematical analysis through a process called *data reduction*.

The use of newly defined "factors" or "principal components" in an analysis can be a bit intimidating to those who do not have a strong quantitative background. We remember a colleague at a Society for American Archaeology Annual meeting who privately confided that she worried that our factors "were made up" and don't correspond to any "real variable that we measured." In a sense this is true, but the factors reflect the correspondence among "the real variables." As a result, they let us look at the underlying structure of the data without trying to remember the exact degree to which each variable corresponds with all other variables at the same time. Consider for example the dissertation research of the senior author, which included measuring 21 metric traits of 424 points recovered by Emil Haury (1975) from Ventana Cave, Arizona (Table 15.1 and Figure 15.1). Many of these variables are

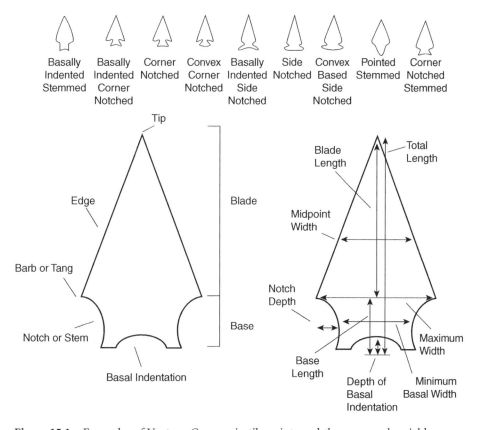

Figure 15.1   Examples of Ventana Cave projectile points and the measured variables

Table 15.1   Metric variables measured for projectile points
from Ventana Cave, Arizona

Weight
Total length
Maximum width
Maximum thickness
Blade length
Maximum blade width
Width at blade midpoint
Maximum blade thickness
Thickness at blade midpoint
Basal length
Maximum basal width
Minimum basal width
Maximum basal thickness
Width of widest notch
Depth of widest notch
Width of thinnest notch
Depth of thinnest notch
Depth of largest stem indentation
Depth of smallest stem indentation
Depth of basal indentation
Average number of serrations

Note: Notch measurements were recorded for corner-notched and
side-notched projectile points whereas stem indentation depths
were recorded for stemmed projectile points.
Source: VanPool, Todd L. (2003). *Explaining Changes in Projectile
Point Morphology: A Case Study from Ventana Cave, Arizona*. Ph.D.
Dissertation, Department of Anthropology. University of New
Mexico, Albuquerque.

directly linked (e.g., base length has an obvious correspondence with total point length), but trying to remember how each variable corresponds with all of the other variables is an impossible task. Using factor analysis, however, it is possible to quantify and easily communicate such relationships.

Factor and principal component analysis may superficially sound like regression analysis, where an independent variable is identified and used to predict the state of one or more dependent variables, but they are quite different. In principal component and factor analysis, the dimensions do not correspond with any single variable within a matrix, but reflect an underlying variable that may be quite specific (e.g., projectile point size), more general (e.g., pot use), and possibly unknown (e.g., unrecognized ethnic or idiosyncratic differences). In effect, all of the variables are treated as dependent variables, and the dimensions are newly defined independent variables. Further, factor analysis and principal component analysis do not seek to maximize the ability to allow the variates of an individual variable to be predicted as is the case with regression, but instead seek to maximize the amount of the data that can be *extracted* (effectively summarized) in the entire assemblage, irrespective of the predictive power for any given variable. They are thus *interdependence* (identification of structure) techniques as opposed to *dependence* (predictive) techniques.

Principle component and factor analysis are very similar but differ in the way they measure variation (Jolliffe, 2002: 180–96). From the perspective of both techniques, there are three possible "types" of variation in a data set: common, specific, and error. Common is defined as variation that is shared among multiple variables in a data set (e.g., all of the measured variables increase as artifact size increases). Specific variance is limited to a single variable within the data set and typically reflects the operation of other influences that do not impact the other variables (e.g., wall thickness in a pottery assemblage may increase with temper size, an unmeasured variable, but it is the only variable that reflects this variation). Error is variation within a single variable resulting from measurement errors. Both factor analysis and principal component analysis are excellent means of measuring common variance, but they differ in their treatment of specific variance and error. Because of the way that factor analysis organizes the underlying data matrix, it measures and only measures common variance (Hair *et al.*, 1995: 375–6). This is why factors are commonly called "common factors". As a result, factor analysis structurally cannot account for specific or error variation, and reflects *only* the underlying dimensions that structure the common variation within the data. All "common" factors must reflect "common variance".

In contrast, principal component analysis doesn't mathematically discard the specific variance and error as factor analysis does. The components can reflect any of these sources, meaning that they can reflect relationships specific to a variable as opposed to characteristics of the entire data set. This can be useful, in that a relationship contained within a single variable that might be obscured in factor analysis can become evident when using principal component analysis. However, the analyst must be careful to correctly determine what variation is being reflected in the results, or face the possibility of incorrectly concluding that some influence is a general influence instead of a specific relationship (Jolliffe, 2002:180–196). This can be a significant error when using principal component analysis for dimensional scaling, so significant in fact that various authors have concluded that, "… principal component analysis should not be used if a researcher wishes to obtain parameters reflecting latent constructs or factors" and "at best, PCA [principal component analysis] provides an approximation to what is truly required" (Jolliffe 2002: 161). Perhaps it isn't necessary to go so far as to say one should never use principal component analysis, but it is fair to say that it should only be used when the researcher can be reasonably sure that specific variance and error is small. We discuss the application of both methods, but focus our discussion primarily on factor analysis.

Mechanically, factor analysis and principal component analysis both allow *variance maximizing rotation*, a fancy way of saying that they allow the data to be structured around the components that explain the most variation. (In truth, variance maximization rotation in principal component analysis isn't really part of the method, but is based on a "borrowing" of the technique from factor analysis (Jolliffe, 2002: 166). Still, it is mathematically defensible and virtually all statistical software packages will perform variance maximizing rotation for components.) Consider Figure 15.2, a simple scatter plot organized using X and Y axes. A regression line drawn through the scatter plot is still defined according to the two original axes, despite the fact that the original axes do not explain any of the variation in and of themselves. To the contrary, the axes are really arbitrary structures used to organize the data, typically defined by the value zero for each variable. Zero is a great number with intuitive meaning, especially when dealing with ratio scale data, but in regards *to the relationship being considered*, it is no more meaningful than any other number. Instead of using an arbitrary number like zero that produces arbitrary axes in regards to any relationship between variables that might be present, we could use the regression line itself as an axis. This would shift the focus from arbitrarily defined axes to a meaningful data structure in which the variation in the scatter plot is reflected by the variation in variates around the regression line (Figure 15.3). It would also simplify the data, in that two axes (one for each variable) are replaced by the regression line. Continuing this process for a multidimensional scatter plot composed of many additional variables, new regression lines could be created that effectively replace the arbitrary x and y axes with axes that are organized according to the relationships within the data. This is exactly what happens with variance maximizing rotation. The factors/components reflect the new axes and are rotated so as to maximize the difference with other factors while minimizing the spread of the variates around them.

The first factor/component defined is the one that summarizes the most variation in the entire data set. The second is the one that maximizes the amount of the

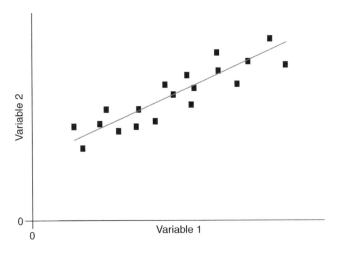

Figure 15.2   A traditional regression line organized using an x- and y-axis value of 0

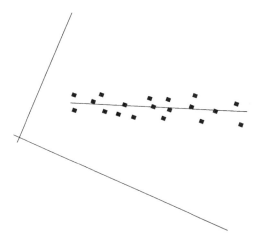

Figure 15.3   The scatter plot illustrated in Figure 15.1 reorganized using the regression line as the axis

remaining variation that is summarized. The third, fourth, and subsequent factors continue to maximize the amount of the remaining variation they summarize until all of the variation is accounted for. Thus, the "placement" of each previous factor impacts the placement of all of the subsequent factors. The first factor is the single best summary of linear relationships in the data; the second is the single best summary of the linear relationships remaining after the first factor, and so forth. The results of the factor analysis or principal component analysis will then report the amount of the variation in each variable accounted for by each factor/component, so that we can easily determine which meaningfully summarize the various variables and how the variables relate with one another. We will illustrate this using the Ventana Cave data previously mentioned, but must first consider a few additional issues before we are ready to "number crunch".

Objectives of Principal Component and Factor Analysis

As with any quantitative method, the starting point for factor analysis and principal component analysis is articulating an appropriate research objective using appropriate theoretical and methodological approaches. This is sometimes difficult, given that it is easy to lose sight of what these methods do. Their general utility is to reduce a data set into a smaller number of composite dimensions (components) assumed to underlie the original variables (i.e., to reduce the number of relationships we must consider to characterize and potentially explain the variation within a larger data set). If the research objective doesn't focus on this goal, then these methods aren't the right techniques and a more appropriate statistical tool should be selected (e.g., use multiple linear regression if you really want to predict the values of certain variates). Further, there are several types of principal component and factor analyses that are used to analyze data in different ways. This is especially true for factor analysis, which typically is divided into two "types" (e.g., Hair *et al.*, 1995). The most common variant is *R factor analysis*, which analyzes a set of variables to identify the dimensions that underlie their correspondences. This approach has been the implicit focus of our discussion to this point.

The results of the R factor analysis can be used for several purposes. First, the dimensions can form the basis for explaining the variation within a matrix. For example, the dimension "artifact size" might control the variation in many variables. Identifying this as an underlying factor might focus the analyst on explaining the general forces operating on the size of a set of artifacts, as opposed to trying to explain changes in artifact length, width, thickness, weight, etc. independently. Second, the identified dimensions can be used as a means of reducing the number of variables considered for presenting the research results. Instead of having to discuss the variation in a whole host of variables, the analyst can simply identify the underlying dimensions and describe the variation in them. Finally, the factor analysis can be used to replace the larger number of variables in the original data with a smaller number of variables (the factors) that can then be compared with one another using additional statistical techniques.

The other type of factor analysis is *Q factor analysis*, which divides the population of objects being considered into different groups (factors) based on the correspondence in the data. This approach is used in archaeology, especially in sourcing studies, but it is computationally difficult and often inferior to the various sorts of cluster analyses that are available (Hair *et al.*, 1995: 371). We bring it up here only in the interest of completeness, but will not discuss it further.

Designing the Principal Component/Factor Analysis

In setting up the principal component or factor analysis, the researcher will need to consider both the number of variables to be included, and their structure. Variables used in these analyses are generally assumed to be metric measurements measured at a ratio or interval scale, although there are special methods useful for interval and even ordinal scale data (e.g., Boolean factor analysis). The number of variables depends on the principal component's or factor analysis's purpose. If the researcher wishes to evaluate some proposed structure that may characterize the data, then the pertinent variables should be included. If instead, the analyst is conducting exploratory data analysis, the list of variables may be more expansive. However, the inclusion of additional variables increases the likelihood of including variables that do not meaningfully correspond with any other variables, meaning that the principal component or factor analysis will not provide useful data reduction for them. Sample size is also important, with samples of 50 or more preferable. Smaller samples will not allow the identification of meaningful but weak correspondences while also increasing the likelihood of identifying spurious correlations.

A common general rule of thumb is to have at least five times as many observations as there are variables to be analyzed. Some propose this should be increased to a ratio of 10 to 1 or even 20 to 1 when possible (e.g., Hair *et al.*, 1995: 373), but such large samples may not always be possible in archaeological contexts.

Assumptions and Conceptual Considerations of Factor Analysis

Principal component and factor analysis both have the assumptions of normality, homoscedasticity, and the presence of linear relationships for the same reasons these assumptions are required in regression analysis. Distributions don't have to be perfectly normal, but outliers will significantly impact the location of components and factors. Heteroscedastic data and non-linear relationships will result in weak components/factors that don't summarize the data well. Departure from these assumptions doesn't necessarily undermine an analysis, but it does weaken its effectiveness. The critical underlying assumptions of principal component and factor analysis are more conceptual than mathematical, however.

To begin with, we ought to have a reason to expect some interconnectedness within the data. As a practical matter, visual inspections of scatter plots of the various variables ought to suggest at least some meaningful correlations. Some software packages even have specific tests such as *Bartlett's test of sphericity* to evaluate whether there are enough apparent correlations to justify using factor analysis (see also Vierra and Carlson, 1981). Even better is if there is some theoretical, methodological, or empirical reason to propose relationships. This will make interpreting the results easier and more interesting.

The analyst should also seek to have a homogeneous population relative to the research question. For example, performing a factor analysis of characteristics for archaeological sites from different time periods when it is previously known that the pertinent characteristics differ between them is inappropriate. The resulting analysis of the combined data will be a poor summary of the underlying structure of each group relative to two separate analyses that divide between them. Thus, when it is known that there are two or more subgroups within a sample that differ in regards to the relationships among variables being analyzed, it is better to separate them and perform multiple principal component or factor analyses as opposed to combining them into a single analysis.

An Example of Factor Analysis

Let's return to our projectile point example. As previously stated we have a series of metric measurements for 424 points. We are interested in understanding the structure of variation in point morphology, which we are certain is mechanically linked in many ways, and communicating this effectively to others. Principal component analysis and, especially, factor analysis are ideal means of doing this.

In keeping with our recommendations above, we start by considering the appropriateness of each variable. We eliminated largest and smallest stem depth and the average number of serrations from the analysis because of stem depths' small sample size and because the lack of direct links between the average number of serrations and other variables cause us to suspect it will be dominated by specific, as opposed to common, variance. Table 15.2 indicates the amount of the variation in each variable accounted for in the factors defined using the factor analysis procedure generated using SPSS, a common statistical software package. (Other packages will present the results in nearly identical manners.) This represents the amount of variation in each variable held in common within the data set, as opposed to reflecting specific variation or error. These results please us, in that none of the variables are wholly or even largely independent of the others. Our factors should provide a useful description of the correspondence among the variables. Still, not all of the variation of each variable is accounted for in the factor analysis (e.g., 27% of the variation in midpoint blade width is not accounted for by the factors). The remaining variation is a product of specific variance caused by influences not shared with the other 17 variables considered here and measurement error.

Table 15.3 presents the actual information for each factor. *Eigenvalues* (also called the *latent root*) are mathematical measures computed using matrix algebra that reflect the amount of variation each factor describes. Each variable contributes variation to a data set such that the total variation in a data set is a product of the

Table 15.2 The amount of the variation in each variable extracted through the factor analysis

Variable	Initial	Extraction
Weight	1	0.95
Total length	1	0.96
Maximum width	1	0.88
Maximum thickness	1	0.90
Blade length	1	0.92
Maximum blade width	1	0.85
Maximum blade thickness	1	0.92
Width at blade midpoint	1	0.73
Thickness at blade midpoint	1	0.94
Basal length	1	0.63
Maximum basal width	1	0.93
Minimum basal width	1	0.89
Maximum basal thickness	1	0.81
Width of widest notch	1	0.78
Depth of widest notch	1	0.83
Width of thinnest notch	1	0.73
Depth of thinnest notch	1	0.96
Depth of basal indentation	1	0.77

Table 15.3   Total variance explained in the factor analysis of the Ventana Cave projectile points

Factor	Eigenvalues	% of variance	Cumulative %
1	6.72	37.34	37.34
2	2.63	14.60	51.94
3	2.11	11.73	63.67
4	1.56	8.68	72.35
5	1.27	7.07	79.42
6	1.11	6.14	85.56
7	0.80	4.45	90.01
8	0.54	3.01	93.02
9	0.43	2.40	95.43
10	0.33	1.85	97.27
11	0.23	1.27	98.54
12	0.14	0.75	99.30
13	0.06	0.35	99.65
14	0.04	0.20	99.85
15	0.02	0.10	99.95
16	0.01	0.05	100
17	Less than .005	0	100
18	Less than .005	0	100

variation in each variable. An eigenvalue of 1 reflects all of the variation contained within a single variable such that the total of all eigenvalues added together will equal the total number of variables, in this case 18. The eigenvalues on Table 15.3 thus reflect the proportion of the total amount of variation in all of the variables described by each factor. The first factor corresponds with an eigenvalue of 6.73, which is roughly 37% of the total variation among all of the variables. Adding all of the eigenvalues together produces a sum of 18, which represents the total common variation within all of the variables. It is always the case that the maximum number of factors that can be defined is the same as the number of variables considered, in this case 18. The first factor has the largest eigenvalue, indicating that it is the single best line summarizing the variation in the data set. The next factor will be the best line summarizing the remaining variation. It will also be *orthogonal* to the first, meaning that in addition to being derived from the proportion of the variance remaining after the first factor is extracted, it must be kept at an angle of 90 degrees to the first factor. This ensures that the two factors are independent of (as opposed to being correlated with) each other. The third factor will be orthogonal to the first two factors, and so on until all of the variance in the data set has been extracted. The first factor can be viewed as the single best summary of a linear relationship within the data, but all subsequent factors are defined only in relationship with the preceding factors.

As is likely evident from Table 15.3, not all of the factors will be analytically or descriptively meaningful. Each factor is the best description of the variation remaining in the data set after the previous factors, but it is possible, in fact almost certain, that the meaningful relationships will be summarized in the first handful of the factors, with the remaining factors effectively describing weak or accidental relationships that are limited to a small number of variables. In our example, the first seven factors summarize 90% of the common variation in the data set, with the final eleven factors summarizing the remaining 10% (Table 15.3). This suggests that the first seven factors may usefully describe the most meaningful and analytically useful relationships among the variables but that the last eleven will be far less useful. Unfortunately, there is no cut and dry means of determining which factors are significant. Most approaches for evaluating their significance use the eigenvalues. A common rule of thumb is the *latent root criterion*, which holds that factors with eigenvalues greater than one are likely significant while those less than one probably are not. The underlying rationale is that a factor should account for at least the variation in one variable to be considered significant. However, this rule of thumb can be overly conservative when dealing with small numbers of variables, say 18 or fewer as is the case here. Still it is a good, and perhaps generally the best, starting point. Applied to our example, this would indicate that first six factors are significant.

Another common approach is the *a priori* method of determining the number of significant factors before the analysis is performed. This is most common when there is a theoretical or empirical reason to expect that a limited number of relationships are present. The results of the factor analysis (or principal component analysis) can then be used to evaluate if the limited number of factors (or components) does really capture most of the variation. Most software packages allow one to specify the number of factors/components extracted.

Other approaches for determining significant factors include the *percentage of variance criterion*, the *scree plot*, and the *heterogeneity of respondents*. The percentage of variance criterion employs the cumulative percentage of the variation as the basis for evaluating factor significance. A common cutoff is 95%, in which all of the factors necessary to account for 95% of the total variation are considered significant. In our example, the first nine factors would be considered significant, despite the fact that factors 8 and 9 both account for only about 3% of the total variation. There is nothing magical about the 95% cutoff, though. Much smaller cumulative percentages can be used (e.g., a 75% cutoff would consider only the first four factors significant). There is no generally accepted demarcation that should be used slavishly.

The scree plot is a simple graphical means of measuring the specific variation characterized by a factor. Remember that factor analysis is designed to measure common variation, as opposed to specific variation or error. Each factor reflects a relationship among the variables, but the later components will likely be dominated by relationships between two variables, as opposed to general relationships among larger variable groups. A scree plot helps determine at what point this happens. It

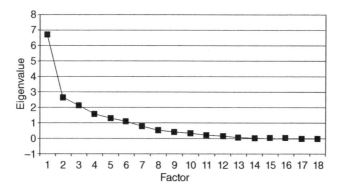

Figure 15.4 Scree plot for the factor analysis evaluating the Ventana Cave projectile point data

is constructed by making a plot of the eigenvalues against the number of the component (Figure 15.4). The initial drop in eigenvalues is likely to be sharp, but at some point the eigenvalues will plateau, causing the plot to approximate a straight line. The point where the curve begins to level is typically considered the spot where most of the common variation has been summarized and more restrictive relationships become dominant. In our case, it seems to us that factor 8 is a good dividing line, although one could defensibly go as low as factor 12. In general, the scree plot tends to include more significant factors that the latent root criterion.

The heterogeneity of respondents approach is less a means of determining how many components should be considered significant, and more of a means of determining if there are variables that are diluting the utility of the factor or principal component analysis. This is especially true for principal component analysis. Given that the principal component analysis includes specific variation, a variable that does not correlate with other variables will be poorly reflected in the initial series of components but will likely be largely or even entirely summarized by one of the ending components. Examining the *factor loadings* (which we will discuss in a moment) to detect if there are later components that account for nearly all of the variation in a single variable will help detect these "rogue" variables (Table 15.4). They can then be eliminated from the analysis, which may help clarify trends and simplify the selecting of meaningful factors/components. This will be less significant for factor analysis, given that it is limited to considering variation among the variables, but it is also likely that two variables that correspond with each other, but not the other variables in the data set, will have poor loadings on early factors while being almost entirely summarized by a later factor. The analyst may want to consider whether these variables should be kept in the analysis.

Ultimately, it is probably best to consider more than one criterion for selecting significant factors, although it seems that the latent root criterion is most common. There are negative consequences to selecting too many or too few components as

Table 15.4 Factor loadings for the first six factors

Variable	Factor					
	1	*2*	*3*	*4*	*5*	*6*
Weight	0.90	−0.31	−0.13	0.09	−0.03	0.13
Total length	0.77	−0.44	−0.19	0.24	−0.01	0.30
Maximum width	0.74	0.55	0.12	−0.01	−0.07	−0.10
Maximum thickness	0.66	−0.27	0.00	−0.29	0.10	−0.54
Blade length	0.72	−0.40	−0.34	0.13	0.03	0.32
Maximum blade width	0.62	0.33	0.15	0.12	−0.56	0.05
Maximum blade thickness	0.78	−0.39	−0.09	−0.30	0.05	−0.25
Width at blade midpoint	0.41	0.38	−0.29	−0.30	0.30	0.39
Thickness at blade midpoint	0.71	0.57	0.19	−0.03	−0.25	0.10
Basal length	0.47	−0.30	0.36	0.41	−0.10	0.06
Maximum basal width	0.47	0.11	0.75	−0.12	0.25	−0.26
Minimum basal width	0.42	0.04	0.80	0.03	−0.05	0.27
Maximum basal thickness	0.46	−0.48	0.19	−0.55	0.09	0.13
Width of widest notch	0.62	−0.01	0.02	0.26	0.58	−0.01
Depth of widest notch	0.72	0.36	−0.28	0.23	0.03	−0.22
Width of thinnest notch	−0.17	0.23	0.16	0.59	0.53	−0.04
Depth of thinnest notch	0.62	0.44	−0.56	0.07	−0.05	−0.25
Depth of basal indentation	0.05	0.60	−0.02	−0.48	0.29	0.30

significant. Selecting too few can obscure the correct structure of the data causing important dimensions to be left out of the analysis. Too many components, though, make the results hard to interpret and summarize for others. In our case, it seems to us that the first six factors summarize most of the variation with the remaining factors being less useful in achieving our goal of characterizing the trends in our data. While this is a defensible decision in our opinion, it remains arbitrary in that other researchers might have chosen to limit the significant components to the first four or might have expanded them to the first 10 or so.

Interpreting the meaning of a factor is based on the previously mentioned *factor loadings*, which reflect the correlation of each variable and the factor (Table 15.4). These values are identical to Pearson's correlation coefficients and reflect the proportion of the variation in a variable summarized by the factor. They can range from 1 to −1, with positive numbers indicating a positive relationship between the factor and the variable and negative values reflecting a negative relationship (i.e., a variable with the negative loading gets smaller as the other variables get larger). As loadings increase to 1 or decrease to −1, the amount of the variation in a variable summarized by the factor increases, whereas values close to zero indicate that the variable and the factor are largely independent. Squaring them creates the equivalent of a regression coefficient (r^2), which identifies the portion of the total variation in the variable described by the factor (e.g., factor 1 describes $(.9)^2$ or .81 of the common variation in weight). The squared factor loadings are directly reflected in the eigenvalues we used to evaluate the significance of the factors (Table 15.3). The eigenvalues are the sum of the squared factor loadings for each factor. This relationship is illustrated in Equation (15.1), where λ_i is the eigenvalue for a particular factor and L_{ij}^2 is the squared factor loading for a given variable.

The eigenvalue

$$\lambda_i = \sum L_{ij}^2 \qquad\qquad\qquad\qquad (15.1)$$

Table 15.5 presents the squared factor loadings from Table 15.4. The sum for each factor produces the eigenvalues listed on Table 15.3. This connection between the factor loadings and the eigenvalues explains why we are able to use the eigenvalues to determine the amount of variation in the dataset a factor describes.

Let's turn our attention to the significance of each of the six factors we identified as meaningful. The first factor in Table 15.4 is a good summary of much of the variation in most of the variables. Only seven variables – width at blade midpoint, basal length, maximum basal width, minimum basal width, maximum basal thickness, width of the thinnest notch, and depth of basal indentation – have factor loadings below .5. This single dimension, which we could call point size, indicates

Table 15.5 The squared factor loadings of each variable for the first six factors

Variable	Factor					
	1	2	3	4	5	6
Weight	0.81	0.10	0.02	0.01	0.00	0.02
Total length	0.59	0.19	0.04	0.06	0.00	0.09
Maximum width	0.55	0.30	0.01	0.00	0.00	0.01
Maximum thickness	0.44	0.07	0.00	0.08	0.01	0.29
Blade length	0.52	0.16	0.12	0.02	0.00	0.10
Maximum blade width	0.38	0.11	0.02	0.01	0.31	0.00
Maximum blade thickness	0.61	0.15	0.01	0.09	0.00	0.06
Width at blade midpoint	0.17	0.14	0.08	0.09	0.09	0.15
Thickness at blade midpoint	0.50	0.32	0.04	0.00	0.06	0.01
Basal length	0.22	0.09	0.13	0.17	0.01	0.00
Maximum basal width	0.22	0.01	0.56	0.01	0.06	0.07
Minimum basal width	0.18	0.00	0.64	0.00	0.00	0.07
Maximum basal thickness	0.21	0.23	0.04	0.30	0.01	0.02
Width of widest notch	0.38	0.00	0.00	0.07	0.34	0.00
Depth of widest notch	0.52	0.13	0.08	0.05	0.00	0.05
Width of thinnest notch	0.03	0.05	0.03	0.35	0.28	0.00
Depth of thinnest notch	0.38	0.19	0.31	0.00	0.00	0.06
Depth of basal indentation	0.00	0.36	0.00	0.23	0.08	0.09
Sum (eigenvalues)	6.72	2.63	2.11	1.56	1.27	1.11

that most of the variables are linked with each other to some degree and that many of them are closely connected. However, many of the variables reflecting basal morphology have comparatively poor loading scores suggesting that they reflect the operation of attributes other than point size. The second factor has high loadings for the depth of basal indentation, maximum width, and thickness at blade mid-point. The factor also has a comparatively high negative loading for maximum basal thickness, indicating that it gets smaller as the others increase. This factor indicates to us that there is a relationship between the width of points and the depth of basal indentations such that wider points tend to have deeper indentations, but that points with deep indentations often have thinner bases. The third factor has com-paratively high loadings for maximum basal width, minimum basal width, and, to a lesser degree, basal length. It also has a high negative loading for the depth of the smaller notch on notched points and a weaker negative loading for maximum blade length. These variables reflect the size of the point base, indicating that points with long, wide bases tend to be associated with less pronounced notching elements and shorter blades relative to other points with similar point lengths. It certainly makes sense that the minimum width of a point's base will be larger for a point with shallow notches relative to a similarly sized point with deep notches. Likewise, the larger the base, the smaller the blade must be for similarly sized points.

The remaining three factors all show positive or negative loadings on a limited number of variables (factor 4, positive loading with the width of the thinnest notch and negative loadings with basal indentation depth and maximum basal thickness; factor 5, positive loadings with the width of both notches and a negative loading with maximum blade width; factor 6, negative correspondence with maximum thickness but weak positive correspondences with width at blade midpoint, maximum blade length, the depth of the basal indentation, total length, and minimum basal width). Factor 4 appears to reflect a relationship between notch depth such that deeper notches correspond with thinner bases and more shallow basal indentations. These traits are mechanically linked such that a point with deep notches can't have a deep basal indentation without breaking (or at least greatly weakening) the base. Factor 5 summarizes a relationship between the width of the two notches for notched projectile points such that the width of both notches tend to get larger together, but (and somewhat counterintuitively) that points with wide bases tend to have narrower notches. It would seem intuitively that points with wide bases could have more pronounced notches without harming the structural integrity of the point. Factor 6 seems to reflect a negative relationship between maximum point thickness and various size measurements, such that larger points tend to be thinner than smaller points. If you know anything about flint knapping, the rela-tionships reflected in factors 5 and 6 are perhaps counterintuitive. It is much more difficult to make a large, thin biface than a large, thick biface, because the knapper has to make longer but thinner flakes to thin the large biface. Doing so takes con-siderable skill, suggesting that both factors 5 and 6 reflect the skill of the flint knapper, in that skilled knappers can make large, thin points with narrow, well-defined notches, whereas less skilled artisans tend to make smaller, thicker points with more pronounced (less well-defined) notches. These six dimensions effectively summarize the most significant portion of the variation reflected in the data for the original 18 variables.

Discussing the six dimensions would be easier and likely more meaningful to our audience than trying to talk about the interrelationships among the 18 original variables. In short we have learned that the overall size of the points, the size and morphology of the basal elements, and the skill of the flint knapper appear to be the major determinants of the variation in the Ventana Cave assemblage. Of the total common variation in all of the variables, the eigenvalues on Table 15.3 indicate that roughly 37% is controlled by general point size (factor 1), 35% is controlled by the various relationships in basal morphology (factors 2, 3, and 4), and 13% appears to be controlled by the skill of the flint knapper (factors 5 and 6). Such insight would have been difficult to derive using a series of ANOVAs, regression analyses, and correlations, which would require tens if not hundreds of direct comparisons to begin to illustrate. It is certainly much easier to communicate to others using the factor analysis. When evaluating the significance of the factors, please don't forget common sense. If the factors don't make sense, don't force your theoretical and analytic framework to accommodate them. Just as there can be spurious correlations, there can be spurious factors even in randomly generated data (Vierra and Carlson, 1981). Inadequate sample sizes can cause meaningful relationships to be obscured and incidental relationships to appear meaningful. As always, the more representative your sample is, the more effective your quantitative methods will be.

In addition to simplifying the presentation of trends in the data, the factor analysis can serve as a useful tool for structuring additional analyses through the process of data reduction. For example, surrogate variables can serve as effective summaries for the trends the factors reflect. This reduces the number of variables that must be considered in further analyses. A variable used as a surrogate should have very high (positive or negative) loadings for the factor, preferably over .9. For factor 1 in Table 15.3, weight, which has a loading of .901, could serve as a surrogate variable for point size, meaning that the close correspondence between this variable and the factor allows the variable to be used to characterize the factor for additional analyses. In other words, the correspondences in variation summarized by factor 1, which is the single best summary of variation within the data set, can be estimated using weight alone. Correspondence analyses such as correlation can consequently use weight to reflect this underlying pattern for subsequent comparisons. Of course, surrogate variables should make sense to you in that there is some empirical or theoretical link between what the underlying structure the component represents and the specific variable. We identified factor 1 as reflecting artifact size, so weight (which is a direct product of artifact size) does seem intuitively to be a reasonable surrogate variable. In contrast even if a variable such as depth of basal indentation had a good factor loading value, we wouldn't use it as a surrogate because of the lack of an empirical basis for expecting basal indentation depth to be tightly linked to point size.

It is not uncommon, though, for a meaningful factor to not have any single variable with an adequately large factor loading to justify its use as a surrogate. In such cases, it is possible to use a *summated scale* to develop a composite value for a set of variables that do correspond reasonably well with the factor. A summated scale is simply an average value for the variables, calculated as (Variable 1 + Variable 2 + ... Variable *A*)/*N*, where *N* is the number of variables. For factor 3, maximum basal width and minimum basal width both show a good, but not ideal, factor loading. It is consequently possible to create a composite surrogate value by averaging the values for the variables for each individual point (i.e., [Max. Basal Width + Min. Basal Width] / 2). This composite will likely better approximate the factor than either variable alone will. Note though that variables used for creating summated scales must be uniformly positively or negatively correlated with the factor, not both.

If no group of variates has particularly high loadings, it may not be possible to create surrogates using a summated scale. For example, none of the variables seem to correspond satisfactorily with factor 6 to calculate a composite variable. Here instead of using summated scales, we can calculate *factor scores* in which values derived using the factor are used. We won't go through the process of calculating these, but instead note that any software package that allows the calculation of principal components and factor analysis will also calculate component and factor scores.

We could of course always calculate factor scores for all of the factors, instead for bothering with surrogate variables. Using surrogate variables and factor scores are alternatives that each have their advantages and disadvantages. Factor scores: (i) can always be calculated, even when there are no clear choices for surrogate variables; and (ii) reflect the composite loading of all variables on the component, whereas surrogate variables reflect only a single variable (or when using summated scales, a handful of variables). As a result, factor scores are generally applicable and reflect more information than does a surrogate variable, *but* this includes information about variables that are poorly correlated with the component. They can therefore be conceptually poor representations of the underlying relationships. Factor scores are also difficult to compare between different analyses, in that the composition of the factors will change with different data sets. For example, factor 1 for our analysis will certainly be different than factor 1 for another projectile point assemblage analyzed using the same variables and factor analysis, because each factor is a synthetic summary reflecting the specific values of each sample.

In comparison, surrogate variables may continue to be useful surrogates for significant factors in different analyses, even if the factors themselves change somewhat (e.g., weight may continue to be a good proxy for factors reflecting point size in multiple data sets). This can facilitate comparisons based on multiple factor/principal component analyses. Also, the relationship reflected by the surrogate variable is often conceptually clear (e.g., weight has an obvious connection to point size, whereas factor scores do not). However, surrogate variables do not reflect the component perfectly (unless the factor loading is 1 or −1). Thus, if you wish to perfectly describe the factor/component, then factor scores are ideal. If instead you wish to maintain conceptual clarity and/or facilitate comparisons with other variables/data sets, then surrogate variables are typically better choices.

Factor Analysis vs. Principal Component Analysis

As previously mentioned, the single most significant difference between factor analysis and principal component analysis is how they consider specific variation and error. Factor analysis does not consider it at all, being limited instead to the common variance by its mathematical structure. The subsequent factors therefore do not describe all of the variation in a data set. We could even see this by adding the squared factor loadings for a given variable across all of the factors. Given that these values are analogous with the regression coefficient, their sums across all of the factors should equal 1, if all of the variation was held in common with other variables. However, many, perhaps all, variables will not share all of their variation in common. The proportion of the specific variation and error in the variable will be reflected by the difference in the summed square loadings and 1, which can be considerable if there are distinct independent variables impacting only a single variable in a data set. In contrast, principle component analysis includes this variation, such that the components reflect all of the variation in all of the variables, even if that variation is unique to only one of them.

As mentioned previously, many authors see this as a drawback, given that the point of the analysis is describing the common variation for data reduction and descriptive purposes. However, insight can also be gained through principal component analysis, in that components that have high component loadings for only one variable reflect this specific variation in the context of the common variation held among the other variables. This will allow the researcher to identify which variables are largely or completely independent of the other variables (i.e., do not share common variation). Of course this same insight could be gained (although perhaps not as easily) using factor analysis: variables reflecting specific variation will have poor factor loadings across all of the factors. Most statistical packages that perform a factor analysis will also allow you to calculate *communality scores*, which is just the sum of the squared factor loadings for a single variable. Communality scores for our example are reflected in Table 15.2. In our case, we do not see any variables that do not have substantial common variation. We might wish to consider whether variables with communality scores less than .60 should be included in the factor analysis at all. Such variables have a high level of specific variation and typically are not well represented by any of the factors.

This intentionally brief introduction to factor analysis and principal component analysis will give you the basis for using and evaluating these quantitative tools. They are very useful, and can save you time, money, and effort in helping to identify which variables are descriptively and explanatorily meaningful. If you find them to be central to your analyses, we encourage you to investigate these methods further. An excellent choice for this is Hair *et al.* (1995). We now turn our attention to some final consideration related to sampling in the next chapter.

Practice Exercises

1 Define the following terms:

    (a) dimensional scaling and data reduction.

    (b) factor analysis and principal component analysis.

    (c) factor loading score, eigenvalues, and communality scores.

    (d) common variance and specific variance.

2 Imagine an archaeological data set that you might find useful to study any subject matter. What ratio and interval scale variables might you include in your data set? Which of these might share common variation with each other? Which would be appropriate for inclusion in a factor or principal component analysis?

3 Following are factor loadings for five factors used to analyze a pottery assemblage.

    (a) Calculate eigenvalues from the factor loading scores.

    (b) Determine which factors likely provide useful information using the latent root criterion, a percentage of variance criterion using a percentage of 85%, and a scree plot.

    (c) Calculate communality scores using the factor loading scores. What are the implications of the communality scores? Do any of the variables seem to have a significant amount of specific variation?

    (d) Describe in the most general terms which variables load with the factors you identified as being most likely useful. Can you see any connection that might explain the associations identified in the factors?

Variable	1	2	3	4	5
Vessel height	.86	−.38	.33	.00	.000
Vessel diameter	.94	−.03	.02	−.12	−.215
Rim thickness	.66	−.37	−.52	.00	.376
Rim angle	.30	.72	.41	−.35	.308
Opening diameter	.47	.63	.02	.57	.125
Maximum wall thickness	.92	.10	.01	−.14	−.219

16

Sampling, Research Designs, and the Archaeological Record

We have noted throughout the previous chapters that properly integrating our intellectual frameworks with our data is central to effectively using quantitative methods. In this concluding chapter we wish to return to this theme and discuss the importance of samples. Samples have been central to most of the material we have presented. We have illustrated how sample statistics can be used to estimate parameters such as population means, regression coefficients, and standard deviations. We have discussed the improved accuracy of large samples relative to small samples, and we have even discussed means of quantifying this using confidence intervals. Sampling at its most basic form is a means of using information from a part of a population to make inferences about the whole population. It has been central to archaeology from its inception. From the earliest days of culture historical research, archaeologists extrapolated information about artifacts, sites, and regions from smaller subsets of the archaeological record. Often the population being characterized was broadly and vaguely defined (e.g., the Desert Culture, which was a general term for the pre-agricultural people living in the Arid West and Southwest of North America (Jennings 1956)), and the sample of sites/artifacts/areas considered was arbitrarily drawn based on the archaeologist's interests, perception of what was important, logistical capabilities, and access. The use of arbitrary samples mixed with often vague classification systems of various phases, periods, cultures, components, etc. would seem to call into question the adequacy of the samples and their interpretations, yet the knowledge created from this research continues to frame and facilitate archaeological research. However, increasing attention to the importance and structure of sampling has caused it to become a topic of intense interest and occasional controversy (Binford 1964, Hole 1980, Mueller 1975, Nance 1983, Orton 2000, Wobst 1983). This increased emphasis reflects in part the increasing importance of explicitly quantitative analysis in archaeology and the desire to ensure sample selection doesn't bias the subsequent statistical analyses.

As mentioned back in Chapter 1, the materials that enter into the archaeological record are just a portion of the material culture used and produced by humans. The artifacts and other things archaeologists collect are a sample of those that originally entered into the archaeological record. Further, because of time and money constraints (as well as the effectiveness of sampling to describe the materials we collect), archaeologists often only analyze a portion of the artifacts and other objects they discover during their research. Sampling is simply part of the archaeological reality.

Hopefully the materials that enter the archaeological record are satisfactory representations of the materials produced through human action. Likewise, the materials that are preserved enough to allow their recovery are hopefully satisfactory representations of the materials entering the archaeological record. And, the materials recovered by archaeologists are hopefully representative of the materials that were preserved. We know that the archaeological record does not always meet these ideal situations, though. No matter how much we might wish it were otherwise, for example, the sample of recovered organic materials from Pleistocene contexts will never adequately reflect those used and produced by humans. Despite our best efforts to develop increasingly effective means of recovering information (e.g., phytolith analysis, residue analysis), too much time has passed for these materials to be preserved in large abundances. Worrying about the unsatisfactory nature of our sample of Pleistocene age organics is a good thing, in that it can help prevent mistaking the absence of perishable materials in the archaeological record as evidence that they were not used in the past, but rectifying this problem is ultimately out of our control as archaeologists. Aside from our inventiveness in finding new lines of evidence, such inadequacies in the archaeological record can only be dealt with in a general, conceptual way. Simply put, we can postulate what sorts of artifacts, features, ecofacts, etc. *should* be present in the archaeological record, but there is no way to quantify what we can't recover. Thus, our sampling strategies cannot account for whether the extant archaeological record allows reliable inferences about past behavior, the degree to which the archaeological record reflects various aspects of past behavior, or even if it is representative of the materials that originally entered into the archaeological record. These are ultimately theoretical and methodological issues.

When "taking a sample" the only issues we typically hope to control quantitatively are: (i) how we can use a portion of the archaeological record to make reliable inferences about the whole of the extant archaeological record; and (ii) how we can use a portion of the material collected to characterize all of the material we have collected. The first is important in contexts such as regional surveys and limited excavation of a site, in which archaeologists wish to use knowledge of a portion of an area to characterize a larger area. The second is important in contexts in which archaeologists don't have the time or money to analyze everything they recovered, or when destructive analysis might discourage the archaeologist from submitting all possible samples. Although much smaller issues than worrying about the adequacy of the archaeological record as a whole, trying to create adequate samples of the extant archaeological record and the materials we recovered are daunting tasks. The principles underlying sampling are fortunately the same in both of these cases. We outline these issues in the following discussion, beginning with the methods of sample selection.

How to Select a Sample

Imagine that we have 300 obsidian artifacts recovered from five sites that we wish to submit for chemical sourcing analysis using Neutron Activation Analysis (NAA). We could potentially submit all of them, but NAA costs money and is a destructive technique that requires a sample to be cut from each artifact. We might not wish to invest so much money and damage all of the artifacts if a smaller sample could provide the information we desire. The crux of the issue is how can we select a sample that is *representative*, that is, that reflects the variation within the entire population without causing a particular part of it to be overrepresented or underrepresented?

One approach would be to select an *arbitrary sample*, say the 30 largest artifacts. Many archaeologists are suspicious of arbitrary samples, but there is nothing inherently wrong with them. Arbitrary samples can accurately reflect the variation in the underlying distribution, especially when the attributed used to select the sample is random *relative to the underlying population of interest*. Would a sample of the 30 largest artifacts be random relative to the obsidian source? Or would artifact size and source be correlated in some way such that certain sources might be overrepresented among the largest artifacts? There is no obvious connection between artifact size and raw material source, so the sample might be representative of the underlying population. But then again, perhaps not. Different raw material sources might have different initial cobble sizes, which in turn could impact the size of the subsequent artifacts. Or, differences in artifact types among the site might cause the large artifacts to come from a single site. If the sites' inhabitants differed in their access to the various sources, this could cause certain raw materials to be overrepresented relative to the entire population of projectile points. We could of course select the sample based on different criteria (the smallest 30 artifacts, artifacts that are already broken and won't be significantly damaged when the NAA samples are removed, a particular class of artifacts such as scrapers, etc.). These samples might even be representative, but issues similar to those that might cause the 30 largest artifacts to be a biased sample might also be present, causing some portion of the population to be systematically overrepresented or underrepresented.

An alternative to arbitrary sampling that would alleviate the inherent difficulties would be a *simple random sample*. Simple random samples are defined as ones in which each member of the population has the same probability of being selected as all other members. To take a random sample of the 300 artifacts, we would merely number them from 1 to 300 and then randomly select 30 of them. The numbers could be drawn randomly from a hat, the "old-fashioned" way of generating random numbers that is still often used to select bingo numbers and lottery results. Another possibility is to use a random number table, which can be found on the World Wide Web and in many statistics books. For illustrative purposes, we have reproduced a small table of random number (Table 16.1). To use these values to select the 30 artifacts in our hypothetical sample, randomly select a starting place. An easy way

Table 16.1   A random numbers table

22268	71761	18011	00658	67574	84537	29338	64595	67104
18416	60755	65910	91590	42890	18510	10364	80384	07485
32048	11960	66409	57622	42147	68268	12145	03582	09146
69243	23553	35280	40193	27188	19606	22088	99136	70600
90923	76504	53468	85859	64682	60698	98085	60288	53213
30468	58549	02576	41781	74755	82953	59022	36830	72286
75297	43662	44592	87076	81593	21495	22719	69933	29484
67996	45174	59578	45151	31894	93965	04713	95535	91965
92513	58291	08398	91912	80389	39105	57454	86535	34165
56305	56188	99235	81324	88307	94990	70924	86842	73895
79306	62686	59479	12683	15883	54232	87962	28155	36024
27024	58768	01949	83543	80139	66129	67276	67582	67697
70276	08662	26334	90350	95587	78588	45862	62337	23125
84328	42326	67073	83655	55504	73612	06662	76554	91215
01469	72013	68684	31619	32289	46940	30176	74260	21114
65775	41648	91735	58567	09016	35829	23547	69881	78301
48165	13611	10467	55997	54010	02011	03100	73460	43651
94951	42869	04277	90338	32820	17824	78402	28547	91627
10340	13066	03775	76505	63356	68357	70351	49734	65705
42297	62018	13920	08372	17592	58600	23638	92161	75515
82858	71966	49603	47021	16882	37970	04715	48698	24357
07998	53398	05186	09661	82667	05934	77813	13877	70076
90875	11985	00079	28194	93309	91229	70132	11486	54909
42918	82016	14801	77746	47762	62511	20903	75600	12072
43740	53042	88570	26717	30481	11051	30050	68592	37181
13076	03243	29885	38339	09791	19946	54667	69698	33908
79558	05394	26016	80053	49841	20830	91366	32718	52280
47277	13243	53327	60390	33789	92623	25503	16098	76501
36833	79218	38867	25493	16801	20787	37938	62283	30631
61642	10812	27497	45842	92763	66612	39565	18694	77063
35974	06817	35969	13173	83546	41030	10425	05215	31376
10481	06768	05114	81421	24515	17531	46839	06659	93158
15808	23336	40326	67604	92158	90336	31693	57999	07730
65186	38277	18546	47617	91716	62183	09983	46484	39435
31494	72562	01877	28028	71765	51632	06221	85816	27831

to do this is by closing your eyes and putting your finger on a number. Start with the number you select, and then move consistently up, down, to the left, or to the right. Look at the first or last three numbers in the random number sequence. In this case, we are interested in numbers between 001 and 300. As we encounter a value within this range, we select the artifact corresponding to the number. All other numbers are discounted. When we have selected 30 artifacts, we stop. Be sure to use a different starting place when obtaining two or more samples as opposed to always starting from the same location, say the upper left corner. Otherwise, your numbers are arbitrary, not truly random. Of course they may be random *in respect to the underlying distribution*, so the samples could still be representative. But the samples become arbitrary, as opposed to random, samples just the same.

Random numbers can also be generated using computers, but be careful with this method. Depending on the software package, the numbers can reflect a pre-programmed random number sequence (and thus always be an arbitrary sample of the same numbers) or be based on an algorithm that gives the appearance of ran-domness but truly isn't. Often these non-random "random" algorithms are based on factors such as how long the computer has been on, so that the "start point" for multiple sequences may be non-random with respect to each other. This is a common problem for those who use agent-based modeling and similar methods, in that a series of multiple runs may not be random at all.

An issue that arises when taking random samples is whether we should sample *with replacement* or *without replacement*. The difference between these sampling techniques lies in whether the same individual can be selected more than once in the sample. Sampling with replacement places each selected individual back into the sampling pool before each new sample member is drawn, whereas sampling without replacement deliberately withholds previously selected individuals from the sampling pool. Sampling with replacement is the conceptually and mathematically easier system, in that the probability of selecting any individual always remains the same throughout the entire selection process. In contrast, sampling without replace-ment causes the probabilities of selecting an individual to change during the sam-pling process. Consider for example that when we select the very first artifact for our obsidian artifact sample, each artifact in our population of 300 has a 1 in 300 chance of being selected. If we are sampling with replacement, the artifact first selected will again be placed into the sample pool, meaning that each artifact will have a 1 in 300 chance of being selected when the second, third, fourth, and sub-sequent members of the sample are selected. The probability of an individual arti-fact being selected will consequently always remain 1 in 300 (or .0033) throughout the selection of the entire sample. In contrast, if we sample without replacement, the first artifact selected is not returned to the sampling pool when drawing subse-quent sample members. The probability associated with selecting any given artifact next is now only 1 in 299. Obviously this isn't a significant change with such a large population, but by the 30th artifact, the probability of selecting a given artifact has risen to 1 in 271. The difference between the probabilities of 1 in 300 (.0033) and 1 in 271 (.0036) is small, but still meaningful. Taking large samples of small populations would result in even more drastic changes in probability (e.g., the probabilities associated with a sample of 30 individuals without replacement from a population of 50 individuals changes from 1 in 50 (.02) to 1 in 21 (.05) by the end of the sampling process).

Sampling with replacement is fairly common in ethnology, but archaeologists and physical anthropologists tend to sample without replacement. Ethnologists studying a variable such as "hours a day spent hunting" may randomly select a hunter to accompany each day, and will not be bothered by the reoccurrence of a particular hunter multiple times. The amount of effort invested in hunting might change day to day, and including the same hunter will provide valuable information about such variation. In archaeology and physical anthropology, though, we typically deal with items that do not change over the course of a study. Including the same artifact in our sample of 30 obsidian artifacts seems odd, given that the artifact's determined source shouldn't differ between measurement events. If it does, then we ought to reconsider the validity, accuracy, and precision of our measurements.

Now there is nothing wrong *per se* with using sampling with replacement in archaeological contexts, even when the same set of measurements will be included twice. The goal of random sampling is to provide a characterization of the underlying population without systematically over representing or under representing some part of it. Including the same set of measurements more than once doesn't hurt this goal at all, so long as the redundancy is random, as opposed to systematic. Further, there are various quantitative methods such as the binomial that require consistent probabilities through multiple events. Some methods even require sampling with replacement, because of the consistency in probabilities (see Orton 2000 for a discussion of many of these). However, we and most other archaeologists find it odd and unnecessarily redundant to include the same measurements of a pot, skull, house, shell, or any other object more than once in a sample when additional items could be selected and analyzed.

Another alternative to an arbitrary sample is a *systematic sample*. The systematic sample starts from a given spot/item and then selects the remaining members of a sample at consistent increments. For the obsidian artifacts, a systematic sample might start by numbering each of the artifacts from 1 to 300 as was done when creating the random sample, and then selecting number 10, number 20, number 30, and so forth at increments of 10 until 30 artifacts are selected. This will produce a useful sample of 30 artifacts that is random *relative to the underlying population*, so long as the underlying population is randomly distributed relative to the assigned numbers.

Systematic samples can also be used when the underlying population is not randomly distributed relative to the assigned numbers, though. Let's say that in our example of 300 artifacts, Site A produced 100 artifacts and the remaining sites each produced 50 artifacts. By happenstance, a random sample might include many more (or too few) artifacts from Site A than expected. If the sites differ in regards to raw material sources, this "accidental" overrepresentation of Site A might bias the sample. A systematic sample could help prevent this, if the artifacts from each site are numbered consecutively (i.e., Site A's artifacts are numbered 1 through 100, Site B's artifacts are numbered 101 through 150, etc.). The systematic sample would ensure that ten artifacts were selected from Site A, and five artifacts were selected from each of the remaining sites, thereby eliminating one possible form of bias. Of course, the sample will not be representative if the numbers are assigned so as to favor some part of the population (e.g., whole, large projectile points are intentionally assigned numbers corresponding to those that will be selected during the sampling). If a researcher is going to do this, he or she would be better off just selecting an arbitrary sample and defending it on that basis.

Systematic samples are perhaps most commonly used in the United States for initial site and area surveys to reduce the cost and time for infield analysis. For example, archaeologists starting surveys along linear tracks such as roads or rail lines will sometimes conduct a systematic pedestrian survey to get an idea of site density before scheduling crews and making logistical arrangements. The systematic survey can be as simple as surveying 100 m along a road right of way, driving 1 km down the road, surveying another 100 m, and then repeating the process. This "leapfrog" survey will help the field director figure out where significant archaeological remains are likely to be encountered. Similar systematic sampling strategies are sometimes used during surveys in which infield artifact analysts walk a certain number of paces, analyze or collect the artifacts within 5 m or some other set distance of their stopping point, and then repeat the process by walking the same number of paces. In our experience, these "dog-leash surveys" (so named because the archaeologist collecting and/or analyzing artifacts in a circumscribed area is like a dog on a leash at each stopping spot) are especially common in areas where plowing and other disturbances make it difficult to define site boundaries and identify spatially distinct features.

Systematic surveys do have the advantage that they are often easy to select (at the very least, we don't have to bother with a random numbers table or similar device). However, we must also be careful to ensure that the underlying population does not have some regularity that corresponds with the systematic unit we are using. If, for example, the geology of an area has led to a consistent ridge and floodplain pattern in which the ridges are roughly 1 km apart, a systematic survey focusing on 100 m every 1 km will not provide an unbiased reflection of the area's archaeology. Depending on where we start, we will instead consistently survey ridge tops or floodplains, without covering the entire range of topographic locations, which in turn can lead to a skewed view of the area's archaeology.

Another sampling alternative is the *stratified random sample*, which combines the best features of simple random samples and systematic samples in some ways. With stratified random samples, we identify some underlying influence that is expected to structure the population, divide the population according to this influence, and then randomly sample within each group. In our obsidian artifact example, we identified several variables that might affect the sources represented at each site (e.g., access to sources, requisite artifact typology, differences in cobble size). Randomly selecting

30 artifacts from the combined sample of 300 artifacts might lead to the overrepresentation of a particular site(s), simply by chance. If so, the sourcing study might not adequately reflect the variation in sources among the sites and the population as a whole. To control for the possibility of over representing or under representing any particular site in the sourcing study, divide the 300 artifacts into five groups according to site, and then randomly select the same proportion of points from each group. Thus, we would randomly select ten artifacts from the 100 obsidian objects from Site A, and five artifacts each from Sites B through E, which would help ensure that the influence of variables creating differences and similarities between sites can be detected, while also helping to prevent systematic bias caused by factors that might influence the variation within sites. (A systematic sampling strategy would also produce a proportionally consistent sample from each site, assuming they were numbered consecutively, but the sample for each site would be selected systematically as opposed to randomly as is the case here.)

So which sampling strategy should archaeologists use? All four of the possibilities we outline here have been and will continue to be useful in archaeological research. Arbitrary samples do not need to be rejected out of hand. Arbitrary samples can be representative when they are selected using criteria that is not causally tied to the variables of interest. For example, researchers working with museum collections often are in truth dealing with arbitrary samples reflecting other peoples' perceptions of what is "worth" keeping/donating, but they generally treat their collections as representative samples of more broadly defined populations based on the assumption that the museum's collection reflects the general population of artifacts. This is often a reasonable assumption, but it should be evaluated. Museum collections often include donations from collectors and researchers who might have a specific interest in certain artifacts or gathered their materials from specific areas, which could introduce significant bias.

Arbitrary samples can also be ideal when the archaeologist has detailed knowledge about the underlying sources of variation. For example, obsidians that are macroscopically distinct can be well represented in an arbitrary sample, which might be preferable to hoping that members of each distinct obsidian group are selected by random chance. By the same token, both New World and Old World culture historians often chose samples to reflect the variation in populations of interest based on their perceptions of that variation, and the continued support of many of their conclusions and characterization indicates their samples were not inordinately biased. In fact, an archaeologist with a good understanding of the material record *may* be as effective in determining a representative sample as any of the methods presented here. However, arbitrary samples can improperly reinforce an archaeologist's preconceived notions when the variation is constrained around that which the archaeologist already has identified. The fundamental issue is that it is generally impossible to *ensure* that an arbitrary sample isn't biased. This becomes a very significant issue when we begin to describe samples visually or use various methods to quantify measures of central tendency and dispersion. Sometimes the "typical" is not well reflected in arbitrary samples because researchers become fascinated by the atypical and unusual. The inordinate concentration on outliers can greatly influence measures such as the mean and the standard deviation, as well as graphical methods of illustrating distributions. We recommend in general that researchers focus on using one of the other sampling methods when they can be applied. Using these approaches, you may not necessarily get all of the cool or unusual stuff, but the sample should be a good reflection of the general characteristics of the population.

By the same token, though, do not consider results derived using random samples as "right." Random samples can, by chance, produce a less than perfect reflection of the entire population. This possibility was in fact implicitly acknowledged in our discussion of confidence intervals, in which we discussed that there is a possibility that a parameter is farther away from our statistic than we expect. Using 95% confidence intervals around a sample mean only ensures that the true mean is within the confidence intervals 95% of the time. For the remaining 5% of the cases, our sample mean is farther away from μ than we expect. The reason for this is because, by happenstance, the underlying population was not properly reflected in the sample. If the sample is derived randomly, this means that some portion of the population has been overrepresented relative to the other portions by chance. Be sure to remember that evaluating the adequacy of the sample is a continual process, as opposed to something performed only when initially choosing the sample. Accidental bias may not be evident until we learn more about our materials. We can of course always increase the size of the sample to improve the certainty that the statistic is a good estimate of the population parameter. This brings us to our next topic, sample size.

How Big a Sample is Necessary?

In general, larger samples are much better representations of the underlying populations than are smaller samples. Assuming that they are drawn in a way that allows the creation of a representative sample, large samples will better reflect the entire variation within a population, including the rare or unusual members. However, even small samples can provide excellent estimates of many population parameters. Exactly what sample size is acceptable for characterizing the underlying population but not so excessively large that it wastes time and money is not an easy issue to determine. Various rules of thumb have been suggested, including samples between 5% and 10% are often adequate to accurately reflect measures of central tendency and dispersion, but these are just guesses. In a large, homogeneous population with little variation, a 10% sample may be unnecessarily large. In a very diverse population, it may not adequately reflect the underlying variation that might interest the researcher. Further, in many archaeological cases we may not be certain exactly how large the population is (e.g., the number of Middle Woodland ceramic vessels curated in museums throughout the world), making it impossible to determine what a 10% sample would be.

When estimating means, one possible approach is to determine the desired error range *a priori*. For example, imagine that we were taking metric measurements of 300 obsidian points, and we wished to determine the likely range of the mean using 95% confidence intervals that are no larger than 2 cm from the upper to the lower confidence interval. Using Equation (8.3) and the information presented in Chapter 8, we know that $L_1 = \bar{Y} - t_{[.05,df]}\dfrac{s}{\sqrt{n}}$ and $L_2 = \bar{Y} + t_{[.05,df]}\dfrac{s}{\sqrt{n}}$ for 95% confidence intervals. In this case we want the value provided by $t_{[.05,df]}\dfrac{s}{\sqrt{n}}$ to equal 1 cm, which would provide confidence intervals 2 cm wide. To solve this, we must have at least a rough estimate of the standard deviation. We could just guess what the standard deviation might be, but that seems unsatisfactory. There are several more secure means of getting estimates. We could use previously calculated standard deviations from other assemblages as estimates, based on the assumption that the range of variation will likely be similar across assemblages. We could also estimate the standard deviation using the range and Table 4.4, if we could quickly identify the largest and smallest individuals. A third alternative, which we prefer, is to start by taking a small sample of a population to determine a preliminary estimate of the standard deviation. Of course, these are not incompatible alternatives, in that they all can be applied to determine a consistent estimate. Let's pretend that we do take a 5% sample of the points, and find a standard deviation of 2.3 cm. Using this value and starting with a sample of 15 (the 5% sample we already collected), we can create a table illustrating the anticipated spread of the confidence intervals (Table 16.2). The table indicates that we reach our desired level of certainty at a sample of 23 points. We can then evaluate whether this sample is in fact adequate using the improved estimate of the standard deviation derived from the larger sample of $n = 23$. We can continue resampling as necessary until we have the confidence intervals we desire.

Table 16.2 Changing confidence intervals as sample size increases

Sample size	t-value	Standard error	Size of confidence interval (cm)
15	2.13	0.59	1.27
16	2.12	0.58	1.22
17	2.11	0.56	1.18
18	2.10	0.54	1.14
19	2.09	0.53	1.10
20	2.09	0.51	1.07
21	2.08	0.50	1.04
22	2.07	0.49	1.02
23	2.07	0.48	0.99
24	2.06	0.47	0.97
25	2.06	0.46	0.95

Some Concluding Thoughts

At this point, we are confident that you, the reader, have the requisite background and knowledge to successfully use quantitative methods in your own research, and to evaluate quantitative applications encountered in other researchers' work. We have not discussed every possible quantitative method that has been or could be applied in archaeology. However, with the knowledge you have gained through this text, you do have the background to research, learn, and implement all of the additional methods that are available. We do want to end with a few parting cautionary notes and words of encouragements.

To begin with, please trust your instincts when applying and interpreting quantitative methods. The approaches outlined here are outstanding ways to evaluate ideas about how the archaeological record is structured and to help us identify relationships we might otherwise overlook. Still, the likelihood of incorrectly identifying a difference when there isn't one (a Type I error) or failing to detect a difference when there is (a Type II error) is always present. Statistical methods are not an infallible means of getting the *right* answer, but are instead useful tools for gauging the likelihood of certain relationships. If your quantitative analysis identifies a difference when you think there shouldn't be one, based on previous research and theoretical grounds, don't necessarily reject your previous ideas. Stop and investigate what could be driving the aberrant result. Remember, when using a critical level of .05, we expect to incorrectly reject a null hypothesis once for every 20 tests we complete. By the same token, when there isn't a statistically significant difference when you think there ought to be one, stop and seriously consider the probability of committing a Type II error. Could you really differentiate between alternate distributions if they were present? If not, then your results likely provide little or no insight into the archaeological record. Additional lines of evidence, including the statistical analyses of other materials, may be used to support or refute the statistical result. Regardless, remember that statistical methods are expected to provide spurious results at least some of the time. If the results don't make sense, investigate the issue more before discarding potentially useful ideas and theories.

On a related note, the .05 critical level is arbitrary and may not be appropriate in all contexts. .05 represents good betting odds, but there may be cases where we are more concerned about committing a Type II error, and need to increase it to .10 or even more. By the same token, there may be cases when a researcher feels that .05 is too high a likelihood of committing a Type I error, and would prefer to lower it to .01 or even .001. You should carefully think about the implications of committing Type I and Type II errors, and adjust your critical value accordingly.

Also, don't confuse statistical significance with some sort of strength of association. When evaluating a null hypothesis using a critical level of .05, a resulting probability of .004 is no more significant than a probability of .04. Both indicate a statistically significant difference. It would be improper to consider the difference reflected by .004 as stronger than that reflected by .04. Both results could be Type I errors at the selected level of significance. Further, you shouldn't change the level of rejection after the analysis is complete so as to appear to minimize the probability of making a Type I error. Using a critical value corresponding with .05 to reject a hypothesis when the probability is .04 but then switching to a critical value corresponding with .01 for a probability of .004 misstates the likelihood of committing a Type I error. In truth, the critical value is really .05, which indicates you will likely make an error once every 20 comparisons. An erroneous result corresponding with a probability of .004 is still an error. Changing the probability after the fact to make the error appear less likely doesn't change the fact. Remember Gregg Easterbrook's famous warning that, "Torture numbers, and they'll confess to anything." Be honest about what you are comparing, how and why the data are organized as they are, and the likelihood of being wrong, and you will be far more likely to produce useful analyses that will withstand the scrutiny and continued evaluation that goes along with archaeology as a growing science.

Finally, remember that quantitative methods rely on our theoretical and methodological frameworks to be meaningful. Without a strong conceptual framework, the importance of any quantitative analysis will be limited to vacuous descriptions of the world (e.g., Variables 1 and 2 might be correlated). Although such relationships can be useful for inductively deriving propositions for future consideration, they do not and cannot provide satisfactory explanations or even descriptions of the archaeological record. The use of quantitative methods is essential to current archaeology, but nothing can replace a rigorous theoretical and analytic framework.

Practice Exercises

1   Define and differentiate between a sample, a statistic, a population, and a population parameter (see Chapter 4). Define and differentiate between arbitrary samples, random samples, systematic samples, and stratified random samples.

2   Identify three examples in the archaeological literature where archaeologists use samples. Answer each of the following questions for the examples you found.

    (a)   What is the underlying population being estimated by the sample?

    (b)   How was the sample selected? How large was the sample?

    (c)   To the best of your ability, evaluate whether the sample seems adequate in terms of its size and its composition for the author's arguments.

3   Following are 60 measurements (mm) of the anterior breadth of the distal trochlea of the distal humeri of deer (*Odocoileus* sp.) presented by Lyman (2006: 1258).

(a) Calculate the mean and standard error using all of the measurements.

(b) Select an arbitrary sample of 12 measurements based on whatever means you care to use. Calculate a mean and standard error.

(c) Randomly select a sample of 12 measurements. Calculate a mean and standard error.

(d) Select a systematic sample of every fifth measurement. Calculate a mean and standard error.

(e) Select a stratified random sample on the basis of deer species so that you have three individuals from *O. virginianus* and nine individuals from *O. hemionus*. Combine the samples and calculate a mean and standard error.

(f) How do each of the sample means you calculated from your samples compare to the population mean you determined from all 60 variates? Which method seems better to you in this case? Defend your answer to the best of your ability.

Taxon	Anterior breadth	Taxon	Anterior breadth	Taxon	Anterior breadth
O. virginianus	25.36	*O. hemionus*	29.56	*O. hemionus*	26.08
O. virginianus	26.28	*O. hemionus*	27.36	*O. hemionus*	28.46
O. virginianus	28.04	*O. hemionus*	28.76	*O. hemionus*	28.06
O. virginianus	24.18	*O. hemionus*	29.04	*O. hemionus*	25.32
O. virginianus	23.92	*O. hemionus*	26.80	*O. hemionus*	28.28
O. virginianus	28.90	*O. hemionus*	27.24	*O. hemionus*	25.50
O. virginianus	24.84	*O. hemionus*	29.74	*O. hemionus*	26.34
O. virginianus	25.70	*O. hemionus*	25.56	*O. hemionus*	25.90
O. virginianus	25.26	*O. hemionus*	26.28	*O. hemionus*	27.90
O. virginianus	26.70	*O. hemionus*	24.80	*O. hemionus*	27.26
O. virginianus	26.32	*O. hemionus*	24.48	*O. hemionus*	25.66
O. virginianus	26.76	*O. hemionus*	26.46	*O. hemionus*	28.12
O. virginianus	27.30	*O. hemionus*	27.62	*O. hemionus*	25.70
O. virginianus	25.12	*O. hemionus*	28.40	*O. hemionus*	27.06
O. virginianus	26.68	*O. hemionus*	26.40	*O. hemionus*	27.70
O. virginianus	26.16	*O. hemionus*	28.46	*O. hemionus*	24.78
O. virginianus	26.68	*O. hemionus*	27.76	*O. hemionus*	29.60
O. hemionus	27.22	*O. hemionus*	26.80	*O. hemionus*	28.52
O. hemionus	25.70	*O. hemionus*	24.86	*O. hemionus*	29.80
O. hemionus	28.28	*O. hemionus*	26.78	*O. hemionus*	28.04

References

Abbott, D. R., S. E. Ingram, and B. G. Kober (2006). Hohokam exchange, and Early Classic Period organization in Central Arizona: Focal villages or linear communities? *Journal of Field Archaeology*, **31**: 285–306.

Alzualde, A., N. Izagirre, S. Alonso, N. Rivera, A. Alonso, A. Azkarate, and C. de la Rúa (2007). Influence of the European kingdoms of late antiquity on the Basque Country: An ancient-DNA study. *Current Anthropology*, **48**: 155–62.

Andrefsky, W., Jr. (2005). *Lithics: Macroscopic Approaches to Analysis*. Cambridge Manuals in Archaeology. Cambridge: Cambridge University Press.

Arnold, D. and A. L. Nieves (1992). Factors affecting ceramic standardization. In G.J. Bey III and C.A. Pool (eds.), *Ceramic Production and Distribution: An Integrated Approach* (pp. 93–113). Boulder, CO: Westview Press.

Baker, G. and C.M. Hazel (2007). Ryan (40RD77), a Late Middle Archaic Benton culture cemetery in Tennessee's Central Basin. *Journal of Alabama Archaeology*, **53**: 1–84.

Baxter, M.J. (1994). *Exploratory Multivariate Analysis in Archaeology*. Edinburgh: Edinburgh University Press.

Bayman, J.M. and J.J. Moniz Nakamura (2001). Craft specialization and adze production on Hawaii Island. *Journal of Field Archaeology*, **28**: 239–52.

Benco, N.L. (1988). Morphological standardization: An approach to the study of craft specialization. In C.C. Kolb and L.M. Lackey (eds.), *A Pot for All Reasons: Ceramic Ecology Revisited* (pp. 57–72). Philadelphia: Temple University.

Benfer, R.A. and A.N. Benfer (1981). Automatic classification of inspectional categories: Multivariate theories of archaeological data. *American Antiquity*, **46**: 381–96.

Binford, L.R. (1964). A consideration of archaeological research design. *American Antiquity*, **29**: 425–41.

Black, S.L. and K. Jolly (2002). *Archaeology by Design*. Walnut Creek, CA: AltaMira Press.

Bustard, W. (1996). *Space as Place: Small and Great House Spatial Organization in Chaco Canyon, New Mexico, A.D. 1000–1150*. Unpublished Ph.D. dissertation, Department of Anthropology, University of New Mexico, Albuquerque, New Mexico.

Chippindale, C. (2000). Capta and data: On the true nature of archaeological information. *American Antiquity*, **65**: 605–12.

Christenson, A. and D.W. Read (1977). Numerical taxonomy, R-mode factor analysis and archaeological classification. *American Antiquity*, **42**: 163–79.

Cochran, W.G. (1954). Some methods for strengthening the common chi-square tests. *Biometrics*, **10**: 417–51.

Conover, W.J. (1980). *Practical Nonparametric Statistics*, 2nd edition. New York: John Wiley & Sons.

Cordell, L.S. (1997). *Archaeology of the Southwest*. New York: Academic Press.

Costin, C.L. and M.B. Hagstrum (1995). Standardization, labor investment, skill, and organization of ceramic production in Late Pre-Hispanic Highland Peru. *American Antiquity*, **60**: 619–39.

Cowgill, G. (1982). Clusters of objects and associations between variables: Two approaches to archaeological classification. In R. Whallon and J.A. Brown (eds.), *Essays on Archaeological Typology* (pp. 30–55). Evanston, IL: Center for American Archeology Press.

Cowgill, G. (1990). Artifact classification and archaeological purposes. In A. Voorrips (ed.), *Mathematics and Information Science in Archaeology: A Flexible Framework* (pp. 61–78). Bonn: Holos Verlag.

Crawford, G.T. (1993). *A Quantitative Analysis and Catalog of British and Irish Iron-Age Swords*. Unpublished MA Thesis, Department of Anthropology, University of Missouri, Columbia.

Crown, P. (1994). *Ceramics and Ideology: Salado Polychrome Pottery*. Albuquerque: University of New Mexico Press.

Crown, P. (1995). The production of the Salado polychromes in the American Southwest. In B.J. Mills and P.L. Crown (eds.), *Ceramic Production in the American Southwest* (pp. 142–66). Tucson: University of Arizona Press.

Crown, P. and S.K. Fish (1996). Gender and status in the Hohokam Pre-Classic to Classic transition. *American Anthropologist*, **98**: 803–17.

Cruz Antillón, R., R.D. Leonard, T.D. Maxwell *et al.* (2004). Galeana, Villa Ahumada, and Casa Chica: Diverse sites in the Casas Grandes region. In G.E. Newell and E. Gallaga (eds.), *Surveying the Archaeology of Northwest Mexico* (pp. 149–76). Salt Lake City: University of Utah Press.

Di Peso, C.C., J.B. Rinaldo, and G.J. Fenner (1974). *Casas Grandes: A Fallen Trading Center of the Gran Chichimeca*, Volumes 2 and 5. Dragoon, AZ: Amerind Foundation.

Dohm, K. (1990). Effect of population nucleation on house size for pueblos in the American Southwest. *Journal of Anthropological Archaeology*, **9**: 201–39.

Donahue, R.E., M.L. Murphy, and L.H. Robbins (2002–2004). Lithic microwear analysis of Middle Stone Age artifacts from White Painting Rock Shelter, Botswana. *Journal of Field Archaeology*, **29**: 155–64.

Ehrenberg, A.S. (1981). The problem of numeracy. *The American Statistician*, **35**: 67–71.

Everitt, B.S. (1992). The analysis of contingency tables. *Monographs on Statistics and Applied Probability* 45. New York: Chapman and Hall.

Fenenga, F. (1953). The weights of chipped stone points: A clue to their functions. *Southwestern Journal of Anthropology*, **9**: 309–23.

Fish, P.R. (1978). Consistency in archaeological measurement and classification: A pilot study. *American Antiquity*, **43**: 86–9.

Ford, J.A. (1938). A chronological method applicable to the Southeast. *American Antiquity*, **3**: 260–4.

Ford, J.A. (1962). A quantitative method for deriving cultural chronology. *Pan American Union, Technical Manual*, 1.

Foster, H.T. and A.D. Cohen (2007). Palynological evidence of the effects of the deerskin trade on forest fires during the eighteenth century in southwestern North America. *American Antiquity*, **72**: 35–51.

Gibbons, J.D. (1985). *Nonparametric Statistical Inference*. New York: M. Dekker.

Gnaden, D. and S. Holdaway (2000). Understanding observer variation when recording stone artifacts. *American Antiquity*, **65**: 739–47.

Good, I.J. (1973). What are degrees of freedom? *The American Statistician*, **27**: 227–8.

Good, I.J. (1985). The paleontology of Hidden Cave: Birds and mammals. In D.H. Thomas (ed.), *The Archaeology of Hidden Cave, Nevada* (pp. 125–61). American Museum of Natural History Anthropological Papers 66(1).

Grayson, D.K. (1984). *Quantitative Zooarchaeology: Topics in the Analysis of Archaeological Faunas*. Studies in Archaeological Science. Orlando: Academic Press.

Hair, J.F., Jr., R.E. Anderson, R.L. Tatham, and W.C. Black (1995). *Multivariate Data Analysis*. Englewood Cliffs, NJ: Prentice Hall.

Hart, J.P., R.A. Daniels, and C.J. Sheviak (2004). Do *Cucurbita Pepo* gourds float fishnets? *American Antiquity*, **69**: 141–8.

Hasenstab, R.J. and W.C. Johnson (2001). Hilltops of the Allegheny Plateau: A preferred microenvironment for Late Prehistoric horticulturalists. In L.P. Sullivan and S.C. Prezzano (eds.), *Archaeology of the Appalachian Highlands* (pp. 3–18). Knoxville: University of Tennessee Press.

Hassan, F.A. (1981). *Demographic Archaeology*. New York: Academic Press.

Haury, E.W. (1975). *The Stratigraphy and Archaeology of Ventana Cave*. Tucson: University of Arizona Press.

Hoffman, C. (1993). Close-interval core sampling: Tests of a method for predicting internal site structure. *Journal of Field Archaeology*, **20**: 461–74.

Hole, B.L. (1980). Sampling in archaeology: A critique. *Annual Review of Anthropology*, **9**: 217–34.

Ireland, S. (2000). *Greek, Roman, and Byzantine Coins in the Museum of Amasya (Ancient Amaseia), Turkey*. Royal Numismatic Society Special Publication No. 33/British Institute of Archaeology at Ankara Monograph No. 27, London.

Hudson, C.M. and M.T. Boulanger (2008). *Assessing the "Grip-ability" of Ceramic Surface Treatments in the American Northeast*. Poster presented at the 73rd Annual Meeting of the Society for American Archaeology, April 2008. Vancouver, British Columbia.

Jefferies, R.W. and B.M. Butler (1982). *The Carrier Mills Archaeological Project: Human Adaptations in the Saline Valley, Illinois*. Carbondale, IL: Center for Archaeological Investigations, Southern Illinois University.

Jennings, J.D. (1956). The Desert Culture. In R. Wauchope (ed.), *The American Southwest: A Problem in Cultural Isolation* (pp. 59–127). Memoirs of the Society for American Archaeology 11. Salt Lake City.

Jolliffe, I.T. (2002). *Principal Component Analysis*. New York: Springer.

Knecht, H.D. (1991). *Technological Innovation and Design during the Early Upper Paleolithic: A Study of Organic Projectile Technologies*. Ph.D. Dissertation, Department of Anthropology, New York University.

Larntz, K. (1978). Small-sample comparison of exact levels for chi-square goodness-of-fit statistics. *Journal of the American Statistical Association*, **73**: 253–63.

Lewis, R.B. (1986). The analysis of contingency tables in archaeology. In M.B. Schiffer (ed.), *Advances in Archaeological Method and Theory*, Volume 9 (pp. 277–310). Orlando: Academic Press.

Lewontin, R.C. and J. Felsenstein (1965). The robustness of homogeneity tests in $2 \times N$ tables. *Biometrics*, **21**: 19–33.

Longacre, W.A., K.L. Kvamme and M. Kobayashi (1988). Southwestern pottery standardization: An ethnoarchaeological view from the Philippines. *The Kiva*, **53**: 101–12.

Lyman, R.L. (2004). Identification and paleoenvironmental significance of late-quaternary ermine (*Mustela erminea*) in the central Columbia Basin, Washington, Northwestern USA. *The Holocene*, **14**: 553–62.

Lyman, R.L. (2006). Identifying bilateral pairs of deer (Odocoileus sp.) bones: How symmetrical is symmetrical enough? *Journal of Archaeological Science*, **33**: 1256–65.

Lyman, R.L. (2008). *Quantitative Paleozoology*. Cambridge: Cambridge University Press.

Lyman, R.L. and T.L. VanPool (2009). Metric data in archaeology: A study of intra-analyst and inter-analyst variation. *American Antiquity*, **74**: 485–504.

McKenzie, D.H. (1970). Statistical analysis of Ohio fluted points. *The Ohio Journal of Science*, **70**: 352–64.

Miller Wieberg, D.A. (2006). *Establishing the Perimortem Interval: Correlation between Bone Moisture Content and Blunt Force Trauma Characters*. MA Thesis, Department of Anthropology, University of Missouri, Columbia.

Mills, B.J. (1995). The organization of Protohistoric Zuni ceramic production. In B.J. Mills and P.L. Crown (eds.), *Ceramic Production in the American Southwest* (pp. 200–30). Tucson: University of Arizona Press.

Mueller, J.W. (1975). *Sampling in Archaeology*. Tucson: University of Arizona Press.

Nance, J.D. (1983). Regional sampling in archaeological survey: The statistical perspective. In M.B. Schiffer (ed.), *Advances in Archaeological Method and Theory*, Volume 6 (pp. 289–356). New York: Academic Press.

Nelson, M.C., M. Hegmon, S. Kulow, and K.G. Schollmeyer (2006). Archaeological and ecological perspectives on reorganization: A case study from the Mimbres region of the U.S. Southwest. *American Antiquity*, **71**: 403–32.

O'Brien, M.J. and R.L. Lyman (1999). *Seriation, Stratigraphy, and Index Fossils: The Backbone of Archaeological Dating*. New York: Kluwer Academic Press.

O'Connell, J.F. (1987). Alyawara site structure and its archaeological implications. *American Antiquity*, **52**: 74–108.

Orton, C.R. (2000). *Sampling in Archaeology*. Cambridge: Cambridge University Press.

Ottaway, B.S. and Q. Wang (2004). *Casting Experiments and Microstructure of Archaeologically Relevant Bronzes*. BAR International Series 1331. Oxford: Archaeopress.

Paine, R.R. (ed.) (1997). *Integrating Archaeological Demography: Multidisciplinary Approaches to Prehistoric Populations*. Carbondale, IL: Center for Archaeological Investigations, Southern Illinois University at Carbondale.

Perttula, T.K. and R. Rogers (2007). The evolution of a Caddo community in Northeastern Texas: The Oak Hill village site (41RK214), Rusk County, Texas. *American Antiquity*, **72**: 71–94.

Petrie, W.M.F. (1899). Sequences in prehistoric remains. *The Journal of the Anthropological Institute of Great Britain and Ireland*, **29**(3/4): 295–301.

Pfeiffer, S. (1977). The skeletal biology of archaic populations of the Great Lakes Region. *Archaeological Survey of Canada*, No. 64. Ottawa: National Museums of Canada.

Phagan, C.J. (1988). Projectile point analysis, part I: Production of statistical type and sub-types. In E. Blinman, C.J. Phagan, and R.H. Wilshusen (eds.), *Dolores Archaeological Program: Supporting Studies: Additive and Reductive Technologies* (pp. 9–86). Denver, CO: United States Department of the Interior, Bureau of Reclamation, Engineering and Research Center.

Prasciunas, M.M. (2007). Bifacial cores and flake production efficiency: An experimental test of technological assumptions. *American Antiquity*, **72**: 334–48.

Rakita, G.F.M., and J.E. Buikstra (2008). Introduction. In *Interacting with the Dead: Perspectives on Mortuary Archaeology for the New Millennium*, edited by G.F.M. Rakita, J.E. Buikstra, L.A. Beck, and S.R. Williams, pp. 1–11. University Press of Florida, Gainsville.

Redmond, B.G. and K.B. Tankersley (2005). Evidence of Early Paleoindian bone modification and use at the Sheriden Cave Site (33WY252), Wyandot County, Ohio. *American Antiquity*, **70**: 503–26.

Riordan, T.B. (2009). "Carry me to Yon Kirk Yard": An investigation of changing burial practices in the seventeenth-century cemetery at St. Mary's City, Maryland. *Historical Archaeology*, **43**: 81–92.

Roscoe, J.T. and J.A. Byars (1971). An investigation of the restraints with respect to sample size commonly imposed on the use of the chi-square test. *Journal of the American Statistical Association*, **66**: 755–9.

Roux, V. (2003). Ceramic standardization and intensity of production: Quantifying degrees of specialization. *American Antiquity* **68**: 768–82.

Runnels, C. and M. Özdoğan (2001). The Palaeolithic of the Bosphorus region, NW Turkey. *Journal of Field Archaeology*, **28**: 69–92.

Ruscavage-Barz, S.M. (1999). *Knowing Your Neighbor: Coalition Period Community Dynamics on the Pajarito Plateau, New Mexico*. Ph.D. Dissertation, Department of Anthropology, Washington State University.

Sahlins, M.D. and E.R. Service (1960). Evolution: Specific and general. In M.D. Sahlins and E.R. Service (eds), *Evolution and Culture* (pp. 12–44). Ann Arbor: University of Michigan Press.

Shott, M.J. (1994). Size and form in the analysis of flake debris: Review and recent approaches. *Journal of Archaeological Method and Theory*, **1**(1): 69–110.

Shott, M.J. and P. Sillitoe (2004). Modeling use-life distributions in archaeology using New Guinea Wola ethnographic data. *American Antiquity*, **69**: 339–55.

Slakter, M.J. (1966). Comparative validity of the chi-square and two modified chi-square goodness of fit tests for small but equal expected frequencies. *Biometrika*, **53**: 619–23.

Sokal, R.R. and F.J. Rohlf (1981). *Biometry*, 2nd edition. San Francisco: W.H. Freeman & Co.

Sokal, R.R. and F.J. Rohlf (1996). *Biometry*, 3rd edition. San Francisco: W.H. Freeman & Co.

Spaulding, A.C. (1953). Statistical techniques for the discovery of artifact types. *American Antiquity*, **19**: 391–3.

Sprehn, M.S. (2003). *Social Complexity and the Specialist Potters of Casas Grandes inNorthern Mexico*. Ph.D. Dissertation, Department of Anthropology, University of New Mexico, Albuquerque.

Styles, B.W. (1985). Reconstruction of availability and utilization of food resources. In R.I. Gilbert, Jr. and J.H. Mielke (eds.), *The Analysis of Prehistoric Diets* (pp. 21–60). Orlando, FL: Academic Press.

Tukey, J.W. (1977). *Exploratory Data Analysis*. Reading, MA: Addison-Wesley.

Van Bueren, T.M. and K. Wooten (2009). Making the most of uncertainties at the Sanderson Farm. *Historical Archaeology*, **43**: 108–34.

VanPool, T.L. (2003). *Explaining Changes in Projectile Point Morphology: A Case Study from Ventana Cave, Arizona*. Ph.D. Dissertation, Department of Anthropology. University of New Mexico, Albuquerque.

VanPool, C.S. and T.L. VanPool (2006). Gender in middle range societies: A case study in Casas Grandes iconograpy. *American Antiquity*, **71**: 53–75.

VanPool, T.L., C.S. VanPool, R. Cruz Antillón, R.D. Leonard, and M. Harmon (2000). Flaked stone and social interaction in the Casas Grandes Region, Chihuahua, Mexico. *Latin American Antiquity*, **11**: 163–74.

Vierra, R.K. and D.L. Carlson (1981). Factor analysis, random data and patterned results. *American Antiquity*, **46**: 272–83.

Watson, P.J., S.A. LeBlanc, and C.L. Redman (1971). *Explanation in Archaeology: An Explicitly Scientific Approach*. New York: Columbia University Press.

Whittaker, J.C., D. Cauklines, and K.A. Kamp (1998). Evaluating consistency in typology and classification. *Journal of Archaeological Method and Theory*, **5**: 129–64.

Wobst, M. (1983). We can't see the forest for the trees: Sampling and the shapes of archaeological distributions. In J.A. Moore and A. Keene (eds.), *Archaeological Hammers and Theories* (pp. 37–85). New York: Academic Press.

Yarnold, J.K. (1970). The minimum expectations in χ^2 goodness-of-fit tests and the accuracy of approximations for the null distribution. *Journal of the American Statistical Association*, **65**: 864–8.

Appendix A

Areas under a Standardized Normal Distribution

Values reflect the area from the mean to the specific Z-score, calculated as $Z = \dfrac{(Y_i - \mu)}{\sigma}$.

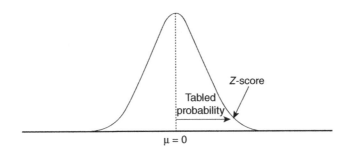

Standard deviation units	0	0.01	0.02	0.03	0.04	0.05	0.06	0.07	0.08	0.09
0.0	0.0000	0.0040	0.0080	0.0120	0.0160	0.0199	0.0239	0.0279	0.0319	0.0359
0.1	0.0398	0.0438	0.0478	0.0517	0.0557	0.0596	0.0636	0.0675	0.0714	0.0753
0.2	0.0793	0.0832	0.0871	0.0910	0.0948	0.0987	0.1026	0.1064	0.1103	0.1141
0.3	0.1179	0.1217	0.1255	0.1293	0.1331	0.1368	0.1406	0.1443	0.1480	0.1517
0.4	0.1554	0.1591	0.1628	0.1664	0.1700	0.1736	0.1772	0.1808	0.1844	0.1879
0.5	0.1915	0.1950	0.1985	0.2019	0.2054	0.2088	0.2123	0.2157	0.2190	0.2224
0.6	0.2257	0.2291	0.2324	0.2357	0.2389	0.2422	0.2454	0.2486	0.2517	0.2549
0.7	0.2580	0.2611	0.2642	0.2673	0.2704	0.2734	0.2764	0.2794	0.2823	0.2852
0.8	0.2881	0.2910	0.2939	0.2967	0.2995	0.3023	0.3051	0.3078	0.3106	0.3133
0.9	0.3159	0.3186	0.3212	0.3238	0.3264	0.3289	0.3315	0.3340	0.3365	0.3389
1.0	0.3413	0.3438	0.3461	0.3485	0.3508	0.3531	0.3554	0.3577	0.3599	0.3621
1.1	0.3643	0.3665	0.3686	0.3708	0.3729	0.3749	0.3770	0.3790	0.3810	0.3830
1.2	0.3849	0.3869	0.3888	0.3907	0.3925	0.3944	0.3962	0.3980	0.3997	0.4015
1.3	0.4032	0.4049	0.4066	0.4082	0.4099	0.4115	0.4131	0.4147	0.4162	0.4177
1.4	0.4192	0.4207	0.4222	0.4236	0.4251	0.4265	0.4279	0.4292	0.4306	0.4319
1.5	0.4332	0.4345	0.4357	0.4370	0.4382	0.4394	0.4406	0.4418	0.4429	0.4441
1.6	0.4452	0.4463	0.4474	0.4484	0.4495	0.4505	0.4515	0.4525	0.4535	0.4545
1.7	0.4554	0.4564	0.4573	0.4582	0.4591	0.4599	0.4608	0.4616	0.4625	0.4633
1.8	0.4641	0.4649	0.4656	0.4664	0.4671	0.4678	0.4686	0.4693	0.4699	0.4706
1.9	0.4713	0.4719	0.4726	0.4732	0.4738	0.4744	0.4750	0.4756	0.4761	0.4767

Standard deviation units	0	0.01	0.02	0.03	0.04	0.05	0.06	0.07	0.08	0.09
2.0	0.4772	0.4778	0.4783	0.4788	0.4793	0.4798	0.4803	0.4808	0.4812	0.4817
2.1	0.4821	0.4826	0.4830	0.4834	0.4838	0.4842	0.4846	0.4850	0.4854	0.4857
2.2	0.4861	0.4864	0.4868	0.4871	0.4875	0.4878	0.4881	0.4884	0.4887	0.4890
2.3	0.4893	0.4896	0.4898	0.4901	0.4904	0.4906	0.4909	0.4911	0.4913	0.4916
2.4	0.4918	0.4920	0.4922	0.4925	0.4927	0.4929	0.4931	0.4932	0.4934	0.4936
2.5	0.4938	0.4940	0.4941	0.4943	0.4945	0.4946	0.4948	0.4949	0.4951	0.4952
2.6	0.4953	0.4955	0.4956	0.4957	0.4959	0.4960	0.4961	0.4962	0.4963	0.4964
2.7	0.4965	0.4966	0.4967	0.4968	0.4969	0.4970	0.4971	0.4972	0.4973	0.4974
2.8	0.4974	0.4975	0.4976	0.4977	0.4977	0.4978	0.4979	0.4979	0.4980	0.4981
2.9	0.4981	0.4982	0.4982	0.4983	0.4984	0.4984	0.4985	0.4985	0.4986	0.4986
3.0	0.4987	0.4987	0.4987	0.4988	0.4988	0.4989	0.4989	0.4989	0.4990	0.4990
3.1	0.4990	0.4991	0.4991	0.4991	0.4992	0.4992	0.4992	0.4992	0.4993	0.4993
3.2	0.4993	0.4993	0.4994	0.4994	0.4994	0.4994	0.4994	0.4995	0.4995	0.4995
3.3	0.4995	0.4995	0.4995	0.4996	0.4996	0.4996	0.4996	0.4996	0.4996	0.4997
3.4	0.4997	0.4997	0.4997	0.4997	0.4997	0.4997	0.4997	0.4997	0.4997	0.4998
3.5	0.4998	0.4998	0.4998	0.4998	0.4998	0.4998	0.4998	0.4998	0.4998	0.4998
3.6	0.4998	0.4998	0.4999	0.4999	0.4999	0.4999	0.4999	0.4999	0.4999	0.4999
3.7	0.4999	0.4999	0.4999	0.4999	0.4999	0.4999	0.4999	0.4999	0.4999	0.4999
3.8	0.4999	0.4999	0.4999	0.4999	0.4999	0.4999	0.4999	0.4999	0.4999	0.4999
3.9	0.5000	0.5000	0.5000	0.5000	0.5000	0.5000	0.5000	0.5000	0.5000	0.5000

Appendix B

Critical Values for the Student's t-Distribution

This table assumes a two-tailed test. Critical values for one-tailed tests would correspond to the listed value for an α twice as large. For example, the critical value for a one-tailed test for $\alpha = .01$ corresponds to the listed value here for $\alpha = .02$. Values not listed on the table can be *interpolated* using the values above and below the desired degrees of freedom. The process is not particularly intuitive but works as illustrated in the following examples. The critical value for $\alpha = .05$ for 43 degrees of freedom can be determined using the tabled values of $t_{[.05,40]} = 2.021$ and $t_{[.05,60]} = 2.000$. Begin by dividing 120 by the degrees of freedom $(120/43 = 2.791)$. Focus only on the portion of value to the right of the decimal point (.791 in this case). Place the value into the following equation $t_{[.05,43]} = (.791 \times 2.021) + [(1 - .791) \times 2.000] = 2.017$. To illustrate this process again, the critical value of $t_{[.05,68]}$ is calculated using the crucial values of $t_{[.05,60]} = 2.000$ and $t_{[.05,120]} = 1.980$. Dividing 120 by 68 degrees of freedom produces $(120/68 = 1.765)$. Placing .765 into the interpolation formula produces $t_{[.05,68]} = (.765 \times 2.000) + [(1 - .765) \times 1.980] = 1.995$.

Critical Values for the Student's t-Distribution

Degrees of freedom	α										
	0.4	0.35	0.3	0.25	0.2	0.15	0.1	0.05	0.02	0.01	0.001
1	1.376	1.632	1.963	2.414	3.078	4.165	6.314	12.706	31.821	63.657	636.619
2	1.061	1.210	1.386	1.604	1.886	2.282	2.920	4.303	6.965	9.925	31.599
3	0.978	1.105	1.250	1.423	1.638	1.924	2.353	3.182	4.541	5.841	12.924
4	0.941	1.057	1.190	1.344	1.533	1.778	2.132	2.776	3.747	4.604	8.610
5	0.920	1.031	1.156	1.301	1.476	1.699	2.015	2.571	3.365	4.032	6.869
6	0.906	1.013	1.134	1.273	1.440	1.650	1.943	2.447	3.143	3.707	5.959
7	0.896	1.001	1.119	1.254	1.415	1.617	1.895	2.365	2.998	3.499	5.408
8	0.889	0.993	1.108	1.240	1.397	1.592	1.860	2.306	2.896	3.355	5.041
9	0.883	0.986	1.100	1.230	1.383	1.574	1.833	2.262	2.821	3.250	4.781
10	0.879	0.980	1.093	1.221	1.372	1.559	1.812	2.228	2.764	3.169	4.587
11	0.876	0.976	1.088	1.214	1.363	1.548	1.796	2.201	2.718	3.106	4.437
12	0.873	0.972	1.083	1.209	1.356	1.538	1.782	2.179	2.681	3.055	4.318
13	0.870	0.969	1.079	1.204	1.350	1.530	1.771	2.160	2.650	3.012	4.221
14	0.868	0.967	1.076	1.200	1.345	1.523	1.761	2.145	2.624	2.977	4.140
15	0.866	0.965	1.074	1.197	1.341	1.517	1.753	2.131	2.602	2.947	4.073
16	0.865	0.963	1.071	1.194	1.337	1.512	1.746	2.120	2.583	2.921	4.015
17	0.863	0.961	1.069	1.191	1.333	1.508	1.740	2.110	2.567	2.898	3.965
18	0.862	0.960	1.067	1.189	1.330	1.504	1.734	2.101	2.552	2.878	3.922
19	0.861	0.958	1.066	1.187	1.328	1.500	1.729	2.093	2.539	2.861	3.883
20	0.860	0.957	1.064	1.185	1.325	1.497	1.725	2.086	2.528	2.845	3.850

21	0.859	0.956	1.063	1.183	1.323	1.494	1.721	2.080	2.518	2.831	3.819
22	0.858	0.955	1.061	1.182	1.321	1.492	1.717	2.074	2.508	2.819	3.792
23	0.858	0.954	1.060	1.180	1.319	1.489	1.714	2.069	2.500	2.807	3.768
24	0.857	0.953	1.059	1.179	1.318	1.487	1.711	2.064	2.492	2.797	3.745
25	0.856	0.952	1.058	1.178	1.316	1.485	1.708	2.060	2.485	2.787	3.725
26	0.856	0.952	1.058	1.177	1.315	1.483	1.706	2.056	2.479	2.779	3.707
27	0.855	0.951	1.057	1.176	1.314	1.482	1.703	2.052	2.473	2.771	3.690
28	0.855	0.950	1.056	1.175	1.313	1.480	1.701	2.048	2.467	2.763	3.674
29	0.854	0.950	1.055	1.174	1.311	1.479	1.699	2.045	2.462	2.756	3.659
30	0.854	0.949	1.055	1.173	1.310	1.477	1.697	2.042	2.457	2.750	3.646
31	0.853	0.949	1.054	1.172	1.309	1.476	1.696	2.040	2.453	2.744	3.633
32	0.853	0.948	1.054	1.172	1.309	1.475	1.694	2.037	2.449	2.738	3.622
33	0.853	0.948	1.053	1.171	1.308	1.474	1.692	2.035	2.445	2.733	3.611
34	0.852	0.948	1.052	1.170	1.307	1.473	1.691	2.032	2.441	2.728	3.601
35	0.852	0.947	1.052	1.170	1.306	1.472	1.690	2.030	2.438	2.724	3.591
36	0.852	0.947	1.052	1.169	1.306	1.471	1.688	2.028	2.434	2.719	3.582
37	0.851	0.947	1.051	1.169	1.305	1.470	1.687	2.026	2.431	2.715	3.574
38	0.851	0.946	1.051	1.168	1.304	1.469	1.686	2.024	2.429	2.712	3.566
39	0.851	0.946	1.050	1.168	1.304	1.468	1.685	2.023	2.426	2.708	3.558
40	0.851	0.946	1.050	1.167	1.303	1.468	1.684	2.021	2.423	2.704	3.551

Critical Values for the Student's t-Distribution

Degrees of freedom	α										
	0.4	0.35	0.3	0.25	0.2	0.15	0.1	0.05	0.02	0.01	0.001
45	0.850	0.944	1.049	1.165	1.301	1.465	1.679	2.014	2.412	2.690	3.520
50	0.849	0.943	1.047	1.164	1.299	1.462	1.676	2.009	2.403	2.678	3.496
55	0.848	0.943	1.046	1.163	1.297	1.460	1.673	2.004	2.396	2.668	3.476
60	0.848	0.942	1.045	1.162	1.296	1.458	1.671	2.000	2.390	2.660	3.460
65	0.847	0.941	1.045	1.161	1.295	1.457	1.669	1.997	2.385	2.654	3.447
70	0.847	0.941	1.044	1.160	1.294	1.456	1.667	1.994	2.381	2.648	3.435
75	0.846	0.940	1.044	1.159	1.293	1.454	1.665	1.992	2.377	2.643	3.425
80	0.846	0.940	1.043	1.159	1.292	1.453	1.664	1.990	2.374	2.639	3.416
85	0.846	0.940	1.043	1.158	1.292	1.453	1.663	1.988	2.371	2.635	3.409
90	0.846	0.939	1.042	1.158	1.291	1.452	1.662	1.987	2.368	2.632	3.402
95	0.845	0.939	1.042	1.157	1.291	1.451	1.661	1.985	2.366	2.629	3.396
100	0.845	0.939	1.042	1.157	1.290	1.451	1.660	1.984	2.364	2.626	3.390
105	0.845	0.939	1.042	1.157	1.290	1.450	1.659	1.983	2.362	2.623	3.386
110	0.845	0.939	1.041	1.156	1.289	1.450	1.659	1.982	2.361	2.621	3.381
115	0.845	0.938	1.041	1.156	1.289	1.449	1.658	1.981	2.359	2.619	3.377
120	0.845	0.938	1.041	1.156	1.289	1.449	1.658	1.980	2.358	2.617	3.373
∞	0.842	0.935	1.036	1.150	1.282	1.440	1.645	1.960	2.326	2.576	3.291

Appendix C

Critical Values for the F-distribution

To find a critical value for F, select the number corresponding to the degrees of freedom for the numerator (across the top of the tables), the degrees of freedom for the denominator (along the left side of the table), and the specific α value. Critical values not listed can be interpolated using the processes illustrated in Appendix B. To interpolate the critical value for $F_{[.05,3,116]}$, you must interpolate between the critical values of $F_{[.05,3,100]} = 2.70$ and $F_{[.05,3,120]} = 2.68$. The value for 120/116 is 1.034. Placing this into the formula introduced in Appendix B produces the following value $F_{[.05,3,116]} = (.034 \times 2.70) + [1 - .034) \times 2.68] = 2.68$.

Critical Values for the F-distribution

df	α						Degrees of freedom (numerator mean squares)										
		1	2	3	4	5	6	7	8	9	10	11	12	13	14	15	16
1	0.50	1.00	1.50	1.71	1.82	1.89	1.94	1.98	2.00	2.03	2.04	2.06	2.07	2.08	2.09	2.09	2.10
1	0.25	5.83	7.50	8.20	8.58	8.82	8.98	9.10	9.19	9.26	9.32	9.37	9.41	9.44	9.47	9.49	9.52
1	0.10	39.86	49.50	53.59	55.83	57.24	58.20	58.91	59.44	59.86	60.19	60.47	60.71	60.90	61.07	61.22	61.35
1	0.05	161.45	199.50	215.71	224.58	230.16	233.99	236.77	238.88	240.54	241.88	242.98	243.91	244.69	245.36	245.95	246.46
1	0.01	4052.18	4999.50	5403.35	5624.58	5763.65	5858.99	5928.36	5981.07	6022.47	6055.85	6083.32	6106.32	6125.86	6142.67	6157.28	6170.10
1	0.001	405284.07	499999.50	540379.20	562499.58	576404.56	585937.11	592873.29	598144.16	602283.99	605620.97	608367.68	610667.82	612622.01	614302.75	615763.66	617045.18
2	0.50	0.67	1.00	1.13	1.21	1.25	1.28	1.30	1.32	1.33	1.35	1.35	1.36	1.37	1.37	1.38	1.38
2	0.25	2.57	3.00	3.15	3.23	3.28	3.31	3.34	3.35	3.37	3.38	3.39	3.39	3.40	3.41	3.41	3.41
2	0.10	8.53	9.00	9.16	9.24	9.29	9.33	9.35	9.37	9.38	9.39	9.40	9.41	9.41	9.42	9.42	9.43
2	0.05	18.51	19.00	19.16	19.25	19.30	19.33	19.35	19.37	19.38	19.40	19.40	19.41	19.42	19.42	19.43	19.43
2	0.01	98.50	99.00	99.17	99.25	99.30	99.33	99.36	99.37	99.39	99.40	99.41	99.42	99.42	99.43	99.43	99.44
2	0.001	998.50	999.00	999.17	999.25	999.30	999.33	999.36	999.37	999.39	999.40	999.41	999.42	999.42	999.43	999.43	999.44
3	0.50	0.59	0.88	1.00	1.06	1.10	1.13	1.15	1.16	1.17	1.18	1.19	1.20	1.20	1.21	1.21	1.21
3	0.25	2.02	2.28	2.36	2.39	2.41	2.42	2.43	2.44	2.44	2.44	2.45	2.45	2.45	2.45	2.46	2.46
3	0.10	5.54	5.46	5.39	5.34	5.31	5.28	5.27	5.25	5.24	5.23	5.22	5.22	5.21	5.20	5.20	5.20
3	0.05	10.13	9.55	9.28	9.12	9.01	8.94	8.89	8.85	8.81	8.79	8.76	8.74	8.73	8.71	8.70	8.69
3	0.01	34.12	30.82	29.46	28.71	28.24	27.91	27.67	27.49	27.35	27.23	27.13	27.05	26.98	26.92	26.87	26.83
3	0.001	167.03	148.50	141.11	137.10	134.58	132.85	131.58	130.62	129.86	129.25	128.74	128.32	127.96	127.64	127.37	127.14
4	0.50	0.55	0.83	0.94	1.00	1.04	1.06	1.08	1.09	1.10	1.11	1.12	1.13	1.13	1.13	1.14	1.14
4	0.25	1.81	2.00	2.05	2.06	2.07	2.08	2.08	2.08	2.08	2.08	2.08	2.08	2.08	2.08	2.08	2.08
4	0.10	4.54	4.32	4.19	4.11	4.05	4.01	3.98	3.95	3.94	3.92	3.91	3.90	3.89	3.88	3.87	3.86
4	0.05	7.71	6.94	6.59	6.39	6.26	6.16	6.09	6.04	6.00	5.96	5.94	5.91	5.89	5.87	5.86	5.84
4	0.01	21.20	18.00	16.69	15.98	15.52	15.21	14.98	14.80	14.66	14.55	14.45	14.37	14.31	14.25	14.20	14.15
4	0.001	74.14	61.25	56.18	53.44	51.71	50.53	49.66	49.00	48.47	48.05	47.70	47.41	47.16	46.95	46.76	46.60
5	0.50	0.53	0.80	0.91	0.96	1.00	1.02	1.04	1.05	1.06	1.07	1.08	1.09	1.09	1.09	1.10	1.10
5	0.25	1.69	1.85	1.88	1.89	1.89	1.89	1.89	1.89	1.89	1.89	1.89	1.89	1.89	1.89	1.89	1.88
5	0.10	4.06	3.78	3.62	3.52	3.45	3.40	3.37	3.34	3.32	3.30	3.28	3.27	3.26	3.25	3.24	3.23
5	0.05	6.61	5.79	5.41	5.19	5.05	4.95	4.88	4.82	4.77	4.74	4.70	4.68	4.66	4.64	4.62	4.60
5	0.01	16.26	13.27	12.06	11.39	10.97	10.67	10.46	10.29	10.16	10.05	9.96	9.89	9.82	9.77	9.72	9.68
5	0.001	47.18	37.12	33.20	31.09	29.75	28.83	28.16	27.65	27.24	26.92	26.65	26.42	26.22	26.06	25.91	25.78
6	0.50	0.51	0.78	0.89	0.94	0.98	1.00	1.02	1.03	1.04	1.05	1.05	1.06	1.06	1.07	1.07	1.08
6	0.25	1.62	1.76	1.78	1.79	1.79	1.78	1.78	1.78	1.77	1.77	1.77	1.77	1.77	1.76	1.76	1.76
6	0.10	3.78	3.46	3.29	3.18	3.11	3.05	3.01	2.98	2.96	2.94	2.92	2.90	2.89	2.88	2.87	2.86
6	0.05	5.99	5.14	4.76	4.53	4.39	4.28	4.21	4.15	4.10	4.06	4.03	4.00	3.98	3.96	3.94	3.92
6	0.01	13.75	10.92	9.78	9.15	8.75	8.47	8.26	8.10	7.98	7.87	7.79	7.72	7.66	7.60	7.56	7.52
6	0.001	35.51	27.00	23.70	21.92	20.80	20.03	19.46	19.03	18.69	18.41	18.18	17.99	17.82	17.68	17.56	17.45
7	0.50	0.51	0.77	0.87	0.93	0.96	0.98	1.00	1.01	1.02	1.03	1.04	1.04	1.05	1.05	1.05	1.06
7	0.25	1.57	1.70	1.72	1.72	1.71	1.71	1.70	1.70	1.69	1.69	1.69	1.68	1.68	1.68	1.68	1.68
7	0.10	3.59	3.26	3.07	2.96	2.88	2.83	2.78	2.75	2.72	2.70	2.68	2.67	2.65	2.64	2.63	2.62
7	0.05	5.59	4.74	4.35	4.12	3.97	3.87	3.79	3.73	3.68	3.64	3.60	3.57	3.55	3.53	3.51	3.49
7	0.01	12.25	9.55	8.45	7.85	7.46	7.19	6.99	6.84	6.72	6.62	6.54	6.47	6.41	6.36	6.31	6.28
7	0.001	29.25	21.69	18.77	17.20	16.21	15.52	15.02	14.63	14.33	14.08	13.88	13.71	13.56	13.43	13.32	13.23
8	0.50	0.50	0.76	0.86	0.91	0.95	0.97	0.99	1.00	1.01	1.02	1.02	1.03	1.03	1.04	1.04	1.04
8	0.25	1.54	1.66	1.67	1.66	1.66	1.65	1.64	1.64	1.63	1.63	1.63	1.62	1.62	1.62	1.62	1.62
8	0.10	3.46	3.11	2.92	2.81	2.73	2.67	2.62	2.59	2.56	2.54	2.52	2.50	2.49	2.48	2.46	2.45
8	0.05	5.32	4.46	4.07	3.84	3.69	3.58	3.50	3.44	3.39	3.35	3.31	3.28	3.26	3.24	3.22	3.20
8	0.01	11.26	8.65	7.59	7.01	6.63	6.37	6.18	6.03	5.91	5.81	5.73	5.67	5.61	5.56	5.52	5.48
8	0.001	25.41	18.49	15.83	14.39	13.48	12.86	12.40	12.05	11.77	11.54	11.35	11.19	11.06	10.94	10.84	10.75
9	0.50	0.49	0.75	0.85	0.91	0.94	0.96	0.98	0.99	1.00	1.01	1.01	1.02	1.02	1.03	1.03	1.03
9	0.25	1.51	1.62	1.63	1.63	1.62	1.61	1.60	1.60	1.59	1.59	1.58	1.58	1.58	1.57	1.57	1.57
9	0.10	3.36	3.01	2.81	2.69	2.61	2.55	2.51	2.47	2.44	2.42	2.40	2.38	2.36	2.35	2.34	2.33
9	0.05	5.12	4.26	3.86	3.63	3.48	3.37	3.29	3.23	3.18	3.14	3.10	3.07	3.05	3.03	3.01	2.99
9	0.01	10.56	8.02	6.99	6.42	6.06	5.80	5.61	5.47	5.35	5.26	5.18	5.11	5.05	5.01	4.96	4.92
9	0.001	22.86	16.39	13.90	12.56	11.71	11.13	10.70	10.37	10.11	9.89	9.72	9.57	9.44	9.33	9.24	9.15

17	18	19	20	25	30	35	40	45	50	60	70	80	90	100	110	120
2.11	2.11	2.11	2.12	2.13	2.15	2.15	2.16	2.16	2.17	2.17	2.18	2.18	2.18	2.18	2.18	2.18
9.53	9.55	9.57	9.58	9.63	9.67	9.70	9.71	9.73	9.74	9.76	9.77	9.78	9.79	9.80	9.80	9.80
61.46	61.57	61.66	61.74	62.05	62.26	62.42	62.53	62.62	62.69	62.79	62.87	62.93	62.97	63.01	63.04	63.06
246.92	247.32	247.69	248.01	249.26	250.10	250.69	251.14	251.49	251.77	252.20	252.50	252.72	252.90	253.04	253.16	253.25
6181.43	6191.53	6200.58	6208.73	6239.83	6260.65	6275.57	6286.78	6295.52	6302.52	6313.03	6320.55	6326.20	6330.59	6334.11	6336.99	6339.39
618178.43	619187.70	620092.29	620907.67	624016.83	626098.96	627590.73	628712.03	629585.61	630285.38	631336.56	632088.51	632653.07	633092.54	633444.33	633732.32	633972.40
1.38	1.39	1.39	1.39	1.40	1.41	1.41	1.42	1.42	1.42	1.43	1.43	1.43	1.43	1.43	1.43	1.43
3.42	3.42	3.42	3.43	3.44	3.44	3.45	3.45	3.45	3.46	3.46	3.46	3.46	3.46	3.47	3.47	3.47
9.43	9.44	9.44	9.44	9.45	9.46	9.46	9.47	9.47	9.47	9.47	9.48	9.48	9.48	9.48	9.48	9.48
19.44	19.44	19.44	19.45	19.46	19.46	19.47	19.47	19.47	19.48	19.48	19.48	19.48	19.48	19.49	19.49	19.49
99.44	99.44	99.45	99.45	99.46	99.47	99.47	99.47	99.48	99.48	99.48	99.48	99.49	99.49	99.49	99.49	99.49
999.44	999.44	999.45	999.45	999.46	999.47	999.47	999.47	999.48	999.48	999.48	999.49	999.49	999.49	999.49	999.49	999.49
1.22	1.22	1.22	1.23	1.23	1.24	1.24	1.25	1.25	1.25	1.25	1.26	1.26	1.26	1.26	1.26	1.26
2.46	2.46	2.46	2.46	2.46	2.47	2.47	2.47	2.47	2.47	2.47	2.47	2.47	2.47	2.47	2.47	2.47
5.19	5.19	5.19	5.18	5.17	5.17	5.16	5.16	5.16	5.15	5.15	5.15	5.15	5.15	5.14	5.14	5.14
8.68	8.67	8.67	8.66	8.63	8.62	8.60	8.59	8.59	8.58	8.57	8.57	8.57	8.56	8.56	8.55	8.55
26.79	26.75	26.72	26.69	26.58	26.50	26.45	26.41	26.38	26.35	26.32	26.29	26.27	26.25	26.24	26.23	26.22
126.93	126.74	126.57	126.42	125.84	125.45	125.17	124.96	124.79	124.66	124.47	124.32	124.22	124.14	124.07	124.01	123.97
1.14	1.15	1.15	1.15	1.16	1.16	1.17	1.17	1.17	1.18	1.18	1.18	1.18	1.18	1.18	1.18	1.18
2.08	2.08	2.08	2.08	2.08	2.08	2.08	2.08	2.08	2.08	2.08	2.08	2.08	2.08	2.08	2.08	2.08
3.86	3.85	3.85	3.84	3.83	3.82	3.81	3.80	3.80	3.80	3.79	3.79	3.78	3.78	3.78	3.78	3.78
5.83	5.82	5.81	5.80	5.77	5.75	5.73	5.72	5.71	5.70	5.69	5.68	5.67	5.67	5.66	5.66	5.66
14.11	14.08	14.05	14.02	13.91	13.84	13.79	13.75	13.71	13.69	13.65	13.63	13.61	13.59	13.58	13.57	13.56
46.45	46.32	46.21	46.10	45.70	45.43	45.23	45.09	44.97	44.88	44.75	44.65	44.57	44.52	44.47	44.43	44.40
1.10	1.11	1.11	1.11	1.12	1.12	1.13	1.13	1.13	1.13	1.14	1.14	1.14	1.14	1.14	1.14	1.14
1.88	1.88	1.88	1.88	1.88	1.88	1.88	1.88	1.88	1.88	1.87	1.87	1.87	1.87	1.87	1.87	1.87
3.22	3.22	3.21	3.21	3.19	3.17	3.16	3.16	3.15	3.15	3.14	3.14	3.13	3.13	3.13	3.12	3.12
4.59	4.58	4.57	4.56	4.52	4.50	4.48	4.46	4.45	4.44	4.43	4.42	4.41	4.41	4.41	4.40	4.40
9.64	9.61	9.58	9.55	9.45	9.38	9.33	9.29	9.26	9.24	9.20	9.18	9.16	9.14	9.13	9.12	9.11
25.67	25.57	25.48	25.39	25.08	24.87	24.72	24.60	24.51	24.44	24.33	24.26	24.20	24.15	24.12	24.09	24.06
1.08	1.08	1.08	1.08	1.09	1.10	1.10	1.10	1.11	1.11	1.11	1.11	1.11	1.11	1.11	1.12	1.12
1.76	1.76	1.76	1.76	1.75	1.75	1.75	1.75	1.75	1.75	1.74	1.74	1.74	1.74	1.74	1.74	1.74
2.85	2.85	2.84	2.84	2.81	2.80	2.79	2.78	2.77	2.77	2.76	2.76	2.75	2.75	2.75	2.74	2.74
3.91	3.90	3.88	3.87	3.83	3.81	3.79	3.77	3.76	3.75	3.74	3.73	3.72	3.72	3.71	3.71	3.70
7.48	7.45	7.42	7.40	7.30	7.23	7.18	7.14	7.11	7.09	7.06	7.03	7.01	7.00	6.99	6.98	6.97
17.35	17.27	17.19	17.12	16.85	16.67	16.54	16.44	16.37	16.31	16.21	16.15	16.10	16.06	16.03	16.00	15.98
1.06	1.06	1.06	1.07	1.07	1.08	1.08	1.08	1.09	1.09	1.09	1.09	1.09	1.09	1.10	1.10	1.10
1.67	1.67	1.67	1.67	1.67	1.66	1.66	1.66	1.66	1.66	1.65	1.65	1.65	1.65	1.65	1.65	1.65
2.61	2.61	2.60	2.59	2.57	2.56	2.54	2.54	2.53	2.52	2.51	2.51	2.50	2.50	2.50	2.49	2.49
3.48	3.47	3.46	3.44	3.40	3.38	3.36	3.34	3.33	3.32	3.30	3.29	3.29	3.28	3.27	3.27	3.27
6.24	6.21	6.18	6.16	6.06	5.99	5.94	5.91	5.88	5.86	5.82	5.80	5.78	5.77	5.75	5.75	5.74
13.14	13.06	12.99	12.93	12.69	12.53	12.41	12.33	12.26	12.20	12.12	12.06	12.01	11.98	11.95	11.93	11.91
1.05	1.05	1.05	1.05	1.06	1.07	1.07	1.07	1.07	1.07	1.08	1.08	1.08	1.08	1.08	1.08	1.08
1.61	1.61	1.61	1.61	1.60	1.60	1.60	1.59	1.59	1.59	1.59	1.59	1.59	1.59	1.58	1.58	1.58
2.45	2.44	2.43	2.42	2.40	2.38	2.37	2.36	2.35	2.35	2.34	2.33	2.33	2.32	2.32	2.32	2.32
3.19	3.17	3.16	3.15	3.11	3.08	3.06	3.04	3.03	3.02	3.01	2.99	2.99	2.98	2.97	2.97	2.97
5.44	5.41	5.38	5.36	5.26	5.20	5.15	5.12	5.09	5.07	5.03	5.01	4.99	4.97	4.96	4.95	4.95
10.67	10.60	10.54	10.48	10.26	10.11	10.00	9.92	9.86	9.80	9.73	9.67	9.63	9.60	9.57	9.55	9.53
1.04	1.04	1.04	1.04	1.05	1.05	1.06	1.06	1.06	1.06	1.07	1.07	1.07	1.07	1.07	1.07	1.07
1.57	1.56	1.56	1.56	1.55	1.55	1.55	1.54	1.54	1.54	1.54	1.54	1.54	1.53	1.53	1.53	1.53
2.32	2.31	2.30	2.30	2.27	2.25	2.24	2.23	2.22	2.22	2.21	2.20	2.20	2.19	2.19	2.19	2.18
2.97	2.96	2.95	2.94	2.89	2.86	2.84	2.83	2.81	2.80	2.79	2.78	2.77	2.76	2.76	2.75	2.75
4.89	4.86	4.83	4.81	4.71	4.65	4.60	4.57	4.54	4.52	4.48	4.46	4.44	4.43	4.41	4.41	4.40
9.08	9.01	8.95	8.90	8.69	8.55	8.45	8.37	8.31	8.26	8.19	8.13	8.09	8.06	8.04	8.02	8.00

df	α																
							Degrees of freedom (numerator mean squares)										
		1	2	3	4	5	6	7	8	9	10	11	12	13	14	15	16
10	0.50	0.49	0.74	0.85	0.90	0.93	0.95	0.97	0.98	0.99	1.00	1.01	1.01	1.02	1.02	1.02	1.03
10	0.25	1.49	1.60	1.60	1.59	1.59	1.58	1.57	1.56	1.56	1.55	1.55	1.54	1.54	1.54	1.53	1.53
10	0.10	3.29	2.92	2.73	2.61	2.52	2.46	2.41	2.38	2.35	2.32	2.30	2.28	2.27	2.26	2.24	2.23
10	0.05	4.96	4.10	3.71	3.48	3.33	3.22	3.14	3.07	3.02	2.98	2.94	2.91	2.89	2.86	2.85	2.83
10	0.01	10.04	7.56	6.55	5.99	5.64	5.39	5.20	5.06	4.94	4.85	4.77	4.71	4.65	4.60	4.56	4.52
10	0.001	21.04	14.91	12.55	11.28	10.48	9.93	9.52	9.20	8.96	8.75	8.59	8.45	8.32	8.22	8.13	8.05
11	0.50	0.49	0.74	0.84	0.89	0.93	0.95	0.96	0.98	0.99	0.99	1.00	1.01	1.01	1.01	1.02	1.02
11	0.25	1.47	1.58	1.58	1.57	1.56	1.55	1.54	1.53	1.53	1.52	1.52	1.51	1.51	1.51	1.50	1.50
11	0.10	3.23	2.86	2.66	2.54	2.45	2.39	2.34	2.30	2.27	2.25	2.23	2.21	2.19	2.18	2.17	2.16
11	0.05	4.84	3.98	3.59	3.36	3.20	3.09	3.01	2.95	2.90	2.85	2.82	2.79	2.76	2.74	2.72	2.70
11	0.01	9.65	7.21	6.22	5.67	5.32	5.07	4.89	4.74	4.63	4.54	4.46	4.40	4.34	4.29	4.25	4.21
11	0.001	19.69	13.81	11.56	10.35	9.58	9.05	8.66	8.35	8.12	7.92	7.76	7.63	7.51	7.41	7.32	7.24
12	0.50	0.48	0.73	0.84	0.89	0.92	0.94	0.96	0.97	0.98	0.99	0.99	1.00	1.00	1.01	1.01	1.01
12	0.25	1.46	1.56	1.56	1.55	1.54	1.53	1.52	1.51	1.51	1.50	1.49	1.49	1.49	1.48	1.48	1.48
12	0.10	3.18	2.81	2.61	2.48	2.39	2.33	2.28	2.24	2.21	2.19	2.17	2.15	2.13	2.12	2.10	2.09
12	0.05	4.75	3.89	3.49	3.26	3.11	3.00	2.91	2.85	2.80	2.75	2.72	2.69	2.66	2.64	2.62	2.60
12	0.01	9.33	6.93	5.95	5.41	5.06	4.82	4.64	4.50	4.39	4.30	4.22	4.16	4.10	4.05	4.01	3.97
12	0.001	18.64	12.97	10.80	9.63	8.89	8.38	8.00	7.71	7.48	7.29	7.14	7.00	6.89	6.79	6.71	6.63
13	0.50	0.48	0.73	0.83	0.88	0.92	0.94	0.96	0.97	0.98	0.98	0.99	1.00	1.00	1.00	1.01	1.01
13	0.25	1.45	1.55	1.55	1.53	1.52	1.51	1.50	1.49	1.49	1.48	1.47	1.47	1.47	1.46	1.46	1.46
13	0.10	3.14	2.76	2.56	2.43	2.35	2.28	2.23	2.20	2.16	2.14	2.12	2.10	2.08	2.07	2.05	2.04
13	0.05	4.67	3.81	3.41	3.18	3.03	2.92	2.83	2.77	2.71	2.67	2.63	2.60	2.58	2.55	2.53	2.51
13	0.01	9.07	6.70	5.74	5.21	4.86	4.62	4.44	4.30	4.19	4.10	4.02	3.96	3.91	3.86	3.82	3.78
13	0.001	17.82	12.31	10.21	9.07	8.35	7.86	7.49	7.21	6.98	6.80	6.65	6.52	6.41	6.31	6.23	6.16
14	0.50	0.48	0.73	0.83	0.88	0.91	0.94	0.95	0.96	0.97	0.98	0.99	0.99	1.00	1.00	1.00	1.01
14	0.25	1.44	1.53	1.53	1.52	1.51	1.50	1.49	1.48	1.47	1.46	1.46	1.45	1.45	1.44	1.44	1.44
14	0.10	3.10	2.73	2.52	2.39	2.31	2.24	2.19	2.15	2.12	2.10	2.07	2.05	2.04	2.02	2.01	2.00
14	0.05	4.60	3.74	3.34	3.11	2.96	2.85	2.76	2.70	2.65	2.60	2.57	2.53	2.51	2.48	2.46	2.44
14	0.01	8.86	6.51	5.56	5.04	4.69	4.46	4.28	4.14	4.03	3.94	3.86	3.80	3.75	3.70	3.66	3.62
14	0.001	17.14	11.78	9.73	8.62	7.92	7.44	7.08	6.80	6.58	6.40	6.26	6.13	6.02	5.93	5.85	5.78
15	0.50	0.48	0.73	0.83	0.88	0.91	0.93	0.95	0.96	0.97	0.98	0.98	0.99	0.99	1.00	1.00	1.00
15	0.25	1.43	1.52	1.52	1.51	1.49	1.48	1.47	1.46	1.46	1.45	1.44	1.44	1.43	1.43	1.43	1.42
15	0.10	3.07	2.70	2.49	2.36	2.27	2.21	2.16	2.12	2.09	2.06	2.04	2.02	2.00	1.99	1.97	1.96
15	0.05	4.54	3.68	3.29	3.06	2.90	2.79	2.71	2.64	2.59	2.54	2.51	2.48	2.45	2.42	2.40	2.38
15	0.01	8.68	6.36	5.42	4.89	4.56	4.32	4.14	4.00	3.89	3.80	3.73	3.67	3.61	3.56	3.52	3.49
15	0.001	16.59	11.34	9.34	8.25	7.57	7.09	6.74	6.47	6.26	6.08	5.94	5.81	5.71	5.62	5.54	5.46
16	0.50	0.48	0.72	0.82	0.88	0.91	0.93	0.95	0.96	0.97	0.97	0.98	0.99	0.99	0.99	1.00	1.00
16	0.25	1.42	1.51	1.51	1.50	1.48	1.47	1.46	1.45	1.44	1.44	1.43	1.43	1.42	1.42	1.41	1.41
16	0.10	3.05	2.67	2.46	2.33	2.24	2.18	2.13	2.09	2.06	2.03	2.01	1.99	1.97	1.95	1.94	1.93
16	0.05	4.49	3.63	3.24	3.01	2.85	2.74	2.66	2.59	2.54	2.49	2.46	2.42	2.40	2.37	2.35	2.33
16	0.01	8.53	6.23	5.29	4.77	4.44	4.20	4.03	3.89	3.78	3.69	3.62	3.55	3.50	3.45	3.41	3.37
16	0.001	16.12	10.97	9.01	7.94	7.27	6.80	6.46	6.19	5.98	5.81	5.67	5.55	5.44	5.35	5.27	5.20
17	0.50	0.47	0.72	0.82	0.87	0.91	0.93	0.94	0.96	0.96	0.97	0.98	0.98	0.99	0.99	0.99	1.00
17	0.25	1.42	1.51	1.50	1.49	1.47	1.46	1.45	1.44	1.43	1.43	1.42	1.41	1.41	1.41	1.40	1.40
17	0.10	3.03	2.64	2.44	2.31	2.22	2.15	2.10	2.06	2.03	2.00	1.98	1.96	1.94	1.93	1.91	1.90
17	0.05	4.45	3.59	3.20	2.96	2.81	2.70	2.61	2.55	2.49	2.45	2.41	2.38	2.35	2.33	2.31	2.29
17	0.01	8.40	6.11	5.18	4.67	4.34	4.10	3.93	3.79	3.68	3.59	3.52	3.46	3.40	3.35	3.31	3.27
17	0.001	15.72	10.66	8.73	7.68	7.02	6.56	6.22	5.96	5.75	5.58	5.44	5.32	5.22	5.13	5.05	4.99
18	0.50	0.47	0.72	0.82	0.87	0.90	0.93	0.94	0.95	0.96	0.97	0.98	0.98	0.99	0.99	0.99	1.00
18	0.25	1.41	1.50	1.49	1.48	1.46	1.45	1.44	1.43	1.42	1.42	1.41	1.40	1.40	1.40	1.39	1.39
18	0.10	3.01	2.62	2.42	2.29	2.20	2.13	2.08	2.04	2.00	1.98	1.95	1.93	1.92	1.90	1.89	1.87
18	0.05	4.41	3.55	3.16	2.93	2.77	2.66	2.58	2.51	2.46	2.41	2.37	2.34	2.31	2.29	2.27	2.25
18	0.01	8.29	6.01	5.09	4.58	4.25	4.01	3.84	3.71	3.60	3.51	3.43	3.37	3.32	3.27	3.23	3.19
18	0.001	15.38	10.39	8.49	7.46	6.81	6.35	6.02	5.76	5.56	5.39	5.25	5.13	5.03	4.94	4.87	4.80

17	18	19	20	25	30	35	40	45	50	60	70	80	90	100	110	120
1.03	1.03	1.03	1.03	1.04	1.05	1.05	1.05	1.05	1.06	1.06	1.06	1.06	1.06	1.06	1.06	1.06
1.53	1.53	1.53	1.52	1.52	1.51	1.51	1.51	1.50	1.50	1.50	1.50	1.50	1.49	1.49	1.49	1.49
2.22	2.22	2.21	2.20	2.17	2.16	2.14	2.13	2.12	2.12	2.11	2.10	2.09	2.09	2.09	2.08	2.08
2.81	2.80	2.79	2.77	2.73	2.70	2.68	2.66	2.65	2.64	2.62	2.61	2.60	2.59	2.59	2.58	2.58
4.49	4.46	4.43	4.41	4.31	4.25	4.20	4.17	4.14	4.12	4.08	4.06	4.04	4.03	4.01	4.00	4.00
7.98	7.91	7.86	7.80	7.60	7.47	7.37	7.30	7.24	7.19	7.12	7.07	7.03	7.00	6.98	6.96	6.94
1.02	1.02	1.03	1.03	1.04	1.04	1.04	1.05	1.05	1.05	1.05	1.05	1.05	1.06	1.06	1.06	1.06
1.50	1.50	1.49	1.49	1.49	1.48	1.48	1.47	1.47	1.47	1.47	1.46	1.46	1.46	1.46	1.46	1.46
2.15	2.14	2.13	2.12	2.10	2.08	2.06	2.05	2.04	2.04	2.03	2.02	2.01	2.01	2.01	2.00	2.00
2.69	2.67	2.66	2.65	2.60	2.57	2.55	2.53	2.52	2.51	2.49	2.48	2.47	2.46	2.45	2.45	2.45
4.18	4.15	4.12	4.10	4.01	3.94	3.89	3.86	3.83	3.81	3.78	3.75	3.73	3.72	3.71	3.70	3.69
7.17	7.11	7.06	7.01	6.81	6.68	6.59	6.52	6.46	6.42	6.35	6.30	6.26	6.23	6.21	6.19	6.18
1.02	1.02	1.02	1.02	1.03	1.03	1.04	1.04	1.04	1.04	1.05	1.05	1.05	1.05	1.05	1.05	1.05
1.47	1.47	1.47	1.47	1.46	1.45	1.45	1.45	1.44	1.44	1.44	1.44	1.44	1.43	1.43	1.43	1.43
2.08	2.08	2.07	2.06	2.03	2.01	2.00	1.99	1.98	1.97	1.96	1.95	1.95	1.94	1.94	1.93	1.93
2.58	2.57	2.56	2.54	2.50	2.47	2.44	2.43	2.41	2.40	2.38	2.37	2.36	2.36	2.35	2.34	2.34
3.94	3.91	3.88	3.86	3.76	3.70	3.65	3.62	3.59	3.57	3.54	3.51	3.49	3.48	3.47	3.46	3.45
6.57	6.51	6.45	6.40	6.22	6.09	6.00	5.93	5.87	5.83	5.76	5.71	5.68	5.65	5.63	5.61	5.59
1.01	1.01	1.02	1.02	1.03	1.03	1.03	1.04	1.04	1.04	1.04	1.04	1.04	1.05	1.05	1.05	1.05
1.45	1.45	1.45	1.45	1.44	1.43	1.43	1.42	1.42	1.42	1.42	1.41	1.41	1.41	1.41	1.41	1.41
2.03	2.02	2.01	2.01	1.98	1.96	1.94	1.93	1.92	1.92	1.90	1.90	1.89	1.89	1.88	1.88	1.88
2.50	2.48	2.47	2.46	2.41	2.38	2.36	2.34	2.33	2.31	2.30	2.28	2.27	2.27	2.26	2.26	2.25
3.75	3.72	3.69	3.66	3.57	3.51	3.46	3.43	3.40	3.38	3.34	3.32	3.30	3.28	3.27	3.26	3.25
6.09	6.03	5.98	5.93	5.75	5.63	5.54	5.47	5.41	5.37	5.30	5.26	5.22	5.19	5.17	5.15	5.14
1.01	1.01	1.01	1.01	1.02	1.03	1.03	1.03	1.03	1.04	1.04	1.04	1.04	1.04	1.04	1.04	1.04
1.44	1.43	1.43	1.43	1.42	1.41	1.41	1.41	1.40	1.40	1.40	1.39	1.39	1.39	1.39	1.39	1.39
1.99	1.98	1.97	1.96	1.93	1.91	1.90	1.89	1.88	1.87	1.86	1.85	1.84	1.84	1.83	1.83	1.83
2.43	2.41	2.40	2.39	2.34	2.31	2.28	2.27	2.25	2.24	2.22	2.21	2.20	2.19	2.19	2.18	2.18
3.59	3.56	3.53	3.51	3.41	3.35	3.30	3.27	3.24	3.22	3.18	3.16	3.14	3.12	3.11	3.10	3.09
5.71	5.66	5.60	5.56	5.38	5.25	5.17	5.10	5.04	5.00	4.94	4.89	4.86	4.83	4.81	4.79	4.77
1.01	1.01	1.01	1.01	1.02	1.02	1.03	1.03	1.03	1.03	1.03	1.04	1.04	1.04	1.04	1.04	1.04
1.42	1.42	1.41	1.41	1.40	1.40	1.39	1.39	1.39	1.38	1.38	1.38	1.37	1.37	1.37	1.37	1.37
1.95	1.94	1.93	1.92	1.89	1.87	1.86	1.85	1.84	1.83	1.82	1.81	1.80	1.80	1.79	1.79	1.79
2.37	2.35	2.34	2.33	2.28	2.25	2.22	2.20	2.19	2.18	2.16	2.15	2.14	2.13	2.12	2.12	2.11
3.45	3.42	3.40	3.37	3.28	3.21	3.17	3.13	3.10	3.08	3.05	3.02	3.00	2.99	2.98	2.97	2.96
5.40	5.35	5.29	5.25	5.07	4.95	4.86	4.80	4.74	4.70	4.64	4.59	4.56	4.53	4.51	4.49	4.47
1.00	1.00	1.01	1.01	1.02	1.02	1.02	1.03	1.03	1.03	1.03	1.03	1.03	1.04	1.04	1.04	1.04
1.41	1.40	1.40	1.40	1.39	1.38	1.38	1.37	1.37	1.37	1.36	1.36	1.36	1.36	1.36	1.36	1.35
1.92	1.91	1.90	1.89	1.86	1.84	1.82	1.81	1.80	1.79	1.78	1.77	1.77	1.76	1.76	1.75	1.75
2.32	2.30	2.29	2.28	2.23	2.19	2.17	2.15	2.14	2.12	2.11	2.09	2.08	2.07	2.07	2.06	2.06
3.34	3.31	3.28	3.26	3.16	3.10	3.05	3.02	2.99	2.97	2.93	2.91	2.89	2.87	2.86	2.85	2.84
5.14	5.09	5.04	4.99	4.82	4.70	4.61	4.54	4.49	4.45	4.39	4.34	4.31	4.28	4.26	4.24	4.23
1.00	1.00	1.00	1.01	1.01	1.02	1.02	1.02	1.03	1.03	1.03	1.03	1.03	1.04	1.04	1.04	1.04
1.39	1.39	1.39	1.39	1.38	1.37	1.37	1.36	1.36	1.36	1.35	1.35	1.35	1.34	1.34	1.34	1.34
1.89	1.88	1.87	1.86	1.83	1.81	1.79	1.78	1.77	1.76	1.75	1.74	1.74	1.73	1.73	1.72	1.72
2.27	2.26	2.24	2.23	2.18	2.15	2.12	2.10	2.09	2.08	2.06	2.05	2.03	2.03	2.02	2.02	2.01
3.24	3.21	3.19	3.16	3.07	3.00	2.96	2.92	2.89	2.87	2.83	2.81	2.79	2.78	2.76	2.75	2.75
4.92	4.87	4.82	4.78	4.60	4.48	4.40	4.33	4.28	4.24	4.18	4.13	4.10	4.07	4.05	4.03	4.02
1.00	1.00	1.00	1.00	1.01	1.02	1.02	1.02	1.02	1.02	1.03	1.03	1.03	1.03	1.03	1.03	1.03
1.38	1.38	1.38	1.38	1.37	1.36	1.35	1.35	1.35	1.34	1.34	1.34	1.33	1.33	1.33	1.33	1.33
1.86	1.85	1.84	1.84	1.80	1.78	1.77	1.75	1.74	1.74	1.72	1.71	1.71	1.70	1.70	1.69	1.69
2.23	2.22	2.20	2.19	2.14	2.11	2.08	2.06	2.05	2.04	2.02	2.00	1.99	1.98	1.98	1.97	1.97
3.16	3.13	3.10	3.08	2.98	2.92	2.87	2.84	2.81	2.78	2.75	2.72	2.70	2.69	2.68	2.67	2.66
4.74	4.68	4.63	4.59	4.42	4.30	4.22	4.15	4.10	4.06	4.00	3.95	3.92	3.89	3.87	3.85	3.84

df	α																
		\multicolumn{16}{c}{Degrees of freedom (numerator mean squares)}														
		1	2	3	4	5	6	7	8	9	10	11	12	13	14	15	16
19	0.50	0.47	0.72	0.82	0.87	0.90	0.92	0.94	0.95	0.96	0.97	0.97	0.98	0.98	0.99	0.99	0.99
19	0.25	1.41	1.49	1.49	1.47	1.46	1.44	1.43	1.42	1.41	1.41	1.40	1.40	1.39	1.39	1.38	1.38
19	0.10	2.99	2.61	2.40	2.27	2.18	2.11	2.06	2.02	1.98	1.96	1.93	1.91	1.89	1.88	1.86	1.85
19	0.05	4.38	3.52	3.13	2.90	2.74	2.63	2.54	2.48	2.42	2.38	2.34	2.31	2.28	2.26	2.23	2.21
19	0.01	8.18	5.93	5.01	4.50	4.17	3.94	3.77	3.63	3.52	3.43	3.36	3.30	3.24	3.19	3.15	3.12
19	0.001	15.08	10.16	8.28	7.27	6.62	6.18	5.85	5.59	5.39	5.22	5.08	4.97	4.87	4.78	4.70	4.64
20	0.50	0.47	0.72	0.82	0.87	0.90	0.92	0.94	0.95	0.96	0.97	0.97	0.98	0.98	0.99	0.99	0.99
20	0.25	1.40	1.49	1.48	1.47	1.45	1.44	1.43	1.42	1.41	1.40	1.39	1.39	1.38	1.38	1.37	1.37
20	0.10	2.97	2.59	2.38	2.25	2.16	2.09	2.04	2.00	1.96	1.94	1.91	1.89	1.87	1.86	1.84	1.83
20	0.05	4.35	3.49	3.10	2.87	2.71	2.60	2.51	2.45	2.39	2.35	2.31	2.28	2.25	2.22	2.20	2.18
20	0.01	8.10	5.85	4.94	4.43	4.10	3.87	3.70	3.56	3.46	3.37	3.29	3.23	3.18	3.13	3.09	3.05
20	0.001	14.82	9.95	8.10	7.10	6.46	6.02	5.69	5.44	5.24	5.08	4.94	4.82	4.72	4.64	4.56	4.49
21	0.50	0.47	0.72	0.81	0.87	0.90	0.92	0.94	0.95	0.96	0.96	0.97	0.98	0.98	0.98	0.99	0.99
21	0.25	1.40	1.48	1.48	1.46	1.44	1.43	1.42	1.41	1.40	1.39	1.39	1.38	1.37	1.37	1.37	1.36
21	0.10	2.96	2.57	2.36	2.23	2.14	2.08	2.02	1.98	1.95	1.92	1.90	1.87	1.86	1.84	1.83	1.81
21	0.05	4.32	3.47	3.07	2.84	2.68	2.57	2.49	2.42	2.37	2.32	2.28	2.25	2.22	2.20	2.18	2.16
21	0.01	8.02	5.78	4.87	4.37	4.04	3.81	3.64	3.51	3.40	3.31	3.24	3.17	3.12	3.07	3.03	2.99
21	0.001	14.59	9.77	7.94	6.95	6.32	5.88	5.56	5.31	5.11	4.95	4.81	4.70	4.60	4.51	4.44	4.37
22	0.50	0.47	0.72	0.81	0.87	0.90	0.92	0.93	0.95	0.96	0.96	0.97	0.97	0.98	0.98	0.99	0.99
22	0.25	1.40	1.48	1.47	1.45	1.44	1.42	1.41	1.40	1.39	1.39	1.38	1.37	1.37	1.36	1.36	1.36
22	0.10	2.95	2.56	2.35	2.22	2.13	2.06	2.01	1.97	1.93	1.90	1.88	1.86	1.84	1.83	1.81	1.80
22	0.05	4.30	3.44	3.05	2.82	2.66	2.55	2.46	2.40	2.34	2.30	2.26	2.23	2.20	2.17	2.15	2.13
22	0.01	7.95	5.72	4.82	4.31	3.99	3.76	3.59	3.45	3.35	3.26	3.18	3.12	3.07	3.02	2.98	2.94
22	0.001	14.38	9.61	7.80	6.81	6.19	5.76	5.44	5.19	4.99	4.83	4.70	4.58	4.49	4.40	4.33	4.26
23	0.50	0.47	0.71	0.81	0.86	0.90	0.92	0.93	0.95	0.95	0.96	0.97	0.97	0.98	0.98	0.98	0.99
23	0.25	1.39	1.47	1.47	1.45	1.43	1.42	1.41	1.40	1.39	1.38	1.37	1.37	1.36	1.36	1.35	1.35
23	0.10	2.94	2.55	2.34	2.21	2.11	2.05	1.99	1.95	1.92	1.89	1.87	1.84	1.83	1.81	1.80	1.78
23	0.05	4.28	3.42	3.03	2.80	2.64	2.53	2.44	2.37	2.32	2.27	2.24	2.20	2.18	2.15	2.13	2.11
23	0.01	7.88	5.66	4.76	4.26	3.94	3.71	3.54	3.41	3.30	3.21	3.14	3.07	3.02	2.97	2.93	2.89
23	0.001	14.20	9.47	7.67	6.70	6.08	5.65	5.33	5.09	4.89	4.73	4.60	4.48	4.39	4.30	4.23	4.16
24	0.50	0.47	0.71	0.81	0.86	0.90	0.92	0.93	0.94	0.95	0.96	0.97	0.97	0.98	0.98	0.98	0.99
24	0.25	1.39	1.47	1.46	1.44	1.43	1.41	1.40	1.39	1.38	1.38	1.37	1.36	1.36	1.35	1.35	1.34
24	0.10	2.93	2.54	2.33	2.19	2.10	2.04	1.98	1.94	1.91	1.88	1.85	1.83	1.81	1.80	1.78	1.77
24	0.05	4.26	3.40	3.01	2.78	2.62	2.51	2.42	2.36	2.30	2.25	2.22	2.18	2.15	2.13	2.11	2.09
24	0.01	7.82	5.61	4.72	4.22	3.90	3.67	3.50	3.36	3.26	3.17	3.09	3.03	2.98	2.93	2.89	2.85
24	0.001	14.03	9.34	7.55	6.59	5.98	5.55	5.23	4.99	4.80	4.64	4.51	4.39	4.30	4.21	4.14	4.07
25	0.50	0.47	0.71	0.81	0.86	0.89	0.92	0.93	0.94	0.95	0.96	0.97	0.97	0.98	0.98	0.98	0.98
25	0.25	1.39	1.47	1.46	1.44	1.42	1.41	1.40	1.39	1.38	1.37	1.36	1.36	1.35	1.35	1.34	1.34
25	0.10	2.92	2.53	2.32	2.18	2.09	2.02	1.97	1.93	1.89	1.87	1.84	1.82	1.80	1.79	1.77	1.76
25	0.05	4.24	3.39	2.99	2.76	2.60	2.49	2.40	2.34	2.28	2.24	2.20	2.16	2.14	2.11	2.09	2.07
25	0.01	7.77	5.57	4.68	4.18	3.85	3.63	3.46	3.32	3.22	3.13	3.06	2.99	2.94	2.89	2.85	2.81
25	0.001	13.88	9.22	7.45	6.49	5.89	5.46	5.15	4.91	4.71	4.56	4.42	4.31	4.22	4.13	4.06	3.99
26	0.50	0.47	0.71	0.81	0.86	0.89	0.91	0.93	0.94	0.95	0.96	0.96	0.97	0.97	0.98	0.98	0.98
26	0.25	1.38	1.46	1.45	1.44	1.42	1.41	1.39	1.38	1.37	1.37	1.36	1.35	1.35	1.34	1.34	1.33
26	0.10	2.91	2.52	2.31	2.17	2.08	2.01	1.96	1.92	1.88	1.86	1.83	1.81	1.79	1.77	1.76	1.75
26	0.05	4.23	3.37	2.98	2.74	2.59	2.47	2.39	2.32	2.27	2.22	2.18	2.15	2.12	2.09	2.07	2.05
26	0.01	7.72	5.53	4.64	4.14	3.82	3.59	3.42	3.29	3.18	3.09	3.02	2.96	2.90	2.86	2.81	2.78
26	0.001	13.74	9.12	7.36	6.41	5.80	5.38	5.07	4.83	4.64	4.48	4.35	4.24	4.14	4.06	3.99	3.92
27	0.50	0.47	0.71	0.81	0.86	0.89	0.91	0.93	0.94	0.95	0.96	0.96	0.97	0.97	0.98	0.98	0.98
27	0.25	1.38	1.46	1.45	1.43	1.42	1.40	1.39	1.38	1.37	1.36	1.35	1.35	1.34	1.34	1.33	1.33
27	0.10	2.90	2.51	2.30	2.17	2.07	2.00	1.95	1.91	1.87	1.85	1.82	1.80	1.78	1.76	1.75	1.74
27	0.05	4.21	3.35	2.96	2.73	2.57	2.46	2.37	2.31	2.25	2.20	2.17	2.13	2.10	2.08	2.06	2.04
27	0.01	7.68	5.49	4.60	4.11	3.78	3.56	3.39	3.26	3.15	3.06	2.99	2.93	2.87	2.82	2.78	2.75
27	0.001	13.61	9.02	7.27	6.33	5.73	5.31	5.00	4.76	4.57	4.41	4.28	4.17	4.08	3.99	3.92	3.86

17	18	19	20	25	30	35	40	45	50	60	70	80	90	100	110	120
1.00	1.00	1.00	1.00	1.01	1.01	1.02	1.02	1.02	1.02	1.02	1.03	1.03	1.03	1.03	1.03	1.03
1.37	1.37	1.37	1.37	1.36	1.35	1.34	1.34	1.34	1.33	1.33	1.33	1.32	1.32	1.32	1.32	1.32
1.84	1.83	1.82	1.81	1.78	1.76	1.74	1.73	1.72	1.71	1.70	1.69	1.68	1.68	1.67	1.67	1.67
2.20	2.18	2.17	2.16	2.11	2.07	2.05	2.03	2.01	2.00	1.98	1.97	1.96	1.95	1.94	1.93	1.93
3.08	3.05	3.03	3.00	2.91	2.84	2.80	2.76	2.73	2.71	2.67	2.65	2.63	2.61	2.60	2.59	2.58
4.58	4.52	4.47	4.43	4.26	4.14	4.06	3.99	3.94	3.90	3.84	3.79	3.76	3.73	3.71	3.69	3.68
0.99	1.00	1.00	1.00	1.01	1.01	1.01	1.02	1.02	1.02	1.02	1.02	1.03	1.03	1.03	1.03	1.03
1.37	1.36	1.36	1.36	1.35	1.34	1.33	1.33	1.33	1.32	1.32	1.32	1.31	1.31	1.31	1.31	1.31
1.82	1.81	1.80	1.79	1.76	1.74	1.72	1.71	1.70	1.69	1.68	1.67	1.66	1.65	1.65	1.65	1.64
2.17	2.15	2.14	2.12	2.07	2.04	2.01	1.99	1.98	1.97	1.95	1.93	1.92	1.91	1.91	1.90	1.90
3.02	2.99	2.96	2.94	2.84	2.78	2.73	2.69	2.67	2.64	2.61	2.58	2.56	2.55	2.54	2.53	2.52
4.44	4.38	4.33	4.29	4.12	4.00	3.92	3.86	3.81	3.77	3.70	3.66	3.62	3.60	3.58	3.56	3.54
0.99	0.99	1.00	1.00	1.01	1.01	1.01	1.02	1.02	1.02	1.02	1.02	1.02	1.02	1.03	1.03	1.03
1.36	1.36	1.35	1.35	1.34	1.33	1.33	1.32	1.32	1.32	1.31	1.31	1.30	1.30	1.30	1.30	1.30
1.80	1.79	1.78	1.78	1.74	1.72	1.70	1.69	1.68	1.67	1.66	1.65	1.64	1.63	1.63	1.63	1.62
2.14	2.12	2.11	2.10	2.05	2.01	1.98	1.96	1.95	1.94	1.92	1.90	1.89	1.88	1.88	1.87	1.87
2.96	2.93	2.90	2.88	2.79	2.72	2.67	2.64	2.61	2.58	2.55	2.52	2.50	2.49	2.48	2.47	2.46
4.31	4.26	4.21	4.17	4.00	3.88	3.80	3.74	3.69	3.64	3.58	3.54	3.50	3.48	3.46	3.44	3.42
0.99	0.99	1.00	1.00	1.00	1.01	1.01	1.01	1.02	1.02	1.02	1.02	1.02	1.02	1.02	1.02	1.03
1.35	1.35	1.35	1.34	1.33	1.32	1.32	1.31	1.31	1.31	1.30	1.30	1.30	1.29	1.29	1.29	1.29
1.79	1.78	1.77	1.76	1.73	1.70	1.68	1.67	1.66	1.65	1.64	1.63	1.62	1.62	1.61	1.61	1.60
2.11	2.10	2.08	2.07	2.02	1.98	1.96	1.94	1.92	1.91	1.89	1.88	1.86	1.86	1.85	1.84	1.84
2.91	2.88	2.85	2.83	2.73	2.67	2.62	2.58	2.55	2.53	2.50	2.47	2.45	2.43	2.42	2.41	2.40
4.20	4.15	4.10	4.06	3.89	3.78	3.69	3.63	3.58	3.54	3.48	3.43	3.40	3.37	3.35	3.33	3.32
0.99	0.99	0.99	1.00	1.00	1.01	1.01	1.01	1.01	1.02	1.02	1.02	1.02	1.02	1.02	1.02	1.02
1.35	1.34	1.34	1.34	1.33	1.32	1.31	1.31	1.30	1.30	1.30	1.29	1.29	1.29	1.29	1.28	1.28
1.77	1.76	1.75	1.74	1.71	1.69	1.67	1.66	1.64	1.64	1.62	1.61	1.61	1.60	1.59	1.59	1.59
2.09	2.08	2.06	2.05	2.00	1.96	1.93	1.91	1.90	1.88	1.86	1.85	1.84	1.83	1.82	1.82	1.81
2.86	2.83	2.80	2.78	2.69	2.62	2.57	2.54	2.51	2.48	2.45	2.42	2.40	2.39	2.37	2.36	2.35
4.10	4.05	4.00	3.96	3.79	3.68	3.60	3.53	3.48	3.44	3.38	3.34	3.30	3.28	3.25	3.24	3.22
0.99	0.99	0.99	0.99	1.00	1.01	1.01	1.01	1.01	1.01	1.02	1.02	1.02	1.02	1.02	1.02	1.02
1.34	1.34	1.33	1.33	1.32	1.31	1.31	1.30	1.30	1.29	1.29	1.28	1.28	1.28	1.28	1.28	1.28
1.76	1.75	1.74	1.73	1.70	1.67	1.65	1.64	1.63	1.62	1.61	1.60	1.59	1.58	1.58	1.57	1.57
2.07	2.05	2.04	2.03	1.97	1.94	1.91	1.89	1.88	1.86	1.84	1.83	1.82	1.81	1.80	1.79	1.79
2.82	2.79	2.76	2.74	2.64	2.58	2.53	2.49	2.46	2.44	2.40	2.38	2.36	2.34	2.33	2.32	2.31
4.02	3.96	3.92	3.87	3.71	3.59	3.51	3.45	3.40	3.36	3.29	3.25	3.22	3.19	3.17	3.15	3.14
0.99	0.99	0.99	0.99	1.00	1.00	1.01	1.01	1.01	1.01	1.02	1.02	1.02	1.02	1.02	1.02	1.02
1.33	1.33	1.33	1.33	1.31	1.31	1.30	1.29	1.29	1.29	1.28	1.28	1.28	1.27	1.27	1.27	1.27
1.75	1.74	1.73	1.72	1.68	1.66	1.64	1.63	1.62	1.61	1.59	1.58	1.58	1.57	1.56	1.56	1.56
2.05	2.04	2.02	2.01	1.96	1.92	1.89	1.87	1.86	1.84	1.82	1.81	1.80	1.79	1.78	1.77	1.77
2.78	2.75	2.72	2.70	2.60	2.54	2.49	2.45	2.42	2.40	2.36	2.34	2.32	2.30	2.29	2.28	2.27
3.94	3.88	3.84	3.79	3.63	3.52	3.43	3.37	3.32	3.28	3.22	3.17	3.14	3.11	3.09	3.07	3.06
0.99	0.99	0.99	0.99	1.00	1.00	1.01	1.01	1.01	1.01	1.01	1.02	1.02	1.02	1.02	1.02	1.02
1.33	1.33	1.32	1.32	1.31	1.30	1.29	1.29	1.28	1.28	1.28	1.27	1.27	1.27	1.27	1.26	1.26
1.73	1.72	1.71	1.71	1.67	1.65	1.63	1.61	1.60	1.59	1.58	1.57	1.56	1.56	1.55	1.55	1.54
2.03	2.02	2.00	1.99	1.94	1.90	1.87	1.85	1.84	1.82	1.80	1.79	1.78	1.77	1.76	1.75	1.75
2.75	2.72	2.69	2.66	2.57	2.50	2.45	2.42	2.39	2.36	2.33	2.30	2.28	2.26	2.25	2.24	2.23
3.86	3.81	3.77	3.72	3.56	3.44	3.36	3.30	3.25	3.21	3.15	3.10	3.07	3.04	3.02	3.00	2.99
0.99	0.99	0.99	0.99	1.00	1.00	1.01	1.01	1.01	1.01	1.01	1.02	1.02	1.02	1.02	1.02	1.02
1.33	1.32	1.32	1.32	1.30	1.30	1.29	1.28	1.28	1.28	1.27	1.27	1.26	1.26	1.26	1.26	1.26
1.72	1.71	1.70	1.70	1.66	1.64	1.62	1.60	1.59	1.58	1.57	1.56	1.55	1.54	1.54	1.53	1.53
2.02	2.00	1.99	1.97	1.92	1.88	1.86	1.84	1.82	1.81	1.79	1.77	1.76	1.75	1.74	1.74	1.73
2.71	2.68	2.66	2.63	2.54	2.47	2.42	2.38	2.35	2.33	2.29	2.27	2.25	2.23	2.22	2.21	2.20
3.80	3.75	3.70	3.66	3.49	3.38	3.30	3.23	3.18	3.14	3.08	3.04	3.00	2.98	2.96	2.94	2.92

df	α	1	2	3	4	5	6	7	8	9	10	11	12	13	14	15	16
								Degrees of freedom (numerator mean squares)									
28	0.50	0.47	0.71	0.81	0.86	0.89	0.91	0.93	0.94	0.95	0.96	0.96	0.97	0.97	0.98	0.98	0.98
28	0.25	1.38	1.46	1.45	1.43	1.41	1.40	1.39	1.38	1.37	1.36	1.35	1.34	1.34	1.33	1.33	1.32
28	0.10	2.89	2.50	2.29	2.16	2.06	2.00	1.94	1.90	1.87	1.84	1.81	1.79	1.77	1.75	1.74	1.73
28	0.05	4.20	3.34	2.95	2.71	2.56	2.45	2.36	2.29	2.24	2.19	2.15	2.12	2.09	2.06	2.04	2.02
28	0.01	7.64	5.45	4.57	4.07	3.75	3.53	3.36	3.23	3.12	3.03	2.96	2.90	2.84	2.79	2.75	2.72
28	0.001	13.50	8.93	7.19	6.25	5.66	5.24	4.93	4.69	4.50	4.35	4.22	4.11	4.01	3.93	3.86	3.80
29	0.50	0.47	0.71	0.81	0.86	0.89	0.91	0.93	0.94	0.95	0.96	0.96	0.97	0.97	0.98	0.98	0.98
29	0.25	1.38	1.45	1.45	1.43	1.41	1.40	1.38	1.37	1.36	1.35	1.35	1.34	1.33	1.33	1.32	1.32
29	0.10	2.89	2.50	2.28	2.15	2.06	1.99	1.93	1.89	1.86	1.83	1.80	1.78	1.76	1.75	1.73	1.72
29	0.05	4.18	3.33	2.93	2.70	2.55	2.43	2.35	2.28	2.22	2.18	2.14	2.10	2.08	2.05	2.03	2.01
29	0.01	7.60	5.42	4.54	4.04	3.73	3.50	3.33	3.20	3.09	3.00	2.93	2.87	2.81	2.77	2.73	2.69
29	0.001	13.39	8.85	7.12	6.19	5.59	5.18	4.87	4.64	4.45	4.29	4.16	4.05	3.96	3.88	3.80	3.74
30	0.50	0.47	0.71	0.81	0.86	0.89	0.91	0.93	0.94	0.95	0.96	0.96	0.97	0.97	0.97	0.98	0.98
30	0.25	1.38	1.45	1.44	1.42	1.41	1.39	1.38	1.37	1.36	1.35	1.34	1.34	1.33	1.33	1.32	1.32
30	0.10	2.88	2.49	2.28	2.14	2.05	1.98	1.93	1.88	1.85	1.82	1.79	1.77	1.75	1.74	1.72	1.71
30	0.05	4.17	3.32	2.92	2.69	2.53	2.42	2.33	2.27	2.21	2.16	2.13	2.09	2.06	2.04	2.01	1.99
30	0.01	7.56	5.39	4.51	4.02	3.70	3.47	3.30	3.17	3.07	2.98	2.91	2.84	2.79	2.74	2.70	2.66
30	0.001	13.29	8.77	7.05	6.12	5.53	5.12	4.82	4.58	4.39	4.24	4.11	4.00	3.91	3.82	3.75	3.69
40	0.50	0.46	0.71	0.80	0.85	0.89	0.91	0.92	0.93	0.94	0.95	0.96	0.96	0.97	0.97	0.97	0.97
40	0.25	1.36	1.44	1.42	1.40	1.39	1.37	1.36	1.35	1.34	1.33	1.32	1.31	1.31	1.30	1.30	1.29
40	0.10	2.84	2.44	2.23	2.09	2.00	1.93	1.87	1.83	1.79	1.76	1.74	1.71	1.70	1.68	1.66	1.65
40	0.05	4.08	3.23	2.84	2.61	2.45	2.34	2.25	2.18	2.12	2.08	2.04	2.00	1.97	1.95	1.92	1.90
40	0.01	7.31	5.18	4.31	3.83	3.51	3.29	3.12	2.99	2.89	2.80	2.73	2.66	2.61	2.56	2.52	2.48
40	0.001	12.61	8.25	6.59	5.70	5.13	4.73	4.44	4.21	4.02	3.87	3.75	3.64	3.55	3.47	3.40	3.34
60	0.50	0.46	0.70	0.80	0.85	0.88	0.90	0.92	0.93	0.94	0.94	0.95	0.96	0.96	0.96	0.97	0.97
60	0.25	1.35	1.42	1.41	1.38	1.37	1.35	1.33	1.32	1.31	1.30	1.29	1.29	1.28	1.27	1.27	1.26
60	0.10	2.79	2.39	2.18	2.04	1.95	1.87	1.82	1.77	1.74	1.71	1.68	1.66	1.64	1.62	1.60	1.59
60	0.05	4.00	3.15	2.76	2.53	2.37	2.25	2.17	2.10	2.04	1.99	1.95	1.92	1.89	1.86	1.84	1.82
60	0.01	7.08	4.98	4.13	3.65	3.34	3.12	2.95	2.82	2.72	2.63	2.56	2.50	2.44	2.39	2.35	2.31
60	0.001	11.97	7.77	6.17	5.31	4.76	4.37	4.09	3.86	3.69	3.54	3.42	3.32	3.23	3.15	3.08	3.02
80	0.50	0.46	0.70	0.80	0.85	0.88	0.90	0.91	0.93	0.93	0.94	0.95	0.95	0.96	0.96	0.96	0.97
80	0.25	1.34	1.41	1.40	1.38	1.36	1.34	1.32	1.31	1.30	1.29	1.28	1.27	1.27	1.26	1.26	1.25
80	0.10	2.77	2.37	2.15	2.02	1.92	1.85	1.79	1.75	1.71	1.68	1.65	1.63	1.61	1.59	1.57	1.56
80	0.05	3.96	3.11	2.72	2.49	2.33	2.21	2.13	2.06	2.00	1.95	1.91	1.88	1.84	1.82	1.79	1.77
80	0.01	6.96	4.88	4.04	3.56	3.26	3.04	2.87	2.74	2.64	2.55	2.48	2.42	2.36	2.31	2.27	2.23
80	0.001	11.67	7.54	5.97	5.12	4.58	4.20	3.92	3.70	3.53	3.39	3.27	3.16	3.07	3.00	2.93	2.87
100	0.50	0.46	0.70	0.79	0.84	0.88	0.90	0.91	0.92	0.93	0.94	0.95	0.95	0.96	0.96	0.96	0.97
100	0.25	1.34	1.41	1.39	1.37	1.35	1.33	1.32	1.30	1.29	1.28	1.27	1.27	1.26	1.25	1.25	1.24
100	0.10	2.76	2.36	2.14	2.00	1.91	1.83	1.78	1.73	1.69	1.66	1.64	1.61	1.59	1.57	1.56	1.54
100	0.05	3.94	3.09	2.70	2.46	2.31	2.19	2.10	2.03	1.97	1.93	1.89	1.85	1.82	1.79	1.77	1.75
100	0.01	6.90	4.82	3.98	3.51	3.21	2.99	2.82	2.69	2.59	2.50	2.43	2.37	2.31	2.27	2.22	2.19
100	0.001	11.50	7.41	5.86	5.02	4.48	4.11	3.83	3.61	3.44	3.30	3.18	3.07	2.99	2.91	2.84	2.78
120	0.50	0.46	0.70	0.79	0.84	0.88	0.90	0.91	0.92	0.93	0.94	0.95	0.95	0.95	0.96	0.96	0.96
120	0.25	1.34	1.40	1.39	1.37	1.35	1.33	1.31	1.30	1.29	1.28	1.27	1.26	1.26	1.25	1.24	1.24
120	0.10	2.75	2.35	2.13	1.99	1.90	1.82	1.77	1.72	1.68	1.65	1.63	1.60	1.58	1.56	1.55	1.53
120	0.05	3.92	3.07	2.68	2.45	2.29	2.18	2.09	2.02	1.96	1.91	1.87	1.83	1.80	1.78	1.75	1.73
120	0.01	6.85	4.79	3.95	3.48	3.17	2.96	2.79	2.66	2.56	2.47	2.40	2.34	2.28	2.23	2.19	2.15
120	0.001	11.38	7.32	5.78	4.95	4.42	4.04	3.77	3.55	3.38	3.24	3.12	3.02	2.93	2.85	2.78	2.72

17	18	19	20	25	30	35	40	45	50	60	70	80	90	100	110	120
0.98	0.99	0.99	0.99	1.00	1.00	1.00	1.01	1.01	1.01	1.01	1.01	1.02	1.02	1.02	1.02	1.02
1.32	1.32	1.31	1.31	1.30	1.29	1.28	1.28	1.27	1.27	1.27	1.26	1.26	1.26	1.25	1.25	1.25
1.71	1.70	1.69	1.69	1.65	1.63	1.61	1.59	1.58	1.57	1.56	1.55	1.54	1.53	1.53	1.52	1.52
2.00	1.99	1.97	1.96	1.91	1.87	1.84	1.82	1.80	1.79	1.77	1.75	1.74	1.73	1.73	1.72	1.71
2.68	2.65	2.63	2.60	2.51	2.44	2.39	2.35	2.32	2.30	2.26	2.24	2.22	2.20	2.19	2.18	2.17
3.74	3.69	3.64	3.60	3.43	3.32	3.24	3.18	3.13	3.09	3.02	2.98	2.94	2.92	2.90	2.88	2.86
0.98	0.99	0.99	0.99	1.00	1.00	1.00	1.01	1.01	1.01	1.01	1.01	1.01	1.02	1.02	1.02	1.02
1.32	1.31	1.31	1.31	1.30	1.29	1.28	1.27	1.27	1.27	1.26	1.26	1.25	1.25	1.25	1.25	1.25
1.71	1.69	1.68	1.68	1.64	1.62	1.60	1.58	1.57	1.56	1.55	1.54	1.53	1.52	1.52	1.51	1.51
1.99	1.97	1.96	1.94	1.89	1.85	1.83	1.81	1.79	1.77	1.75	1.74	1.73	1.72	1.71	1.70	1.70
2.66	2.63	2.60	2.57	2.48	2.41	2.36	2.33	2.30	2.27	2.23	2.21	2.19	2.17	2.16	2.15	2.14
3.68	3.63	3.59	3.54	3.38	3.27	3.18	3.12	3.07	3.03	2.97	2.92	2.89	2.86	2.84	2.82	2.81
0.98	0.99	0.99	0.99	1.00	1.00	1.00	1.01	1.01	1.01	1.01	1.01	1.01	1.02	1.02	1.02	1.02
1.31	1.31	1.31	1.30	1.29	1.28	1.28	1.27	1.27	1.26	1.26	1.25	1.25	1.25	1.25	1.24	1.24
1.70	1.69	1.68	1.67	1.63	1.61	1.59	1.57	1.56	1.55	1.54	1.53	1.52	1.51	1.51	1.50	1.50
1.98	1.96	1.95	1.93	1.88	1.84	1.81	1.79	1.77	1.76	1.74	1.72	1.71	1.70	1.70	1.69	1.68
2.63	2.60	2.57	2.55	2.45	2.39	2.34	2.30	2.27	2.25	2.21	2.18	2.16	2.14	2.13	2.12	2.11
3.63	3.58	3.53	3.49	3.33	3.22	3.13	3.07	3.02	2.98	2.92	2.87	2.84	2.81	2.79	2.77	2.76
0.98	0.98	0.98	0.98	0.99	0.99	1.00	1.00	1.00	1.00	1.01	1.01	1.01	1.01	1.01	1.01	1.01
1.29	1.28	1.28	1.28	1.26	1.25	1.25	1.24	1.23	1.23	1.22	1.22	1.22	1.21	1.21	1.21	1.21
1.64	1.62	1.61	1.61	1.57	1.54	1.52	1.51	1.49	1.48	1.47	1.46	1.45	1.44	1.43	1.43	1.42
1.89	1.87	1.85	1.84	1.78	1.74	1.72	1.69	1.67	1.66	1.64	1.62	1.61	1.60	1.59	1.58	1.58
2.45	2.42	2.39	2.37	2.27	2.20	2.15	2.11	2.08	2.06	2.02	1.99	1.97	1.95	1.94	1.93	1.92
3.28	3.23	3.19	3.14	2.98	2.87	2.79	2.73	2.68	2.64	2.57	2.53	2.49	2.47	2.44	2.43	2.41
0.97	0.97	0.98	0.98	0.98	0.99	0.99	0.99	1.00	1.00	1.00	1.00	1.00	1.00	1.00	1.01	1.01
1.26	1.26	1.25	1.25	1.23	1.22	1.21	1.21	1.20	1.20	1.19	1.19	1.18	1.18	1.18	1.17	1.17
1.58	1.56	1.55	1.54	1.50	1.48	1.45	1.44	1.42	1.41	1.40	1.38	1.37	1.36	1.36	1.35	1.35
1.80	1.78	1.76	1.75	1.69	1.65	1.62	1.59	1.57	1.56	1.53	1.52	1.50	1.49	1.48	1.47	1.47
2.28	2.25	2.22	2.20	2.10	2.03	1.98	1.94	1.90	1.88	1.84	1.81	1.78	1.76	1.75	1.74	1.73
2.96	2.91	2.87	2.83	2.67	2.55	2.47	2.41	2.36	2.32	2.25	2.21	2.17	2.14	2.12	2.10	2.08
0.97	0.97	0.97	0.97	0.98	0.99	0.99	0.99	0.99	0.99	1.00	1.00	1.00	1.00	1.00	1.00	1.00
1.25	1.24	1.24	1.23	1.22	1.21	1.20	1.19	1.19	1.18	1.17	1.17	1.16	1.16	1.16	1.15	1.15
1.55	1.53	1.52	1.51	1.47	1.44	1.42	1.40	1.39	1.38	1.36	1.34	1.33	1.33	1.32	1.31	1.31
1.75	1.73	1.72	1.70	1.64	1.60	1.57	1.54	1.52	1.51	1.48	1.46	1.45	1.44	1.43	1.42	1.41
2.20	2.17	2.14	2.12	2.01	1.94	1.89	1.85	1.82	1.79	1.75	1.71	1.69	1.67	1.65	1.64	1.63
2.81	2.76	2.72	2.68	2.52	2.41	2.32	2.26	2.21	2.16	2.10	2.05	2.01	1.98	1.96	1.94	1.92
0.97	0.97	0.97	0.97	0.98	0.98	0.99	0.99	0.99	0.99	1.00	1.00	1.00	1.00	1.00	1.00	1.00
1.24	1.23	1.23	1.23	1.21	1.20	1.19	1.18	1.18	1.17	1.16	1.16	1.15	1.15	1.14	1.14	1.14
1.53	1.52	1.50	1.49	1.45	1.42	1.40	1.38	1.37	1.35	1.34	1.32	1.31	1.30	1.29	1.29	1.28
1.73	1.71	1.69	1.68	1.62	1.57	1.54	1.52	1.49	1.48	1.45	1.43	1.41	1.40	1.39	1.38	1.38
2.15	2.12	2.09	2.07	1.97	1.89	1.84	1.80	1.76	1.74	1.69	1.66	1.63	1.61	1.60	1.58	1.57
2.73	2.68	2.63	2.59	2.43	2.32	2.24	2.17	2.12	2.08	2.01	1.96	1.92	1.89	1.87	1.85	1.83
0.97	0.97	0.97	0.97	0.98	0.98	0.99	0.99	0.99	0.99	0.99	1.00	1.00	1.00	1.00	1.00	1.00
1.23	1.23	1.22	1.22	1.20	1.19	1.18	1.18	1.17	1.16	1.16	1.15	1.14	1.14	1.14	1.13	1.13
1.52	1.50	1.49	1.48	1.44	1.41	1.39	1.37	1.35	1.34	1.32	1.31	1.29	1.28	1.28	1.27	1.26
1.71	1.69	1.67	1.66	1.60	1.55	1.52	1.50	1.47	1.46	1.43	1.41	1.39	1.38	1.37	1.36	1.35
2.12	2.09	2.06	2.03	1.93	1.86	1.81	1.76	1.73	1.70	1.66	1.62	1.60	1.58	1.56	1.55	1.53
2.67	2.62	2.58	2.53	2.37	2.26	2.18	2.11	2.06	2.02	1.95	1.90	1.86	1.83	1.81	1.78	1.77

Appendix D

Critical Values for the Chi-Square Distribution

To determine a critical value identify the number corresponding with the degrees of freedom and the applicable α level.

df	α					
	0.5	0.1	0.05	0.025	0.01	0.001
1	0.455	2.706	3.841	5.024	6.635	10.828
2	1.386	4.605	5.991	7.378	9.210	13.816
3	2.366	6.251	7.815	9.348	11.345	16.266
4	3.357	7.779	9.488	11.143	13.277	18.467
5	4.351	9.236	11.070	12.833	15.086	20.515
6	5.348	10.645	12.592	14.449	16.812	22.458
7	6.346	12.017	14.067	16.013	18.475	24.322
8	7.344	13.362	15.507	17.535	20.090	26.124
9	8.343	14.684	16.919	19.023	21.666	27.877
10	9.342	15.987	18.307	20.483	23.209	29.588
11	10.341	17.275	19.675	21.920	24.725	31.264
12	11.340	18.549	21.026	23.337	26.217	32.909
13	12.340	19.812	22.362	24.736	27.688	34.528
14	13.339	21.064	23.685	26.119	29.141	36.123
15	14.339	22.307	24.996	27.488	30.578	37.697
16	15.338	23.542	26.296	28.845	32.000	39.252
17	16.338	24.769	27.587	30.191	33.409	40.790
18	17.338	25.989	28.869	31.526	34.805	42.312
19	18.338	27.204	30.144	32.852	36.191	43.820

df	α					
	0.5	0.1	0.05	0.025	0.01	0.001
20	19.337	28.412	31.410	34.170	37.566	45.315
21	20.337	29.615	32.671	35.479	38.932	46.797
22	21.337	30.813	33.924	36.781	40.289	48.268
23	22.337	32.007	35.172	38.076	41.638	49.728
24	23.337	33.196	36.415	39.364	42.980	51.179
25	24.337	34.382	37.652	40.646	44.314	52.620
26	25.336	35.563	38.885	41.923	45.642	54.052
27	26.336	36.741	40.113	43.195	46.963	55.476
28	27.336	37.916	41.337	44.461	48.278	56.892
29	28.336	39.087	42.557	45.722	49.588	58.301
30	29.336	40.256	43.773	46.979	50.892	59.703
31	30.336	41.422	44.985	48.232	52.191	61.098
32	31.336	42.585	46.194	49.480	53.486	62.487
33	32.336	43.745	47.400	50.725	54.776	63.870
34	33.336	44.903	48.602	51.966	56.061	65.247
35	34.336	46.059	49.802	53.203	57.342	66.619
36	35.336	47.212	50.998	54.437	58.619	67.985
37	36.336	48.363	52.192	55.668	59.893	69.346
38	37.335	49.513	53.384	56.896	61.162	70.703
39	38.335	50.660	54.572	58.120	62.428	72.055
40	39.335	51.805	55.758	59.342	63.691	73.402
41	40.335	52.949	56.942	60.561	64.950	74.745
42	41.335	54.090	58.124	61.777	66.206	76.084
43	42.335	55.230	59.304	62.990	67.459	77.419
44	43.335	56.369	60.481	64.201	68.710	78.750
45	44.335	57.505	61.656	65.410	69.957	80.077
46	45.335	58.641	62.830	66.617	71.201	81.400
47	46.335	59.774	64.001	67.821	72.443	82.720
48	47.335	60.907	65.171	69.023	73.683	84.037
49	48.335	62.038	66.339	70.222	74.919	85.351
50	49.335	63.167	67.505	71.420	76.154	86.661

Appendix E

Critical Values for the Wilcoxon Two-Sample U-Test

The following table reflects the critical values for the U-distribution for alpha values of .05 and .01. Remember that this is a special case in which the null hypothesis is rejected if the calculated value is *smaller* than the critical value. If the desired critical value is not listed on this table, use the Z-score approximation instead (Equation (14.2)).

$\alpha = .05$ (One-tailed test: $\alpha = .025$)

Smallest sample size	Largest sample size															
	5	6	7	8	9	10	11	12	13	14	15	16	17	18	19	20
3	0	1	1	2	2	3	3	4	4	5	5	6	6	7	7	8
4	1	2	3	4	4	5	6	7	8	9	10	11	11	12	13	14
5	2	3	5	6	7	8	9	11	12	13	14	15	17	18	19	20
6		5	6	8	10	11	13	14	16	17	19	21	22	24	25	27
7			8	10	12	14	16	18	20	22	24	26	28	30	32	34
8				13	15	17	19	22	24	26	29	31	34	36	38	41
9					17	20	23	26	28	31	34	37	39	42	45	48
10						23	26	29	33	36	39	42	45	48	52	55
11							30	33	37	40	44	47	51	55	58	62
12								37	41	45	49	53	57	61	65	69
13									45	50	54	59	63	67	72	76
14										55	59	64	67	74	78	83
15											64	70	75	80	85	90
16												75	81	86	92	98
17													87	93	99	105
18														99	106	112
19															113	119
20																127

α = .01 (One-tailed test: α = .005)

Smallest sample size	Largest sample size															
	5	6	7	8	9	10	11	12	13	14	15	16	17	18	19	20
3	0	0	0	0	0	0	0	1	1	1	2	2	2	2	3	3
4	0	0	0	1	1	2	2	3	3	4	5	5	6	6	7	8
5	0	1	1	2	3	4	5	6	7	7	8	9	10	11	12	13
6		2	3	4	5	6	7	9	10	11	12	13	15	16	17	18
7			4	6	7	9	10	12	13	15	16	18	19	21	22	24
8				7	9	11	13	15	17	18	20	22	24	26	28	30
9					11	13	16	18	20	22	24	27	29	31	33	36
10						16	18	21	24	26	29	31	34	37	39	42
11							21	24	27	30	33	36	39	42	45	46
12								27	31	34	37	41	44	47	51	54
13									34	38	42	45	49	53	56	60
14										42	46	50	54	58	63	67
15											51	55	60	64	69	73
16												60	65	70	4	79
17													70	75	81	86
18														81	87	92
19															93	99
20																105

Index

Printed and bound by CPI Group (UK) Ltd, Croydon, CR0 4YY

27/10/2024

14580381-0001